AF886874

Introduction to Soliton Theory: Applications to Mechanics

Fundamental Theories of Physics

An International Book Series on The Fundamental Theories of Physics: Their Clarification, Development and Application

Editor:
ALWYN VAN DER MERWE, *University of Denver, U.S.A.*

Editorial Advisory Board:
GIANCARLO GHIRARDI, *University of Trieste, Italy*
LAWRENCE P. HORWITZ, *Tel-Aviv University, Israel*
BRIAN D. JOSEPHSON, *University of Cambridge, U.K.*
CLIVE KILMISTER, *University of London, U.K.*
PEKKA J. LAHTI, *University of Turku, Finland*
ASHER PERES, *Israel Institute of Technology, Israel*
EDUARD PRUGOVECKI, *University of Toronto, Canada*
FRANCO SELLERI, *Università di Bara, Italy*
TONY SUDBURY, *University of York, U.K.*
HANS-JÜRGEN TREDER, *Zentralinstitut für Astrophysik der Akademie der Wissenschaften, Germany*

Volume 143

Introduction to Soliton Theory: Applications to Mechanics

by

Ligia Munteanu

Institute of Solid Mechanics,
Romanian Academy, Bucharest, Romania

and

Stefania Donescu

Technical University of Civil Engineering,
Department of Mathematics, Bucharest, Romania

KLUWER ACADEMIC PUBLISHERS
DORDRECHT / BOSTON / LONDON

A C.I.P. Catalogue record for this book is available from the Library of Congress.

ISBN 1-4020-2576-9 (HB)
ISBN 1-4020-2577-7 (e-book)

Published by Kluwer Academic Publishers,
P.O. Box 17, 3300 AA Dordrecht, The Netherlands.

Sold and distributed in North, Central and South America
by Kluwer Academic Publishers,
101 Philip Drive, Norwell, MA 02061, U.S.A.

In all other countries, sold and distributed
by Kluwer Academic Publishers,
P.O. Box 322, 3300 AH Dordrecht, The Netherlands.

Printed on acid-free paper

All Rights Reserved
© 2004 Kluwer Academic Publishers
No part of this work may be reproduced, stored in a retrieval system, or transmitted
in any form or by any means, electronic, mechanical, photocopying, microfilming, recording
or otherwise, without written permission from the Publisher, with the exception
of any material supplied specifically for the purpose of being entered
and executed on a computer system, for exclusive use by the purchaser of the work.

Printed in the Netherlands.

Contents

Preface ix

Part 1. INTRODUCTION TO SOLITON THEORY

1. MATHEMATICAL METHODS 1

1.1 Scope of the chapter 1
1.2 Scattering theory 1
1.3 Inverse scattering theory 12
1.4 Cnoidal method 17
1.5 Hirota method 25
1.6 Linear equivalence method (LEM) 31
1.7 Bäcklund transformation 39
1.8 Painlevé analysis 46

2. SOME PROPERTIES OF NONLINEAR EQUATIONS 53

2.1 Scope of the chapter 53
2.2 General properties of the linear waves 53
2.3 Some properties of nonlinear equations 59
2.4 Symmetry groups of nonlinear equations 62
2.5 Noether theorem 66
2.6 Inverse Lagrange problem 69
2.7 Recursion operators 73

3. SOLITONS AND NONLINEAR EQUATIONS 78

3.1 Scope of the chapter 78
3.2 Korteweg and de Vries equation (KdV) 78
3.3 Derivation of the KdV equation 86

3.4 Scattering problem for the KdV equation 90
3.5 Inverse scattering problem for the KdV equation 95
3.6 Multi-soliton solutions of the KdV equation 101
3.7 Boussinesq, modified KdV and Burgers equations 107
3.8 The sine-Gordon and Schrödinger equations 112
3.9 Tricomi system and the simple pendulum 115

Part 2. APPLICATIONS TO MECHANICS 121

4. STATICS AND DYNAMICS OF THE THIN ELASTIC ROD 121

4.1 Scope of the chapter 121
4.2 Fundamental equations 122
4.3 The equivalence theorem 132
4.4 Exact solutions of the equilibrium equations 134
4.5 Exact solutions of the motion equations 146

5. VIBRATIONS OF THIN ELASTIC RODS 149

5.1 Scope of the chapter 149
5.2 Linear and nonlinear vibrations 149
5.3 Transverse vibrations of the helical rod 155
5.4 A special class of DRIP media 159
5.5 Interaction of waves 163
5.6 Vibrations of a heterogeneous string 166

6. THE COUPLED PENDULUM 173

6.1 Scope of the chapter 173
6.2 Motion equations. Problem E1 173
6.3 Problem E2 177
6.4 LEM solutions of the system E2 180
6.5 Cnoidal solutions 185
6.6 Modal interaction in periodic structures 191

7. DYNAMICS OF THE LEFT VENTRICLE 197

7.1 Scope of the chapter 197
7.2 The mathematical model 198
7.3 Cnoidal solutions 206
7.4 Numerical results 209
7.5 A nonlinear system with essential energy influx 213

8. THE FLOW OF BLOOD IN ARTERIES 220
8.1 Scope of the chapter 220
8.2 A nonlinear model of blood flow in arteries 221
8.3 Two-soliton solutions 228
8.4 A micropolar model of blood flow in arteries 235

9. INTERMODAL INTERACTION OF WAVES 242
9.1 Scope of the chapter 242
9.2 A plate with Cantor-like structure 243
9.3 The eigenvalue problem 248
9.4 Subharmonic waves generation 249
9.5 Internal solitary waves in a stratified fluid 255
9.6 The motion of a micropolar fluid in inclined open channels 259
9.7 Cnoidal solutions 265
9.8 The effect of surface tension on the solitary waves 269

10. ON THE TZITZEICA SURFACES AND SOME RELATED PROBLEMS 273
10.1 Scope of the chapter 273
10.2 Tzitzeica surfaces 273
10.3 Symmetry group theory applied to Tzitzeica equations 276
10.4 The relation between the forced oscillator and a Tzitzeica curve 283
10.5 Sound propagation in a nonlinear medium 285
10.6 The pseudospherical reduction of a nonlinear problem 291

References 298
Index 305

Preface

This monograph is planned to provide the application of the soliton theory to solve certain practical problems selected from the fields of solid mechanics, fluid mechanics and biomechanics. The work is based mainly on the authors' research carried out at their home institutes, and on some specified, significant results existing in the published literature. The methodology to study a given evolution equation is to seek the waves of permanent form, to test whether it possesses any symmetry properties, and whether it is stable and solitonic in nature.

Students of physics, applied mathematics, and engineering are usually exposed to various branches of nonlinear mechanics, especially to the soliton theory. The soliton is regarded as an entity, a quasi-particle, which conserves its character and interacts with the surroundings and other solitons as a particle. It is related to a strange phenomenon, which consists in the propagation of certain waves without attenuation in dissipative media. This phenomenon has been known for about 200 years (it was described, for example, by the Joule Verne's novel *Les histoires de Jean Marie Cabidoulin*, Éd. Hetzel), but its detailed quantitative description became possible only in the last 30 years due to the exceptional development of computers.

The discovery of the physical soliton is attributed to John Scott Russell. In 1834, Russell was observing a boat being drawn along a narrow channel by a pair of horses. He followed it on horseback and observed an amazing phenomenon: when the boat suddenly stopped, a bow wave detached from the boat and rolled forward with great velocity, having the shape of a large solitary elevation, with a rounded well-defined heap of water. The solitary wave continued its motion along the channel without change of form or velocity. The scientist followed it on horseback as it propagated at about eight or nine miles an hour, but after one or two miles he lost it. Russell was convinced that he had observed an important phenomenon, and he built an experimental tank in his garden to continue the studies of what he named *the wave of translation*.

The wave of translation was regarded as a curiosity until the 1960s, when scientists began to use computers to study nonlinear wave propagation. The discovery of mathematical solutions started with the analysis of nonlinear partial differential equations, such as the work of Boussinesq and Rayleigh, independently, in the 1870s. Boussinesq and Rayleigh explained theoretically the Russell observation and later reproduction in a laboratory experiment. Korteweg and de Vries derived in 1895 the equation for water waves in shallow channels, and confirmed the existence of solitons.

An explosion of works occurred when it was discovered that many phenomena in physics, electronics, mechanics and biology might be described by using the theory of solitons. Nonlinear mechanics is often faced with the unexpected appearance of *chaos* or *order*. Within this framework the soliton plays the role of *order*. The discovery of orderly stable pulses as an effect of nonlinearity is surprising. The results obtained in the linear theory of waves, by ignoring the nonlinear parts, are most frequently too far from reality to be useful. The linearisation misses an important phenomenon, solitons, which are waves, which maintain their identity indefinitely just when we most expect that dispersion effects will lead to their disappearance. The soliton as the solution of the completely integrable partial differential equations are stable in collision process even if interaction between the solitons takes place in a nonlinear way.

The unexpected results obtained in 1955 by Fermi, Pasta and Ulam in the study of a nonlinear anharmonic oscillator, generate much of the work on solitons. Their attempt to demonstrate that the nonlinear interactions between the normal modes of vibrations lead to the energy of the system being evenly distributed throughout all the modes, as a result of the equipartition of energy, failed. The energy does not spread throughout all the modes but recollect after a time in the initial mode where it was when the experiment was started.

In 1965, Zabusky and Kruskal approached the Fermi, Pasta and Ulam problem from the continuum point of view. They rederived the Korteweg and de Vries equation and found its stable wave solutions by numerical computation. They showed that these solutions preserve their shape and velocities after two of them collide, interact and then spread apart again. They named such waves *solitons*.

Gardner, Green, Kruskal and Miura introduced in 1974 the Inverse Scattering Transform to integrate nonlinear evolution equations. The conserved features of solitons become intimately related to the notion of symmetry and to the construction of pseudospherical surfaces. The Gauss–Weingarten system for the pseudospherical surfaces yields sine-Gordon equation, providing a bridge to soliton theory.

A privileged surface related to the certain nonlinear equations that admit solitonic solutions, is the Tzitzeica surface (1910). Developments in the geometry of such surface gave a gradual clarification of predictable properties in natural phenomena.

A remarkable number of evolution equations (sine-Gordon, Korteweg de Vries, Boussinesq, Schrödinger and others) considered by the end of the 19th century, radically changed the thinking of scientists about the nature of nonlinearity. These equations admit solitonic behavior characterized by an infinite number of conservation laws and an infinite number of exact solutions.

In 1973, Wahlquist and Estabrook showed that these equations admit invariance under a Bäcklund transformation, and possess multi-soliton solutions expressed as simple superposition formulae relating explicit solutions among themselves.

The theory of soliton stores the information on some famous equations: the Korteweg de Vries equation, the nonlinear Schrödinger equation, the sine-Gordon equation, the Boussinesq equation, and others. This theory provides a fascinating glimpse into studying the nonlinear processes in which the combination of dispersion and nonlinearity together lead to the appearance of solitons.

This book addresses practical and concrete resolution methods of certain nonlinear equations of evolution, such as the motion of the thin elastic rod, vibrations of the initial deformed thin elastic rod, the coupled pendulum oscillations, dynamics of the left ventricle, transient flow of blood in arteries, the subharmonic waves generation in a piezoelectric plate with Cantor-like structure, and some problems of deformation in inhomogeneous media strongly related to Tzitzeica surfaces. George Tzitzeica is a great Romanian geometer (1873–1939), and the relation of his surfaces to the soliton theory and to certain nonlinear mechanical problems has a long history, owing its origin to geometric investigations carried out in the 19th century.

The present monograph is not a simple translation of its predecessor which appeared at the Publishing House of the Romanian Academy in 2002. Major improvements outline the way in which the soliton theory is applied to solve some engineering problems. In each chapter a different problem illustrates the common origin of the physical phenomenon: the existence of solitons in a solitonic medium.

The book requires as preliminaries only the mathematical knowledge acquired by a student in a technical university. It is addressed to both beginner and advanced practitioners interested in using the soliton theory in various topics of the physical, mechanical, earth and life sciences. We also hope it will induce students and engineers to read more difficult papers in this field, many of them given in the references.

Authors

PART 1

INTRODUCTION TO SOLITON THEORY

Chapter 1

MATHEMATICAL METHODS

1.1 Scope of the chapter

This chapter introduces the fundamental ideas underlying some mathematical methods to study a certain class of nonlinear partial differential equations known as evolution equations, which possess a special type of elementary solution. These solutions known as *solitons* have the form of localized waves that conserve their properties even after interaction among them, and then act somewhat like particles. These equations have interesting properties: an infinite number of local conserved quantities, an infinite number of exact solutions expressed in terms of the Jacobi elliptic functions (*cnoidal solutions*) or the hyperbolic functions (*solitonic solutions* or *solitons*), and the simple formulae for nonlinear superposition of explicit solutions. Such equations were considered *integrable* or more accurately, *exactly solvable*. Given an evolution equation, it is natural to ask whether it is integrable, or it admits the exact solutions or solitons, whether its solutions are stable or not. This question is still open, and efforts are made for collecting the main results concerning the analysis of nonlinear equations.

Substantial parts of this chapter are based on the monographs of Dodd *et al.* (1982), Lamb (1980), Drazin (1983), Drazin and Johnson (1989), Munteanu and Donescu (2002), Toma (1995) and on the articles of Hirota (1980) and Osborne (1995).

1.2 Scattering theory

Historically, the scattering theory was fairly well understood by about 1850. It took almost one hundred years before the inverse scattering theory could be applied.

Since 1951, various types of nonlinear equations with a soliton as a solution have been solved by direct and inverse scattering theories. However, given any evolution equation, it is natural to ask whether it can be solved in the context of the scattering theory. This question is related to the Painlevé property. We may say that a nonlinear partial differential equation is solvable by inverse scattering technique if, and only if, every ordinary differential equation derived from it, by exact reduction, satisfies the Painlevé property (Ablowitz *et al.*). The Painlevé property refers to the absence of movable critical points for an ordinary differential equation.

Let us begin with the equation known as a Schrödinger equation, of frequent occurrence in applied mathematics (Lamb)

$$\varphi_{xx} + [\lambda - u(x,t)]\varphi = 0\,, \tag{1.2.1}$$

where $\varphi : \mathrm{R} \to \mathrm{R}$ is a dimensionless scalar field in one space coordinate x. The potential function $u(x,t)$ contains a parameter t, that may be the temporal variable, $t \geq 0$. At this point, t is only a parameter, so that the shape of $u(x,t)$ varies from t. Subscripts that involve x or t are used to denote partial derivatives, for example $u_t = \dfrac{\partial u}{\partial t}$, $u_x = \dfrac{\partial u}{\partial x}$.

If the function u depends only on x, $a \leq x \leq b$, where a and b can be infinity, the equation (1.2.1) for imposed boundary conditions at $x = a$ and b, leads to certain values of the constant λ (the eigenvalues λ_j) for which the equation has a nonzero solution (the eigenfunctions $\varphi_j(x)$).

For a given function $u(x)$, the determination of the dependence of the solution φ on the parameter λ and the dependence of the eigenvalues λ_j on the boundary conditions is known as a Sturm-Liouville problem. The solutions of (1.2.1) exist only if the function $u(x)$ is integrable, that is $\int_a^b | u(x) | \,\mathrm{d}x < \infty$. The spectrum of eigenvalues λ_j is made up of two cases corresponding to $\lambda > 0$ and $\lambda < 0$. The case $\lambda = 0$ does not occur if $u(x) \neq 0$.

In particular, for $u(x) = -2\mathrm{sech}^2 x$, and the boundary conditions $\varphi(\pm\infty) = 0$ leads to the single eigenvalue $\lambda = -1$ with the associated eigenfunction $\varphi = \mathrm{sech}\, x$. The scattering solutions of (1.2.1) are made up of linear combinations of the functions $\varphi_1 = \exp(\mathrm{i}\sqrt{\lambda}\, x)(\mathrm{i}\sqrt{\lambda} - \tanh x)$, and $\varphi_2 = \exp(-\mathrm{i}\sqrt{\lambda}\, x)(\mathrm{i}\sqrt{\lambda} + \tanh x)$.

The solving of the Schrödinger equation (1.2.1) when the potential function $u(x)$ is specified is referred to as *the direct scattering problem*. If u depends on x and t, $u = u(x,t)$, then we expect the values of the λ_j to depend upon t. It is interesting to ask whether or not there are potential functions $u(x,t)$ for which the λ_j remain unchanged as the parameter t is varied.

In particular, if $u = u(x+t)$ satisfies the linear partial differential equation $u_x = u_t$, the variation of t has no effect upon the eigenvalues λ_j. Also, the eigenvalues are invariant to the variation of t, if $u(x,t)$ satisfies the nonlinear partial differential equation

$$u_t + u u_x + u_{xxx} = 0\,, \tag{1.2.2}$$

known as the Korteweg–de Vries equation (KdV).

Therefore, solving the KdV equation is related to finding the potentials in a Sturm-Liouville equation, and vice versa.

The direct scattering problem is concerned with determining of a wave function φ when the potential u is specified. Determination of a potential u from information about the wave function φ is referred to as *the inverse scattering problem.*

THEOREM 1.2.1 *Let S be a pre-hilbertian space of functions $y : \mathrm{R}^2 \to \mathrm{R}$. Let us consider the operators $L : S \to S$, $B : S \to S$ having the properties:*

a) $\langle Ly_1, y_2 \rangle = \langle y_1, Ly_2 \rangle$, $\forall y_1, y_2 \in S$.

b) *L admits only simple eigenvalues, namely $\lambda(t)$ is an eigenvalue for L if there exists the function $\Psi \in S$, so that*

$$\langle L, \Psi \rangle (x,t) = \lambda(t) \Psi(x,t) . \tag{1.2.3}$$

c) $\langle B, a(t)y \rangle = a(t) \langle B, y \rangle$, $\forall y \in S$, *and* $a(t)y \in S$.

It follows that the relations

$$L_t + LB - BL = 0 , \tag{1.2.4}$$

$$L_t : S \to S, \quad \langle L_t, y \rangle (x,t) = \langle L, y \rangle_{,t} (x,t) - \langle L, y_t \rangle (x,t) , \tag{1.2.5}$$

are verified. Also, it follows that

1. *the eigenvalues are constants*

$$\lambda(t) = \lambda \in \mathrm{R}, \quad \forall t \in \mathrm{R} , \tag{1.2.6}$$

2. *the eigenfunctions verify the evolution equation*

$$\Psi_t(x,t) = \langle B + I\alpha(t), \Psi \rangle (x,t) \quad \forall x \in \mathrm{R} , \ \forall t \in \mathrm{R} , \tag{1.2.7}$$

where α is an arbitrary function of t.

Proof. Let $\lambda(t)$ be an eigenvalue so that

$$\langle L, \Psi \rangle (x,t) = \lambda(t) \Psi(x,t) , \ \forall x \in \mathrm{R} .$$

We can write

$$\langle L_t, \Psi \rangle (x,t) + \langle L, \Psi_t \rangle (x,t) = \lambda_t(t) \Psi(x,t) + \lambda(t) \Psi_t(x,t) ,$$

$$\langle L_t, \Psi \rangle (x,t) + \langle L, \Psi_t \rangle (x,t) = \lambda_t(t) \Psi(x,t) + \lambda(t) \Psi_t(x,t) , \ \forall x,t \in \mathrm{R} .$$

From (1.2.4) and (1.2.5) it results

$$\lambda_t(t) \Psi(x,t) = \langle L - \lambda, -B\Psi + \Psi_t \rangle (x,t) ,$$

and multiplying to Ψ, we obtain

$$\langle \lambda_t(t) \Psi \cdot \Psi \rangle_S = \langle \langle L - \lambda, -B\Psi + \Psi_t \rangle \cdot \Psi \rangle_S = \langle \langle L - \lambda, \Psi \rangle \cdot (-B\Psi + \Psi_t) \rangle_S = 0.$$

This implies $\lambda_t = 0$. The function $-B\Psi + \Psi_t$ is also an eigenfunction for L, corresponding to non time-dependent λ. Therefore, there exists an arbitrary function $\alpha(t)$ so that

$$-B\Psi + \Psi_t = \alpha(t)\Psi .$$

Considering a new dependent variable

$$\Psi(x,t) = \varphi(x,t)\exp\left(\int\alpha(t)\mathrm{dt}\right),$$

the equation (1.2.7) yields

$$\varphi_t = \langle B, \varphi\rangle . \tag{1.2.8}$$

To illustrate this, let us consider the example

$$\{y : \mathrm{R}\times\mathrm{R}_+ \to \mathrm{R},\ \ y, y_x \to 0,\ |x| \to \infty\} ,$$

with the scalar product

$$\langle y_1, y_2\rangle_S = \int_{-\infty}^{\infty} y_1(x,t) y_2(x,t)\mathrm{dx} ,$$

and operators $L, B : S \to S$

$$\langle L, y\rangle = -y_{xx} + u(x,t)y , \tag{1.2.9}$$

$$\langle B, y\rangle = -4y_{xxx} + 6u(x,t)y_x + 3u_x(x,t)y . \tag{1.2.10}$$

According to

$$\langle L_t, y\rangle = u_t(x,t)y ,$$

$$\langle L, By\rangle = 4y_{xxxxx} - 10uy_{xxx} - 15u_x y_{xx} - 12u_{xx}y_x + 6u^2 y_x - 3u_{xxx}y + 3uu_x y,$$

$$\langle B, Ly\rangle = 4y_{xxxxx} - 10uy_{xxx} - 15u_x y_{xx} - 12u_{xx}y_x + 6u^2 y_x - 4u_{xxx}y + 9uu_x y,$$

it is found that (1.2.3) can be written under the standard form of the KdV equation

$$u_t + u_{xxx} - 6uu_x = 0 .$$

The operators satisfy the properties mentioned in the theorem 1.2.1. For (1.2.5) we find

$$\Psi_{xx} + (\lambda - u)\Psi = 0 ,$$

and the eigenfunction Ψ corresponding to λ verifies

$$\Psi_t = -4\Psi_{xxx} + 6u\Psi_x + 3u_x\Psi + \alpha(t)\Psi .$$

The new dependent variable, $\Psi(x,t) = \varphi(x,t)\exp\left(\int\alpha(t)\mathrm{d}t\right)$, states that φ is also an eigenfunction corresponding to λ

$$\varphi_{xx} + (\lambda - u)\varphi = 0, \tag{1.2.11}$$

which verifies the equation

$$\varphi_t = -4\varphi_{xxx} + 6u\varphi_x + 3u_x\varphi. \tag{1.2.12}$$

Consequently, finding solutions to the KdV equation is related to solving the Schrödinger equation

$$\varphi_{xx} + [k^2 - u(x,t)]\varphi = 0. \tag{1.2.13}$$

Note that in (1.2.13), t is playing the role of a parameter, k is a real or a pure complex number ik, $k > 0$, and the potential function u has the property $u \to 0$, $|x| \to \infty$.

For localized potentials $u(x)$, all solutions of (1.2.13) will reduce to a linear combination of the functions $\exp(ikx)$, and $\exp(-ikx)$ as $|x| \to \infty$.

Following Faddeev (1967), the solutions of the Schrödinger equation are expressed as linear combinations of a solution $f_1(x,k)$ that reduces to $\exp(ikx)$, as $x \to \infty$, and a solution $f_2(x,k)$ that reduces to $\exp(-ikx)$, as $x \to -\infty$.

By definition, $f_1(x,k)$ and $f_2(x,k)$ are fundamental solutions of (1.2.13) and are exact solutions of (1.2.13) and verify

$$f_1(x,k) \approx \exp(ikx), \quad x \to \infty,$$

$$f_1(x,k)\exp(-ikx) \to 1, \quad |x| \to \infty, \tag{1.2.14a}$$

$$f_2(x,k) \approx \exp(-ikx), \quad x \to -\infty,$$

$$f_2(x,k)\exp(ikx) \to 1, \quad |x| \to -\infty, \tag{1.2.14b}$$

THEOREM 1.2.2 *Fundamental solutions* $f_1(x,k)$ *and* $f_2(x,k)$ *verify the equations*

$$f_1(x,k) = \exp(ikx) - \frac{1}{k}\int_x^{\infty} \sin k(x-\alpha) u(\alpha) f_1(\alpha,k)\,\mathrm{d}\alpha, \tag{1.2.15a}$$

$$f_2(x,k) = \exp(-ikx) - \frac{1}{k}\int_{-\infty}^{x} \sin k(x-\alpha) u(\alpha) f_1(\alpha,k)\,\mathrm{d}\alpha. \tag{1.2.15b}$$

Proof. The homogeneous equation associated to (1.2.13), $\varphi'' + k^2\varphi = 0$, admits the solutions $\varphi(x) = A(x)\exp(ikx) + B(x)\exp(-ikx)$, with A, B arbitrary constants.

By applying the method of variation of constants, we obtain

$$A'(x) = \frac{1}{2ik} u(x)\varphi(x)\exp(-ikx),$$

$$B'(x) = -\frac{1}{2ik} u(x)\varphi(x)\exp(ikx),$$

and then, by integration, we have

$$A(x) = \frac{1}{2ik}\int_0^x u(\alpha)\varphi(\alpha)\exp(-ik\alpha)d\alpha + C_1,$$

$$B(x) = -\frac{1}{2ik}\int_0^x u(\alpha)\varphi(\alpha)\exp(ik\alpha)d\alpha + C_2.$$

The constants C_1, C_2 are found from (1.2.14a)

$$C_1 = 1 - \frac{1}{2ik}\int_0^\infty u(\alpha)\varphi(\alpha)\exp(-ik\alpha)d\alpha,$$

$$C_2 = \frac{1}{2ik}\int_0^\infty u(\alpha)\varphi(\alpha)\exp(ik\alpha)d\alpha.$$

Substitution of these expressions into $\varphi(x)$, leads to

$$\varphi(x,k) = \exp(ikx) - \frac{1}{2ik}\int_x^\infty u(\alpha)\varphi(\alpha)\exp ik(x-\alpha)d\alpha +$$

$$+\frac{1}{2ik}\int_x^\infty u(\alpha)\varphi(\alpha)\exp ik(\alpha - x)d\alpha.$$

The function f_2 is derived in an analogous manner.

Equations (1.2.15) are the Volterra integral equations, which can be solved by an iteration procedure. More specifically, the substitution of $\exp(ikx)$ into (1.2.15a) yields to the conclusion that the resulting integrals converge for $\mathrm{Im}(k) > 0$.

For integral equations of Volterra, the resulting series expansion is always convergent. Hence, the functions f_1, f_2 are analytic in the upper half of the complex k plane. For real $u(x)$ and k, we have $f_i(x,-k) = f_i^*(x,k)$, $i = 1,2$, where "$*$" is the complex conjugate operator.

From (1.2.15) we see that the functions $f_1(x,k)$, $f_1(x,-k)$ are independent. The functions $f_2(x,k)$, $f_2(x,-k)$ are also independent. So, there exist the coefficients $c_{ij}(k)$, $i,j = 1,2$, depending on *k*, so that

$$f_2(x,k) = c_{11}(k)f_1(x,k) + c_{12}(k)f_1(x,-k), \tag{1.2.16a}$$

$$f_1(x,k) = c_{21}(k)f_2(x,k) + c_{22}(k)f_2(x,-k). \tag{1.2.16b}$$

From the limiting form of $f_i(x,\pm k)$, $i = 1,2$, we may write

$$f_2(x,k) \approx \exp(-ikx), \quad x \to -\infty,$$

$$f_2(x,k) \approx c_{11}(k)\exp(ikx) + c_{12}(k)\exp(-ikx), \quad x \to \infty,$$

which means the solution corresponds to a scattering problem in which the incident wave is coming from ∞ with an amplitude $c_{12}(k)$, and is reflected with an amplitude $c_{11}(k)$, and transmitted to $-\infty$ with an amplitude of unity.

In particular, the fundamental solutions for the potential $u(x) = -2\text{sech}^2 x$, are obtained by solving the equation

$$z'' = (k^2 - 2\text{sech}^2 x)z, \quad -\infty < x < \infty.$$

By using the substitution $y = \tanh x$, $-1 < x < 1$, we obtain the associated Legendre equation (Drazin and Johnson)

$$\frac{\mathrm{d}}{\mathrm{d}y}[(1-y^2)\frac{\mathrm{d}z}{\mathrm{d}y}] + (-2 + \frac{k^2}{1-y^2})z = 0,$$

whose general solution is given by

$$z = A\exp(kx)(k-y) - B\exp(-kx)(k+y).$$

From here we obtain the fundamental solutions

$$f_1(x,k) = \frac{1}{ik-1}\exp(ikx)(ik - \tanh x),$$

$$f_2(x,k) = \frac{1}{ik-1}\exp(-ikx)(ik + \tanh x).$$

Let us introduce the reflection and transmission coefficients for an incident wave of unit amplitude (Achenbach). The ratio

$$R_R(k) = \frac{c_{11}(k)}{c_{12}(k)}, \tag{1.2.17}$$

is the reflection coefficient at ∞, and the ratio

$$T_R(k) = \frac{1}{c_{12}(k)}, \tag{1.2.18}$$

is the transmission coefficient at ∞. The subscript R refers to a wave incident from the right. Similarly, we have

$$f_1(x,k) \approx \exp(ikx), \quad x \to \infty,$$

$$f_1(x,k) \approx c_{21}(k)\exp(ikx) + c_{22}(k)\exp(-ikx), \quad x \to -\infty,$$

which means the incident wave from $-\infty$ with an amplitude $c_{21}(k)$ is reflected with an amplitude $c_{22}(k)$ and transmitted to ∞ with an amplitude of unity.

The ratio

$$R_L(k) = \frac{c_{22}(k)}{c_{21}(k)}, \tag{1.2.19}$$

is the reflection coefficient at $-\infty$, and the ratio

$$T_L(k) = \frac{1}{c_{21}(k)}, \tag{1.2.20}$$

is the transmission coefficient al $-\infty$.

The Wronskian of any two functions φ_1 and φ_2, is defined as

$$w[\varphi_1(x), \varphi_2(x)] = \varphi_{1,x}(x)\varphi_2(x) - \varphi_1(x)\varphi_{2,x}(x). \tag{1.2.21}$$

If φ_1 and φ_2 are two linearly independent solutions of (1.2.13), then their Wronskian is a constant

$$w[\varphi_1(x;k), \varphi_2(x;k)] = f(k), \ \forall k \in \mathrm{C}. \tag{1.2.22}$$

The relation (1.2.22) results by adding (1.2.14) written for φ_1 and multiplied by φ_2, to (1.2.13) written for φ_2 and multiplied by $-\varphi_1$. It results

$$\frac{\mathrm{d}}{\mathrm{d}x} w[\varphi_1, \varphi_2] = 0.$$

According to definition of the Wronskian and (1.2.22), the following properties hold

$$w[f_1(x;k), f_1(x;-k)] = 2\mathrm{i}k, \ w[f_2(x;k), f_2(x;-k)] = -2\mathrm{i}k, \tag{1.2.23}$$

where f_1, f_2 are fundamental solutions (1.2.15).

Substituting $f_1(x;k)$ from (1.2.16b) into (1.2.16a), and substituting $f_2(x;k)$ from (1.2.16a) into (1.2.16b), $\forall k \in \mathrm{C}$, and taking account of the independency of $f_2(x;k), f_2(x,-k)$, the following relations are obtained

$$\begin{gathered} c_{11}(k)c_{22}(k) + c_{12}(k)c_{21}(-k) = 1, \\ c_{11}(k)c_{21}(k) + c_{12}(k)c_{22}(-k) = 0, \\ c_{21}(k)c_{12}(-k) + c_{22}(k)c_{11}(k) = 1, \\ c_{21}(k)c_{11}(-k) + c_{22}(k)c_{12}(k) = 0. \end{gathered} \tag{1.2.24}$$

The coefficients c_{ij} may be written in terms of the Wronskian

$$\begin{gathered} c_{11}(k) = \frac{1}{2\mathrm{i}k} w[f_2(x;k), f_1(x;-k)], \\ c_{22}(k) = \frac{1}{2\mathrm{i}k} w[f_2(x;-k), f_1(x;k)], \end{gathered} \tag{1.2.25}$$

$$c_{12}(k) = c_{21}(k) = \frac{1}{2ik} w[f_1(x;k), f_2(x;k)],$$

$\forall k \in C$. That yields $T_R(k) = T_L(k) = T(k)$. The relations (1.2.25) are obtained from (1.2.16) and (1.2.23). Furthermore, from $f_i\ (x;-k) = f_i^*(x;k),\ i = 1,2,$ we have

$$R_R(k)T(-k) + R_L(-k)T(k) = 0\,,\ \forall k \in C\,, \tag{1.2.26}$$

and

$$\begin{gathered} c_{12}(-k) = c_{12}{}^*(k), \\ c_{11}(k) = -c_{22}{}^*(k) = -c_{22}(-k), \\ |c_{12}(k)|^2 = 1 + |c_{11}(k)|^2 = 1 + |c_{22}(k)|^2, \\ |T(k)|^2 + |R_R(k)|^2 = |T(k)|^2 + |R_L(k)|^2 = 1, \\ R_L^*(-k) = R_L(k), \quad R_R^*(-k) = R_R(k), \end{gathered} \tag{1.2.27}$$

$\forall k \in R$. The location of the poles of the transmission and reflection coefficients in the upper half-plane are important to obtain information about the localized or bound-state solutions. Consider now the poles of $T(k)$.

THEOREM 1.2.3 *For real potential functions $u: R \to R$, any poles of the transmission coefficient in the upper half-plane must be on the imaginary axis. More precisely, if $k_0 \in C$ is a pole for $T(k)$, then $k_0 = i\kappa_0,\ \kappa_0 \in R_+$.*

Proof. Let $k_0 \in C$ be a pole for $T(k) = \dfrac{1}{c_{12}(k)}$. Then it is a zero for c_{12}, $c_{12}(k_0) = 0$. According to $(1.2.27)_3$, for $\forall k \in R$ we have $c_{12}(k) \neq 0$, and then $\operatorname{Im} k_0 > 0$.

Writing (1.2.13) for k_0

$$\varphi'' + (k_0^2 - u)\varphi = 0\,,$$

and similarly for k_0^*

$$\varphi^{*}{}'' + (k_0^{*2} - u)\varphi^* = 0\,,$$

and subtracting them, it follows that

$$\varphi''\varphi^* - \varphi\varphi^{*}{}'' = -(k_0^2 - k_0^{*2})\varphi\varphi^*.$$

Integrating then over x, from $-\infty$ to ∞, yields

$$w(\varphi,\varphi^*)\big|_{+\infty} - w(\varphi,\varphi^*)\big|_{-\infty} = -(k_0^2 - k_0^{*2}) \int_{-\infty}^{\infty} |\varphi|^2\, dx\,.$$

The Wronskian of φ, φ^* being a constant, it follows that

$$\operatorname{Re} k_0 \operatorname{Im} k_0 \int_{-\infty}^{\infty} |\varphi|^2 \, dx = 0,$$

and hence, from $\operatorname{Im} k_0 > 0$, it results $\operatorname{Re} k_0 = 0$.

When $c_{12}(k_0) = 0$, $k_0 \in C$, the fundamental solutions are linearly dependent, and then lead to

$$f_2(x, k_0) = c_{11}(k_0) f_1(x, k_0), \tag{1.2.28}$$

and

$$c_{22}(k_0) = \frac{1}{c_{11}(k_0)}. \tag{1.2.29}$$

This property results from (1.2.16a,b) written for $k = k_0$.

Next, we show that the value of the residuum of the function $T(k)$ in every pole $k_l = \mathrm{i}\kappa_l, \kappa_l > 0$ is given by

$$\operatorname{Res}(T(k))(k_l) = \frac{1}{\dot{c}_{12}(k_l)} = \frac{\mathrm{i}}{\int_{-\infty}^{\infty} f_1(x, k_l) f_2(x, k_l) \mathrm{d}x}. \tag{1.2.30}$$

To obtain this, let us differentiate with respect to k, the relation $(1.2.25)_3$ and set $k = k_l$. According to (1.2.28) and (1.2.29) we have

$$\left. \frac{\mathrm{d}}{\mathrm{d}k} c_{12}(k) \right|_{k=k_l} = \frac{1}{2\mathrm{i}k_l} c_{11}(k_l) w_1(x, k_l) + \frac{1}{2\mathrm{i}k_l} c_{22}(k_l) w_2(x, k_l), \tag{1.2.31}$$

where

$$w_i(x, k_l) = w[\left. \frac{\partial f_i}{\partial k}(x, k) \right|_{k=k_l}; f_i(x, k_l)], \quad i = 1, 2.$$

To obtain w_1, let us multiply (1.2.13) written for $f_1(x, k)$ with $f_1(x, k_l)$, then multiply (1.2.13) written for $f_1(x, k_l)$ with $f_1(x, k)$, and add the results. We have

$$\frac{\partial}{\partial x}[f_1(x, k) \frac{\partial}{\partial x} f_1(x, k_l) - f_1(x, k_l) \frac{\partial}{\partial x} f_1(x, k)] - (k^2 - k_l^2) f_1(x, k) f_1(x, k_l) = 0.$$

Differentiating the above relation with respect to $k \in \mathrm{R}$, we have

$$\frac{\partial}{\partial x} w[f_1(x, k_l); \left. \frac{\partial}{\partial k} f_1(x, k) \right|_{k=k_l}] = 2k_l (f_1(x, k_l))^2.$$

Integration from x to ∞, gives

$$A - w[f_1(x, k_l); \frac{\partial}{\partial k} f_1(x, k_l)] = 2k_l \int_x^{\infty} (f_1(\alpha, k_l))^2 \mathrm{d}\alpha,$$

$$A = \lim_{x\to\infty} w[f_1(x,k_l); \frac{\partial}{\partial k} f_1(x,k_l)] = 0 .$$

It follows that

$$w_1 = 2k_l \int_x^{\infty} [f_1(\alpha;k_l)]^2 \, \mathrm{d}\alpha .$$

In a similar way we obtain

$$w_2 = 2k_l \int_{-\infty}^{x} [f_2(\alpha;k_l)]^2 \, \mathrm{d}\alpha .$$

Substitution of w_i, $i = 1,2$, into (1.2.31) yields

$$\dot{c}_{12}(k)\big|_{k=k_l} = \frac{c_{11}(k_l)}{\mathrm{i}} \int_{-\infty}^{\infty} [f_1(\alpha;k_l)]^2 \, \mathrm{d}\alpha =$$

$$= \frac{1}{\mathrm{i}} \int_{-\infty}^{\infty} [f_1(\alpha;k_l)] f_2(\alpha;k_l) \mathrm{d}\alpha = \frac{c_{22}(k_l)}{\mathrm{i}} \int_{-\infty}^{\infty} [f_{21}(\alpha;k_l)]^2 \, \mathrm{d}\alpha.$$

Note that

$$\int_{-\infty}^{\infty} c_{11}(k_l)\gamma_l [f_1(x;k_l)]^2 \, \mathrm{d}x = 1 , \qquad (1.2.32a)$$

$$\int_{-\infty}^{\infty} c_{22}(k_l)\gamma_l [f_2(x;k_l)]^2 \, \mathrm{d}x = 1 . \qquad (1.2.32b)$$

Thus, the quantities $[\gamma_l c_{11}(k_l)]^{1/2}$ and $[\gamma_l c_{22}(k_l)]^{1/2}$ are the normalization constants for the bound-state wave functions $f_i(x,k_l)$, $i = 1,2$.

Using (1.2.30) we may write the normalization constants as

$$m_{Rl} = \gamma_l c_{11}(k_l) = -\mathrm{i}\frac{c_{11}(k_l)}{\dot{c}_{12}(k_l)} = \{ \int_{-\infty}^{\infty} [f_1(x;k_l)]^2 \mathrm{d}x \}^{-1} , \qquad (1.2.33a)$$

$$m_{Ll} = \gamma_l c_{22}(k_l) = -\mathrm{i}\frac{c_{22}(k_l)}{\dot{c}_{12}(k_l)} = \{ \int_{-\infty}^{\infty} [f_2(x;k_l)]^2 \mathrm{d}x \}^{-1} , \qquad (1.2.33b)$$

where $m_{Rl}, m_{Ll} \in \mathrm{R}$, due to the fact that $f_i(x,k_l) \in \mathrm{R}$, $k_l = \mathrm{i}\kappa_l$, $i = 1,2$.

Any poles of the transmission coefficient are simple because, if k_l is a pole for $T(k)$, $k_l = \mathrm{i}\kappa_l$, $\kappa_l > 0$, then it has the properties (1.2.32a), (1.2.33a), and it results

$$\dot{c}_{12}(k_l) = -\mathrm{i}c_{11}(k_l) \int_{-\infty}^{\infty} [f_1(x;k_l)]^2 \, \mathrm{d}x \neq 0 .$$

1.3 Inverse scattering theory

The inverse scattering theory was firstly considered to solve an inverse physical problem of finding the shape of a mechanical object, which vibrates, from the knowledge of the energy or amplitude at each frequencies (Drazin and Johnson).

In our terms, the methods consist in determination of the potential function u from given coefficients $c_{ij}(k)$, that relate the fundamental solutions of the equation

$$\varphi''+(k^2-u)\varphi=0\,. \tag{1.3.1}$$

The fundamental solutions of the Schrödinger equation may be written under the form

$$f_1(x,k)=\exp(\mathrm{i}kx)+\int_x^\infty A_R(x,x')\exp(\mathrm{i}kx')\mathrm{d}x'\,, \tag{1.3.2a}$$

$$f_2(x,k)=\exp(-\mathrm{i}kx)+\int_{-\infty}^x A_L(x,x')\exp(-\mathrm{i}kx')\mathrm{d}x'\,. \tag{1.3.2b}$$

Balanis considered these forms in 1972, by solving the elastically braced vibrating string equation

$$y_{xx}-y_{x'x'}-u(x)y=0\,, \tag{1.3.3}$$

for which the solutions are written as

$$y_1(x,x')=\delta(x'-x)+\theta(x'-x)A_R(x,x')\,, \tag{1.3.4a}$$

$$y_2(x,x')=\delta(x'+x)+\theta(x'+x)A_L(x,-x'), \tag{1.3.4b}$$

where δ is the Dirac function, θ is the Heaviside function, and A_R, A_L are functions that describe the scattering or wake. Applying the Fourier transform

$$F[y(x,x')](x,k)=\int_{-\infty}^\infty y(x,x')\exp(\mathrm{i}kx')\mathrm{d}x'\,,$$

to (1.3.3), we find (Lamb)

$$F[y(x,x')]_{xx}+(k^2-u(x))F[y(x,x')]=0\,. \tag{1.3.5}$$

The equation (1.3.5) admits as solutions the Fourier transform of (1.3.4)

$$F[y_1(x,x')](x,k)=\exp(\mathrm{i}kx)+\int_x^\infty A_R(x,x')\exp(\mathrm{i}kx')\mathrm{d}x'\,, \tag{1.3.6a}$$

$$F[y_2(x,x')](x,k)=\exp(-\mathrm{i}kx)+\int_{-\infty}^x A_L(x,-x')\exp(-\mathrm{i}kx')\mathrm{d}x'\,. \tag{1.3.6b}$$

From (1.3.5), we see that the solutions of (1.3.1) take the form (1.3.6)

$$f_i(x,k)=F[y_i(x,x')](x,k),\quad i=1,2\,.$$

Substituting f_1, f_2 into (1.3.1) we derive the conditions to be verified by A_R, A_L. For this, we write f_1 as

$$f_1(x,k) = \exp(\mathrm{i}kx)[1 + \frac{\mathrm{i}}{k} A_R(x,x') - \frac{1}{k^2}\frac{\partial A_R}{\partial x'}(x,x')]_{x'=x} -$$
$$-\frac{1}{k^2}\int_x^\infty \frac{\partial^2 A_R}{\partial x'^2}(x,x')\exp(\mathrm{i}kx')\mathrm{d}x',$$

then integrate it by parts and introduce into (1.3.1). By imposing the conditions

$$A_R(x,x'),\ A_{Rx'}(x,x') \to 0, \text{ for } x' \to \infty,$$

we find

$$-\exp(\mathrm{i}kx)[2\frac{\mathrm{d}A_R}{\mathrm{d}x}(x,x) + u(x)] +$$
$$+\int_x^\infty [\frac{\partial^2 A_R}{\partial x^2}(x,x') - \frac{\partial^2 A_R}{\partial x'^2}(x,x') - u(x)A_R(x,x')]\exp(\mathrm{i}kx')\mathrm{d}x' = 0.$$

Therefore, the equation (1.3.1) is verified for

$$u(x) = -2\frac{\mathrm{d}A_R}{\mathrm{d}x}(x,x'),$$

$$A_R(x,x') = 0, \quad x' < x, \tag{1.3.7a}$$

$$\frac{\partial^2 A_R}{\partial x^2}(x,x') - \frac{\partial^2 A_R}{\partial x'^2}(x,x') - u(x)A_R(x,x') = 0, \quad x' > x.$$

Similarly, f_2 verifies the equation (1.3.1) for

$$u(x) = 2\frac{\mathrm{d}A_L}{\mathrm{d}x}(x,x'),$$

$$A_L(x,x') = 0, \quad x' > x, \tag{1.3.7b}$$

$$\frac{\partial^2 A_L}{\partial x^2}(x,x') - \frac{\partial^2 A_L}{\partial x'^2}(x,x') - u(x)A_L(x,x') = 0, \quad x' < x,$$

and the Faddeev condition is verified (Faddeev 1958)

$$\int_{-\infty}^{\infty}(1+|x|)u(x)\mathrm{d}x < \infty.$$

From (1.3.7) we see that for given A_R, A_L, we can find the potential function u.

Next, we try to determine the functions A_R, A_L in terms of the coefficients, $c_{ij}(k),\ i,j = 1,2$, considered specified. For this we write (1.2.14a) under the form

$$T(k)f_2(x,k) = R_R(k)f_1(x,k) + f_1(x,-k)\,, \tag{1.3.8}$$

and derive the corresponding relation in the time domain. Taking the Fourier transform does this

$$F^{-1}[f(x,k)](x,x') = \frac{1}{2\pi}\int_{-\infty}^{\infty} f_1(x,k)\exp(-ikx')\mathrm{d}k\,.$$

By noting

$$\Gamma(z) = \frac{1}{2\pi}\int_{-\infty}^{\infty}[T(k)-1]\exp(-ikz)\mathrm{d}k\,, \tag{1.3.9}$$

the Fourier transform on the left-hand side of (1.3.8) yields

$$F^{-1}[T(k)f_2(x,k)](x,x') = \frac{1}{2\pi}\int_{-\infty}^{\infty}(1+\int_{-\infty}^{\infty}\Gamma(z)\exp(ikz)\mathrm{d}z)\times$$

$$\times(\int_{-\infty}^{\infty} y_2(x,x'')\exp(ikx'')\mathrm{d}x'')\exp(-ikx')\mathrm{d}k =$$

$$= y_2(x,x') + \frac{1}{2\pi}\iiint_{-\infty,\infty}\Gamma(z)y_2(x,x'')\exp[ik(z+x''-x')]\mathrm{d}k\mathrm{d}x''\mathrm{d}z =$$

$$= \delta(x'+x) + \theta(x'+x)A_L(x,-x') +$$

$$+\Gamma(x+x') + \int_{-x}^{\infty}\Gamma(x'-x'')A_L(x,-x'')\mathrm{d}x''. \tag{1.3.10}$$

In a similar way, by noting

$$r_R(z) = \frac{1}{2\pi}\int_{-\infty}^{\infty} R_R(k)\exp(ikz)\mathrm{d}k\,, \tag{1.3.11}$$

the Fourier transform on the right-hand side of (1.3.8) leads to

$$F^{-1}[R_R(k)f_1(x,k) + f_1(x,-k)](x,x') =$$

$$= \frac{1}{2\pi}\int_{-\infty}^{\infty} R_R(k)f_1(x,k)\exp(-ikx')\mathrm{d}k + y_1(x,-x') =$$

$$= \frac{1}{2\pi}\int_{-\infty}^{\infty}(\int_{-\infty}^{\infty} r_R(z)\exp(ikz)\mathrm{d}z)(\int_{-\infty}^{\infty} y_1\exp(ikx'')\mathrm{d}x'')\exp(-ikx')\mathrm{d}k + y_1(x,-x') =$$

$$= \iint_{-\infty,\infty} r_R(z)y_1(x,x'')\{\int_{-\infty}^{\infty}\frac{1}{2\pi}\exp[ik(-z+x''-x')]\mathrm{d}k\}\mathrm{d}x''\mathrm{d}z + y_1(x,-x') =$$

$$= \int_{-\infty}^{\infty} r_R(x''-x')y_1(x,x'')\mathrm{d}x'' + y_1(x,-x') =$$

$$= r_R(x-x') + \int_x^{\infty} r_R(x''-x')A_R(x,x'')dx'' +$$

$$+\delta(x'+x) + \theta(-x'-x)A_R(x,-x'). \qquad (1.3.12)$$

Substituting (1.3.10) and (1.3.12) into (1.3.8) we have

$$\theta(x'+x)A_L(x,-x') + \Gamma(x+x') + \int_{-x}^{\infty} \Gamma(x'-x'')A_L(x,-x'')dx'' =$$
$$= r_R(x-x') + \int_x^{\infty} r_R(x''-x')A_R(x,x'')dx'' + \theta(-x'-x)A_R(x,-x'). \qquad (1.3.13)$$

We study now the case $x+x'<0$. To interpret (1.3.13) we must evaluate the function $\Gamma(z)$, $z<0$.

Case 1. When the transmission coefficient $T(k)$ possesses neither poles nor zeros in the upper half-plane, then

$$\Gamma(z) = 0. \qquad (1.3.14)$$

To show this, we consider the closed contour in the complex plane $C = C_R \cup [-R,R]$ where C_R a semicircle of radius R.

According to the Cauchy theorem we have

$$\int_C [T(k)-1]\exp(-ikz)dk = 0,$$

due to the fact that the integrant is an olomorphic function in the simple convex domain enclosed by the contour C.

Then, we can write

$$\int_{C_R} [T(k)-1]\exp(-ikz)dk = -\int_{-R}^{R} [T(k)-1]\exp(-ikz)dk. \qquad (1.3.15)$$

According to the Jordan lemma, if C_R is a semi-circle in the upper half-plane, centered in zero and having the radius R, and the function $G(k)$ satisfies the condition $G(k)\to 0$, $k\to\infty$, in the upper half-plane and on the real axis, and m is a positive real number, then we have

$$\int_{C_R} G(k)\exp(ikm)dk \to 0, \quad R\to\infty.$$

Here $m=-z\in \mathrm{R}_+$, $G(k)=T(k)-1$. If $T(k)\to 1$, $k\to\infty$, we may write

$$\int_{C_R} [T(k)-1]\exp(-ikz)dk \to 0, \quad R\to\infty.$$

From (1.3.15) it results

$$\Gamma(z) = \lim_{R\to\infty} \int_{-R}^{R} [T(k)-1]\exp(-\mathrm{i}kz)\mathrm{d}k = 0\,.$$

When $T(k)$ contains no poles in the upper half-plane, the equation (1.3.13) becomes

$$r_R(x-x') + \int_x^{\infty} r_R(x''-x')A_R(x,x'')\mathrm{d}x'' + A_R(x,-x') = 0, \quad x+x'<0\,,$$

or, denoting $-x'=y$

$$r_R(x+y) + \int_x^{\infty} r_R(x''+y)A_R(x,x'')\mathrm{d}x'' + A_R(x,y) = 0, \quad x<y\,. \tag{1.3.16a}$$

In a similar way we obtain

$$r_L(x+y) + \int_{-\infty}^{x} r_L(x''+y)A_L(x,x'')\mathrm{d}x'' + A_L(x,y) = 0, \quad x>y\,, \tag{1.3.16b}$$

where

$$r_L(z) = \frac{1}{2\pi}\int_{-\infty}^{\infty} R_L(k)\exp(-\mathrm{i}kz)\mathrm{d}k\,.$$

Case 2. If $T(k)$ contains first-order zeros or poles in the upper half-plane, then they are situated on the imaginary axis $k_l = \mathrm{i}\kappa_l,\ \ \kappa_l>0,\ \ l=1,2,...,n$.

From the residuum theorem we find

$$\Gamma(z) = \int_{-\infty}^{\infty} \frac{1}{2\pi}[T(k)-1]\exp(-\mathrm{i}kz)\mathrm{d}k = 2\pi\mathrm{i}\sum_{l=1}^{n}\frac{\exp(-\mathrm{i}k_l z)}{2\pi}\mathrm{Rez}[T(k_l)] =$$

$$= \mathrm{i}\sum_{l=1}^{n}\exp(k_l z)\mathrm{i}\gamma_l = -\sum_{l=1}^{n}\exp(k_l z)\gamma_l, \tag{1.3.17}$$

where γ_l is given by (1.2.31).

The relation (1.3.13) is written as

$$\begin{aligned} -\sum_{l=1}^{n}\exp[k_l(x+x')]\gamma_l - \int_{-x}^{\infty}\sum_{l=1}^{n}\exp[k_l(x-x'')]\gamma_l A_L(x,-x'')\mathrm{d}x'' = \\ = r_R(x-x') + \int_x^{\infty} r_R(x''-x')A_R(x,x'')\mathrm{d}x'' + A_R(x,-x'), \end{aligned} \tag{1.3.18}$$

for $x+x'<0$.

In terms of $-x'=y$, and taking into consideration that

$$f_2(x,\mathrm{i}\kappa_l)=c_{11}(\mathrm{i}\kappa_l)f_1(x,\mathrm{i}\kappa_l)=c_{11}(\mathrm{i}\kappa_l)[\exp(-\kappa_l x)+\int_x^\infty A_R(x,x')\exp(-\kappa_l x'')\mathrm{d}x''],$$

the equation (1.3.18) becomes

$$\Omega_R(x+y)+\int_x^\infty \Omega_R(x''+y)A_R(x,x'')\mathrm{d}x''+A_R(x,y)=0\,,\ x<y\,. \qquad (1.3.19a)$$

Similarly, we obtain

$$\Omega_L(x+y)+\int_{-\infty}^x \Omega_L(x''+y)A_R(x,x'')\mathrm{d}x''+A_L(x,y)=0, \quad x>y\,. \qquad (1.3.19b)$$

The functions $\Omega_R(z)$ and $\Omega_L(z)$ are defined as

$$\Omega_R(z)=r_R(z)+\sum_{l=1}^n \gamma_l c_{11}(\mathrm{i}\kappa_l)\exp(-k_l z) \ = \qquad (1.3.20a)$$
$$=\frac{1}{2\pi}\int_{-\infty}^\infty \frac{c_{11}(k)}{c_{12}(k)}\exp(ikz)\mathrm{d}k+\mathrm{i}\sum_{l=1}^n m_{Rl}(\mathrm{i}\kappa_l)\exp(-\kappa_l z),$$

$$\Omega_L(z)=\frac{1}{2\pi}\int_{-\infty}^\infty \frac{c_{22}(k)}{c_{12}(k)}\exp(-ikz)\mathrm{d}k+\mathrm{i}\sum_{l=1}^n m_{Ll}(\mathrm{i}\kappa_l)\exp(\kappa_l z)\,, \qquad (1.3.20b)$$

with m_{Rl} and m_{Ll} given by (1.2.31)

$$m_{Rl}(\mathrm{i}\kappa_l)=-\mathrm{i}\frac{c_{11}(\mathrm{i}\kappa_l)}{\dot{c}_{12}(\mathrm{i}\kappa_l)}\,,\ m_{Ll}(\mathrm{i}\kappa_l)=-\mathrm{i}\frac{c_{22}(\mathrm{i}\kappa_l)}{\dot{c}_{12}(\mathrm{i}\kappa_l)}\,.$$

In this case we have obtained the same integral equations (1.3.19) as in the first case, with the difference that r_R is replaced to Ω_R.

These equations are known as Marchenko equations (Agranovich and Marchenko), and they can be used to determine A_R or A_L when one of the reflection coefficients r_R or r_L is specified.

Solutions of Marchenko equations are the functions A_R, A_L, which allow the determination of the potential function u. The Marchenko equations are also used to determine the reflection coefficients when the potential and hence the fundamental solutions and the functions A_R or A_L are specified.

We can say that determination of R_R, R_L and r_R, r_L, is made from (1.2.15) and (1.2.17), and determination of Ω_R, Ω_L from (1.3.20).

1.4 Cnoidal method

The inverse scattering theory generally solves certain nonlinear differential equations, which have cnoidal solutions. The mathematical and physical structure of the inverse scattering transform solutions has been extensively studied in both one and two dimensions (Osborne, Drazin and Johnson, Ablowitz and Segur, Ablowitz and Clarkson). The theta-function representation of the solutions is describable as a linear superposition of Jacobi elliptic functions (*cnoidal functions*) and additional terms, which include nonlinear interactions among them.

Osborne is suggesting that the method is reducible to a generalization of the Fourier series with the cnoidal functions as the fundamental basis function. This is because the cnoidal functions are much richer than the trigonometric or hyperbolic functions, that is, the modulus m of the cnoidal function, $0 \le m \le 1$, can be varied to obtain a sine or cosine function $(m \cong 0)$, a Stokes function $(m \cong 0.5)$ or a solitonic function, sech or tanh $(m \cong 1)$ (Nettel).

Since the original paper by Korteweg and DeVries, it remains an open question (Ablowitz and Segur): "*if the KdV linearised equation can be solved by an ordinary Fourier series as a linear superposition of sine waves, can the KdV equation itself be solved by a generalization of Fourier series which uses the cnoidal wave as the fundamental basis function?*"

This method requires brief information necessary to describe the cnoidal waves. The arc length of the ellipse is related to the integral

$$E(z) = \int_0^z \frac{\sqrt{(1-k^2x^2)}\mathrm{d}x}{\sqrt{(1-x^2)}},$$

with $0 < k < 1$. Another elliptical integral is given by

$$F(z) = \int_0^z \frac{\mathrm{d}x}{\sqrt{(1-x^2)(1-k^2x^2)}}.$$

The integrals $E(z)$ and $F(z)$ are Jacobi elliptic integrals of the first and the second kinds. Legendre is the first who works with these integrals, being followed by Abel (1802–1829) and Jacobi (1804–1851). Jacobi inspired by Gauss, discovered in 1820 that the inverse of $F(z)$ is an elliptical double-periodic integral

$$F^{-1}(\omega) = \mathrm{sn}(\omega).$$

Jacobi compares the integral

$$v = \int_0^\varphi \frac{\mathrm{d}\theta}{(1-m\sin^2\theta)^{1/2}}, \tag{1.4.1}$$

where $0 \le m \le 1$, to the elementary integral

$$w = \int_0^\varphi \frac{\mathrm{d}t}{(1-t^2)^{1/2}}, \tag{1.4.2}$$

and observed that (1.4.2) defines the inverse of the trigonometric function sin if we use the notations $t = \sin\theta$ and $\psi = \sin w$. He defines a new pair of inverse functions from (1.4.1)

$$\mathrm{sn}\, v = \sin\varphi, \quad \mathrm{cn}\, v = \cos\varphi. \tag{1.4.3}$$

These are two of the Jacobi elliptic functions, usually written $\mathrm{sn}(v,m)$ and $\mathrm{cn}(v,m)$ to denote the dependence on the parameter m. The angle φ is called the amplitude $\varphi = \mathrm{am}\, u$. We also define the Jacobi elliptic function $\mathrm{dn} v = (1 - m\sin^2\varphi)^{1/2}$.

For $m = 0$, we have

$$v = \varphi\,,\ \mathrm{cn}(v,0) = \cos\varphi = \cos v\,,$$

$$v = \varphi\ \ \mathrm{sn}(v,0) = \sin\varphi = \sin v\,,\ \mathrm{dn}(v,0) = 1\,, \tag{1.4.4}$$

and for $m = 1$

$$v = \mathrm{arcsech}(\cos\varphi)\,,\ \mathrm{cn}(v,1) = \mathrm{sech} v\,,$$

$$\mathrm{sn}(v,1) = \tanh v\,,\ \mathrm{dn}(v,1) = \mathrm{sech} v\,. \tag{1.4.5}$$

The functions $\mathrm{sn}\, v$ and $\mathrm{cn}\, v$ are periodic functions with the period

$$\int_0^{2\pi} \frac{d\theta}{(1 - m\sin^2\theta)^{1/2}} = 4\int_0^{\pi/2} \frac{d\theta}{(1 - m\sin^2\theta)^{1/2}}.$$

The later integral is the complete elliptic integral of the first kind

$$K(m) = \int_0^{\pi/2} \frac{d\theta}{(1 - m\sin^2\theta)^{1/2}}\,. \tag{1.4.6}$$

The period of the function $\mathrm{dn}\, v$ is $2K$. For $m = 0$ we have $K(0) = \pi/2$. For increasing of m, $K(m)$ increases monotonically

$$K(m) \approx \frac{1}{2}\log\frac{16}{1-m}\,.$$

Thus, this periodicity of $\mathrm{sn}(v,1)$ and $\mathrm{cn}(v,1) = \mathrm{sech} v$ is lost for $m = 1$, so $K(m) \to \infty$.

Some important algebraic and differential relations between the cnoidal functions are given below

$$\mathrm{cn}^2 + \mathrm{sn}^2 = 1\,,\ \mathrm{dn}^2 + m\,\mathrm{sn}^2 = 1\,,\ \frac{d}{dv}\mathrm{cn} = -\mathrm{sn}\,\mathrm{dn}\,,$$

$$\frac{d}{dv}\mathrm{sn} = \mathrm{cn}\,\mathrm{dn}\,,\ \frac{d}{dv}\mathrm{dn} = -m\,\mathrm{sn}\,\mathrm{cn}\,, \tag{1.4.7}$$

where the argument v and parameter m are the same throughout relations.

Now, consider the function $\wp(t)$ introduced by Weierstrass (1815–1897) in 1850, which verifies the equation

$$\dot{\wp}^2 = 4\wp^3 - g_2\wp - g_3\,, \tag{1.4.8}$$

where the superimposed point means differentiation with respect to t.

If e_1, e_2, e_3 are real roots of the equation $4y^3 - g_2 y - g_3 = 0$ with $e_1 > e_2 > e_3$, then (1.4.8) can be written under the form

$$\dot{\wp}^2 = 4(\wp - e_1)(\wp - e_2)(\wp - e_3), \tag{1.4.9}$$

with

$$g_2 = 2(e_1^2 + e_2^2 + e_3^2),$$

$$g_3 = 4e_1 e_2 e_3, \quad e_1 + e_2 + e_3 = 0.$$

Introducing

$$\Delta = g_2^3 - 27g_3^2, \tag{1.4.10}$$

when $\Delta > 0$, equation (1.4.9) admits the elliptic Weierstrass function as a particular solution, which is reducing in this case to the Jacobi elliptic function cn

$$\wp(t + \delta'; g_2, g_3) = e_2 - (e_2 - e_3)\mathrm{cn}^2(\sqrt{e_1 - e_3}t + \delta'), \tag{1.4.11}$$

where δ' is an arbitrary real constant.

If we impose initial conditions to (1.4.9)

$$\wp(0) = \theta_0, \quad \wp'(0) = \theta_{p0}, \tag{1.4.12}$$

then a linear superposition of cnoidal functions (1.4.11) is also a solution for (1.4.8)

$$\theta_{lin} = 2\sum_{k=0}^{n} \alpha_k \mathrm{cn}^2[\omega_k t; m_k], \tag{1.4.13}$$

where the angular frequencies ω_k, and amplitudes α_k depend on θ_0, θ_{p0}.

When $\Delta < 0$ the solution of (1.4.9) is

$$\wp = e_2 + H_2 \frac{1 + \mathrm{cn}(2t\sqrt{H_2} + \delta')}{1 - \mathrm{cn}(2t\sqrt{H_2} + \delta')},$$

with

$$m = \frac{1}{2} - \frac{3e_2}{4H_2}, \quad H_2 = 3e_2^2 - \frac{g_2}{4}.$$

When $\Delta = 0$, we have $e_1 = e_2 = c$, $e_3 = -2c$, and the solution of (1.4.9) is

$$\wp = c + \frac{3c}{\sinh^2(\sqrt{3c}t + \delta')}.$$

Since the calculation of the elliptic functions is very important for practical problems, in Chapter 10, the Shen-Ling method to construct a Weierstrass elliptic function from the solutions of the Van der Pol's equation is presented.

Consider now a generalized Weierstrass equation with a polynomial of n degree in $\theta(t)$

$$\dot{\theta}^2 = P_n(\theta)\,. \tag{1.4.14}$$

The functional form of solutions of (1.4.14) is determined by the zeros of the right-hand side polynomial.

For biquadratic polynomial, $n = 4$, we can have four real zeros, two real and two purely imaginary zeros, four purely imaginary zeros or four genuinely complex zeros. For $n = 5$ the functional form of solutions depends also on the zeros of the polynomial.

For all cases the solutions are expressed in terms of Jacobi elliptic functions, the hyperbolic and trigonometric functions.

In the following we present the cnoidal method. Osborne discussed this method for integrable nonlinear equations that have periodic boundary conditions, in particular for the KdV equation. In 2002, Munteanu and Donescu have extended this method to nonlinear partial differential equations that can be reduced to Weierstrass equations of the type (1.4.14).

We present the method in context of the KdV equation

$$\theta_t + c_0\theta_x + \alpha\theta\theta_x + \beta\theta_{xxx} = 0\,, \tag{1.4.15}$$

where c_0, α and β are constants.

For $\alpha = 0$, the linearized KdV equation is solved by the Fourier series. The solutions are expressed as a sum of sine waves using the linear dispersion relation $\omega = c_0 k - \beta k^3$.

The general solution to the KdV equation with periodic boundary conditions may be written in the terms of the *theta function* representation (Dubrovin *et al.*)

$$\theta(x,t) = \frac{2}{\lambda}\frac{\mathrm{d}^2}{\mathrm{d}x^2}\log\Theta_n(\eta_1,\eta_2,\ldots,\eta_n)\,, \tag{1.4.16}$$

where $\lambda = \alpha / 6\beta$, and Θ is the *theta function* defined as

$$\Theta_n(\eta_1,\eta_2,\ldots,\eta_n) = \sum_{M\in(-\infty,\infty)} \exp(\mathrm{i}\sum_{i=1}^{n} M_i\eta_i + \frac{1}{2}\sum_{i,j=1}^{n} M_i B_{ij} M_j)\,, \tag{1.4.17}$$

with n the number of degrees of freedom for a particular solution of the KdV equation, and

$$\eta_j = k_j x - \omega_j t + \phi_j\,,\ 1 \le j \le N\,. \tag{1.4.18}$$

In (1.4.18), k_j are the wave numbers, the ω_j are the frequencies and the ϕ_j are the phases. Let us introduce the vectors of wave numbers, frequencies and constant phases

$$k = [k_1,k_2,\ldots,k_n],\ \ \omega = [\omega_1,\omega_2,\ldots,\omega_n],$$

$$\phi = [\phi_1,\phi_2\ldots,\phi_n]\,,\ \eta = [\eta_1,\eta_2,\ldots,\eta_n]\,. \tag{1.4.19}$$

The vector η can be written as

$$\eta = kx - \omega t + \phi\,. \tag{1.4.20}$$

Also, we can write

$$M\eta = Kx - \Omega t + \Phi ,$$

$$M = [M_1, M_2, ..., M_n],$$

$$K = Mk , \ \Omega = M\omega , \Phi = M\phi .$$

The integer components in M are the integer indices in (1.4.17). The matrix B can be decomposed in a diagonal matrix D and an off-diagonal matrix O, that is

$$B = D + O . \tag{1.4.21}$$

THEOREM 1.4.1 (Osborne) *The solution* $\theta(x,t)$ *of KdV equation (1.4.15) can be written as*

$$\theta(x,t) = \frac{2}{\lambda}\frac{\partial^2}{\partial x^2}\log \Theta_n(\eta) = \theta_{lin}(\eta) + \theta_{int}(\eta) , \tag{1.4.22}$$

where θ_{lin} *represents a linear superposition of cnoidal waves*

$$\theta_{lin}(\eta) = \frac{2}{\lambda}\frac{\partial^2}{\partial x^2}\log G(\eta) , \tag{1.4.23}$$

$$G(\eta) = \sum_M \exp(iM\eta + \frac{1}{2}M^T DM) , \tag{1.4.24}$$

and θ_{int} *represents a nonlinear interaction among the cnoidal waves*

$$\theta_{\text{int}}(\eta) = 2\frac{\partial^2}{\partial t^2}\log(1 + \frac{F\ (\eta, C)}{G\ (\eta)}) , \tag{1.4.25}$$

$$F\ (\eta, C) = \sum_{M^\alpha} C\exp(iM\eta + \frac{1}{2}M^T DM) , \tag{1.4.26}$$

$$C = \exp(\frac{1}{2}M^T OM) - 1 . \tag{1.4.27}$$

Proof. The decomposition (1.4.22) result easily from (1.4.16), (1.4.17) and (1.4.21). Consider the case with no interactions ($O = 0$, and then $C = 0$). The function $G(\eta)$ yields

$$G(\eta) = \prod_{m=1}^{n} G_m(M_m\eta_m) , \tag{1.4.28}$$

where

$$G_m(\eta_m) = \sum_{M_m=-\infty}^{\infty} \exp(iM_m\eta_m + \frac{1}{2}M_m^2 D_{mm}) . \tag{1.4.29}$$

So, the linear term (1.4.23) becomes

$$\theta_{lin}(\eta) = \sum_{m=1}^{n} \frac{2}{\lambda} \frac{\partial^2}{\partial x^2} \log G_m(M_m \eta_m) =$$
$$= \sum_{m=1}^{n} \frac{2}{\lambda} \eta_m \mathrm{cn}^2 \{(K(m_m)/\pi)[x - C_m t, m_m]\}, \tag{1.4.30}$$

where $K(m)$ is the elliptic integral (1.4.6). The relation (1.4.30) provides the interpretation of the first term on the right-hand side of (1.4.22) as a linear superposition of cnoidal waves.

The solution (1.4.30) represents the cnoidal wave solution for the KdV equation. The moduli m_m and the phase speeds C_m, $m = 1, 2, ..., n$, are given by

$$m_m K^2(m_m) = 4\pi^2 U_n,$$

$$C_m = c_0 (1 + \frac{2\eta_m}{h} - 2k_m^2 h^2 \frac{K^2(m_m)}{3\pi^2}),$$

where $U_m = \dfrac{3\eta_m}{8k_m^3 h^3}$ is the Ursell number.

Consider now a nonlinear system of equations that govern the motion of a dynamical system

$$\frac{\mathrm{d}\theta_i}{\mathrm{d}t} = F_i(\theta_1, \theta_2, ..., \theta_n), \quad i = 1, ..., n, \quad n \geq 3, \tag{1.4.31}$$

with $x \in \mathrm{R}^n$, $t \in [0, T]$, $T \in \mathrm{R}$, where F may be of the form

$$F_i = \sum_{p=1}^{n} a_{ip} \theta_p + \sum_{p,q=1}^{n} b_{ipq} \theta_p \theta_q + \sum_{p,q,r=1}^{n} c_{ipqr} \theta_p \theta_q \theta_r +$$
$$+ \sum_{p,q,r,l=1}^{n} d_{ipqrl} \theta_p \theta_q \theta_r \theta_l + \sum_{p,q,r,l,m=1}^{n} e_{ipqrlm} \theta_p \theta_q \theta_r \theta_l \theta_m + ..., \tag{1.4.32}$$

with $i = 1, 2, ..., n$, and $a, b, c...$ constants.

The system of equations has the remarkable property that it can be reduced to Weierstrass equations of the type (1.4.14). In the following, we present the cnoidal method, suitable to be used for equations of the form (1.4.31). To simplify the presentation, let us omit the index i and note the solution by $\theta(t)$.

We introduce the function transformation

$$\theta = 2 \frac{\mathrm{d}^2}{\mathrm{d}t^2} \log \Theta_n(t), \tag{1.4.33}$$

where the theta function $\Theta_n(t)$ are defined as

$$\Theta_1 = 1 + \exp(\mathrm{i}\omega_1 t + B_{11}),$$

$$\Theta_2 = 1 + \exp(\mathrm{i}\omega_1 t + B_{11}) + \exp(\mathrm{i}\omega_2 t + B_{22}) + \exp(\omega_1 + \omega_2 + B_{12}),$$

$$\begin{aligned}\Theta_3 &= 1+\exp(\mathrm{i}\omega_1 t + B_{11}) + \exp(\mathrm{i}\omega_2 t + B_{22}) + \\ &+ \exp(\mathrm{i}\omega_3 t + B_{33}) + \exp(\omega_1+\omega_2+B_{12}) + \\ &+ \exp(\omega_1+\omega_3+B_{13}) + \exp(\omega_2+\omega_3+B_{23}) + \\ &+ \exp(\omega_1+\omega_2+\omega_3+B_{12}+B_{13}+B_{23}),\end{aligned} \tag{1.4.34}$$

and

$$\Theta_n = \sum_{M\in(-\infty,\infty)} \exp(\mathrm{i}\sum_{i=1}^{n} M_i\omega_i t + \frac{1}{2}\sum_{i<j}^{n} B_{ij} M_i M_j), \tag{1.4.35}$$

$$\exp B_{ij} = \left(\frac{\omega_i-\omega_j}{\omega_i+\omega_j}\right)^2, \quad \exp B_{ii} = \omega_i^2. \tag{1.4.36}$$

Further, consider (1.4.21) and write the solution (1.4.33) under the form

$$\theta(t) = 2\frac{\partial^2}{\partial t^2}\log\Theta_n(\eta) = \theta_{lin}(\eta) + \theta_{int}(\eta), \tag{1.4.37}$$

for $\eta = -\omega t + \phi$. The first term θ_{lin} represents, as above, a linear superposition of cnoidal waves. Indeed, after a little manipulation and algebraic calculus, (1.4.23) gives

$$\theta_{lin} = \sum_{l=1}^{n} \alpha_l [\frac{2\pi}{K_l\sqrt{m_l}} \sum_{k=0}^{\infty} [\frac{q_l^{k+1/2}}{1+q_l^{2k+1}} \cos(2k+1)\frac{\pi\omega_l t}{2K_l}]^2] \,. \tag{1.4.38}$$

In (1.4.38) we recognize the expression (Abramowitz and Stegun, Magnus *et al.*)

$$\theta_{lin} = \sum_{l=1}^{n} \alpha_l \mathrm{cn}^2[\omega_l t; m_l], \tag{1.4.39}$$

with

$$q = \exp(-\pi\frac{K'}{K}),$$

$$K = K(m) + \int_0^{\pi/2} \frac{\mathrm{d}u}{\sqrt{1-m\sin^2 u}},$$

$$K'(m_1) = K(m), \quad m + m_1 = 1.$$

The second term θ_{int} represents a nonlinear superposition or interaction among cnoidal waves. We write this term as

$$2\frac{\mathrm{d}^2}{\mathrm{d}t^2}\log[1+\frac{F(t)}{G(t)}] \approx \frac{\beta_k \mathrm{cn}^2(\omega t, m_k)}{1+\gamma_k \mathrm{cn}^2(\omega t, m_k)}. \tag{1.4.40}$$

If m_k take the values 0 or 1, the relation (1.4.40) is directly verified. For $0 \le m_k \le 1$, the relation is numerically verified with an error of $|e| \le 5\times 10^{-7}$. Consequently, we have

$$\theta_{int}(x,t) = \frac{\sum_{k=0}^{n} \beta_k \mathrm{cn}^2[\omega_k t; m_k]}{1+\sum_{k=0}^{n} \lambda_k \mathrm{cn}^2[\omega_k t; m_k]} . \tag{1.4.41}$$

As a result, the cnoidal method yields to solutions consisting of a linear superposition and a nonlinear superposition of cnoidal waves.

1.5 Hirota method

In 1971, Hirota showed that certain evolution equations can be reduced to bilinear differential equations. He introduces a dependent-variable transformation (Drazin and Johnson, Hirota)

$$u(x,t) = -2\frac{\partial^2}{\partial x^2}\ln f(x,t) , \tag{1.5.1}$$

where f has the property

$$f, f_x, f_{xx} \to 0 , \quad \text{as } |x| \to \infty .$$

We shall describe the method in the context of the KdV equation

$$u_t - 6uu_x + u_{xxx} = 0 .$$

Substituting (1.5.1) into the KdV equation we obtain

$$ff_{xt} - f_x f_t + ff_{xxxx} - 4f_x f_{xxx} + 3f_{xx}^{\ 2} = 0 . \tag{1.5.2}$$

This equation can be reduced to a bilinear form by using the *Hirota operator* (Hirota and Satsuma)

$$D_t^{\ m} D_x^{\ n} : V\times V \to V,$$

$$D_t^{\ m} D_x^{\ n}(a,b)(x,t) = (\frac{\partial}{\partial t} - \frac{\partial}{\partial t'})^m (\frac{\partial}{\partial x} - \frac{\partial}{\partial x'})^n a(x,t)b(x',t')\Big|_{\substack{x=x' \\ t=t'}} , \tag{1.5.3}$$

where m, n are positive integers, V is a functions space, in particular $V = \{ f : \mathrm{R}\times\mathrm{R} \to \mathrm{R}, f \in \mathrm{C}^n(\mathrm{R})\times \mathrm{C}^m(\mathrm{R})\}$, and a,b two arbitrary functions in V.

By virtue of the definition (1.5.3), the Hirota operator is

1. Bilinear

$$D_t^{\ m} D_x^{\ n}(\alpha a_1 + a_2, b) = \alpha D_t^{\ m} D_x^{\ n}(a_1,b) + D_t^{\ m} D_x^{\ n}(a_2,b) ,$$

$$D_t^{\ m} D_x^{\ n}(a, \alpha b_1 + b_2) = \alpha D_t^{\ m} D_x^{\ n}(a,b_1) + D_t^{\ m} D_x^{\ n}(a,b_2) ,$$

$\forall a_1, a_2, b_1, b_2 \in V$, $\alpha \in R$.

2. Symmetric

$$D_t^m D_x^n(a,b) = D_t^m D_x^n(b,a),\ D_t^m D_x^n(a,b) = D_x^n D_t^m(a,b),$$

$\forall a,b \in V$. For $m = n = 1$, the Hirota operator (1.5.3) reduces to

$$D_t D_x(a,b)(x,t) = (a_{xt}b - a_t b_x - a_x b_t + ab_{tx})(x,t), \tag{1.5.4}$$

and for $m = 0,\ n = 4$, to

$$D_x^{\ 4}(a,b)(x,t) = (a_{xxxx}b - 4a_{xxx}b_x + 6a_{xx}b_{xx} - 4a_x b_{xxx} + ab_{xxxx})(x,t), \tag{1.5.5}$$

$\forall a,b \in V$.

The equation for f (1.5.2) can be rewritten in a form of a bilinear differential equation

$$D_x D_t (f,f) + D_x^4 (f,f) = 0, \tag{1.5.6}$$

or, by denoting

$$B = D_x (D_t + D_x^3),$$

to

$$B(f,f) = 0. \tag{1.5.7}$$

To solve (1.5.7), we consider the solutions of the form

$$f(x,t) = \sum_{n=0}^{N} \varepsilon^n f_n(x,t), \tag{1.5.8}$$

with $\varepsilon \neq 0$ a positive, real number, and $f_0(x,t) = 1$, $\forall (x,t) \in \mathrm{R} \times \mathrm{R}$.

Substitution of (1.5.8) into (1.5.7) yields

$$\begin{gathered} B(f_0, f_0) + \varepsilon[B(f_1,1) + B(1,f_1)] + \ldots \\ \ldots + \varepsilon^n \sum_{r=0}^{n} B(f_{n-r}, f_r) + \ldots + \varepsilon^N \sum_{r=0}^{N} B(f_{N-r}, f_r) = 0. \end{gathered} \tag{1.5.9}$$

From $B(f_0, f_0) = 0$, we obtain

$$\begin{gathered} B(f_1,1) = 0, \\ 2B(f_2,1) = -B(f_1,f_1), \\ 2B(f_3,1) = -B(f_2,f_1) - B(f_1,f_2), \end{gathered} \tag{1.5.10}$$

$$2B(f_n,1) = -\sum_{r=1}^{n-1} B(f_{n=r}, f_r),\quad n = 2,\ldots,N,$$

that are the sufficient condition for the equation (1.5.9) to be verified, $\forall \varepsilon$ real and positive number. So this time, we calculate some particular expressions for B

$$B(a,b)(x,t) = (a_{xt}b - a_t b_x - a_x b_t + ab_{xt} + a_{xxxx}b_x - 4a_{xxx}b_x + \\ +6a_{xx}b_{xx} - 4a_x b_{xxx} + ab_{xxxx})(x,t),$$

where $a,b \in V$. For $b \equiv 1$ we find

$$B(a,1) = a_{xt} + a_{xxx}.$$

For

$$a(x,t) = \exp(\theta_1), \; \theta_1 = k_1 x + \varpi_1 t + \alpha_1,$$

with k_1, ϖ_1, α_1, real and positive numbers, we have

$$B(\exp(\theta_1),1) = (k_1\varpi_1 + k_1^4)\exp(\theta_1). \tag{1.5.11}$$

Again, for

$$a(x,t) = \exp(\theta_1), \; b(x,t) = \exp(\theta_2), \; \theta_i = k_i x + \varpi_i t + \alpha_i, \quad i = 1,2,$$

it follows that

$$B[\exp(\theta_1), \exp(\theta_2)] = [(k_2 - k_1)(\varpi_2 - \varpi_1) + (k_2 - k_1)^4]\exp(\theta_1 + \theta_2). \tag{1.5.12}$$

The expressions (1.5.11) and (1.5.12) suggest some forms to be chosen for the functions f_n, $n \geq 1$. For $f_1(x,t) = \exp(\theta_1)$, the equation $(1.5.10)_1$ leads to $\varpi_1 = -k_1^3$, and the others equations of (1.5.10) are identically verified if $f_n \equiv 0, \; n \geq 2$.

Thus, we obtain

$$f_1(x,t) = 1 + \varepsilon\exp(k_1 x - k_1^3 t + \alpha_1), \tag{1.5.13}$$

$$u(x,t) = -\frac{k_1^2}{2}\operatorname{sech}^2[\frac{k_1 x - k_1^3 t + \alpha_1 + \ln\varepsilon}{2}]. \tag{1.5.14}$$

An initial condition of the form

$$u(x,0) = -2\operatorname{sech}^2 x, \tag{1.5.15}$$

leads to $\alpha_1 = -\ln\varepsilon, \; k_1 = 2$.

We can say that a solution of the KdV equation with the initial condition (1.5.15) is given by the soliton wave

$$u(x,0) = -2\operatorname{sech}^2(x - 4t). \tag{1.5.16}$$

If we consider the function

$$f_1(x,t) = \exp(\theta_1) + \exp(\theta_2),$$

$$\theta_i = k_i x + \varpi_i t + \alpha_i, \quad i = 1,2,$$

with k_1, ϖ_1, α_1, positive and real numbers, the equation $(1.5.10)_1$ is verified if $\varpi_2 = -k_i^3, \; i = 1,2$. For $B(f_1, f_1)$ we have

$$B(f_1, f_1) = 2[(k_2 - k_1)(\varpi_2 - \varpi_1) + (k_2 - k_1)^4]\exp(\theta_1 + \theta_2).$$

Note that $(1.5.10)_2$ suggests the form

$$f_2(x,t) = A\exp(\theta_1+\theta_2)\,,$$

where A is a real constant which is determined from $(1.5.10)_2$

$$A_2 = (\frac{k_1-k_2}{k_1+k_2})^2\exp(\theta_1+\theta_2)\,.$$

The constants k_1, k_2 are determined so that $k_1+k_2\neq 0$.

Choosing $f_n=0,\quad n\geq 3$, the other equations (1.5.10) are identically verified. We have

$$f(x,t) = 1+\varepsilon[\exp(\theta_1)+\exp(\theta_2)]+\varepsilon^2(\frac{k_1-k_2}{k_1+k_2})^2\exp(\theta_1+\theta_2), \tag{1.5.17}$$

$$u(x,t) = -2\frac{\partial}{\partial x}[\frac{\varepsilon k_1\exp(\theta_1)+\varepsilon k_2\exp(\theta_2)+\varepsilon^2\frac{(k_1-k_2)^2}{k_1+k_2}\exp(\theta_1+\theta_2)}{1+\varepsilon\exp(\theta_1)+\varepsilon\exp(\theta_2)+\varepsilon^2(\frac{k_1-k_2}{k_1+k_2})^2\exp(\theta_1+\theta_2)}], \tag{1.5.18}$$

with $\theta_i = k_i x - k_i^3 t+\alpha_i,\quad i=1,2$.

An initial condition of the form

$$u(x,0) = -6\,\mathrm{sech}^2 x\,, \tag{1.5.19}$$

is verified for $\alpha_1=\alpha_2=\ln\frac{3}{\varepsilon},\quad k_1=-2,\quad k_2=-4$. Consequently, the solution is

$$u(x,t) = -2\frac{\partial}{\partial x}[\frac{-6\exp(-6x+72t)-6\exp(-2x+8t)-12\exp(-4x+64t)}{3\exp(-2x+8t)+3\exp(-4x+64t)+\exp(-6x+72t)}], \tag{1.5.20}$$

and represents the two-soliton solution.

For $f_1(x,t)=\sum_{i=1}^{3}\exp(\theta_i)$, $\theta_i = k_i x+\omega_i t+\alpha_i,\quad i=1,2,3$, the equation $(1.5.10)_1$ leads to $\omega_i=-k_i^3,\quad i=1,2,3$.

Choosing $f_2(x,t)=\sum_{i,j=1}^{3}A_{ij}\exp(\theta_i+\theta_j)$, the equation $(1.5.10)_2$ is verified for

$$A_{ij} = \frac{1}{2}\frac{(k_i-k_j)^2}{(k_i+k_j)^2},\quad i,j=1,2,3\,.$$

For

$$f_3(x,t) = \sum_{i,j,l=1}^{3}A_{ijl}\exp(\theta_i+\theta_j+\theta_l),\quad i,j,l=1,2,3\,.$$

The equation $(1.5.10)_3$ yields

$$A_{ijl} = -\frac{1}{2}\frac{(k_i - k_j)^2}{(k_i + k_j)^2}\frac{(k_i - k_l)(k_j - k_l)}{(k_i + k_l)(k_j + k_l)}\frac{k_i + k_j - k_l}{k_i + k_j + k_l}, \quad i, j = 1, 2, 3 .$$

For $f_n(x,t) = 0, \quad n \geq 4$, equations (1.5.10) are verified and, finally, we have

$$f(x,t) = 1 + \varepsilon \sum_{i=1}^{3} \exp(\theta_i) + \varepsilon^2 \sum_{i,j=1}^{3} A_{ij} \exp(\theta_i + \theta_j) +$$
$$+ \varepsilon^3 \sum_{i,j,l=1}^{3} A_{ijl} \exp(\theta_i + \theta_j + \theta_l), \quad i, j, l = 1, 2, 3.$$

In summary, we have

$$\begin{aligned} f(x,t) &= 1 + \varepsilon[\exp\theta_1 + \exp\theta_2 + \exp\theta_3] + \\ &+ \varepsilon^2[\frac{(k_1 - k_2)^2}{(k_1 + k_2)^2}\exp(\theta_1 + \theta_2) + \frac{(k_3 - k_2)^2}{(k_3 + k_2)^2}\exp(\theta_3 + \theta_2) + \\ &+ \frac{(k_1 - k_3)^2}{(k_1 + k_3)^2}\exp(\theta_1 + \theta_3)] + \varepsilon^3 \frac{(k_1 - k_2)}{(k_1 + k_2)}\frac{(k_1 - k_3)(k_2 - k_3)}{(k_1 + k_3)(k_2 + k_3)} \times \\ &\times \frac{k_3^3(k_1 - k_2) + k_2^3(k_3 - k_1) + k_1^3(k_2 - k_3)}{k_1 + k_2 + k_3}\exp(\theta_3 + \theta_2 + \theta_3). \end{aligned} \quad (1.5.21)$$

Substitution of (1.5.21) into (1.5.1) leads to a solution $u(x,t)$, which is a three-soliton solution. An initial condition of the form

$$u(x,0) = -12 \operatorname{sech}^2 x, \quad (1.5.22)$$

is verified for

$$\alpha_1 = \ln\frac{6}{\varepsilon}, \quad \alpha_2 = \ln\frac{15}{\varepsilon}, \alpha_3 = \ln\frac{10}{\varepsilon}, \quad k_1 = -2, \quad k_2 = -4, \ k_3 = -6 .$$

It follows that

$$\begin{aligned} u(x,t) &= -2\frac{\partial^2}{\partial x^2}[\ln(1 + 6\exp(-2x + 8t) + 15\exp(-4x + 64t) + \\ &+ 10\exp(-6x + 216t) + 10\exp(-6x + 72t) + 15\exp(-8x + 224t) + \\ &+ 6\exp(-10x + 280t) + \exp(-12x + 288t))]. \end{aligned} \quad (1.5.23)$$

In an inductive way, the N-soliton solution is obtained as

$$f_1(x,t) = \sum_{i=1}^{N} \exp(\theta_i) \ , \ \theta_i = k_i x - k_i^3 t + \alpha_i, \quad i = 1, ..., N ,$$

and the solution is

$$f(x,t) = 1+\varepsilon\sum_{i=1}^{N}\exp(\theta_i)+\varepsilon^2\sum_{i,j=1}^{N}A_{ij}\exp(\theta_i+\theta_j)+...+$$
$$+\varepsilon^N\sum_{i_1...i_N=1}^{N}A_{i_1...i_N}\exp(\theta_{i_1}+...+\theta_{i_N}). \tag{1.5.24}$$

We mention that the Hirota bilinear method obtains the exact N-soliton solution for many evolution equations. The bilinear method is effective for completely integrable systems and also for nearly integrable equations. This method is more useful for stability analysis than the ordinary perturbation method or the perturbation treatment of the inverse scattering method.

Next, we are going to investigate the stability of the KdV soliton with respect to wavefront bending, for example. The amplitude of the wave is calculated for a small perturbation of the argument θ. If the amplitude remains constant, the soliton is stable.

Consider a particular solution of the KdV equation in the bilinear form (1.5.6)

$$f = 1+\exp(-2\xi_0),\ \xi_0 = k_0(x-Vt). \tag{1.5.25}$$

Now assume the phase and amplitude of the KdV equation are slowly varying functions of y, perpendicularly to direction x. To study the stability of the soliton against transverse perturbation, it is convenient to introduce the Kadomtsev and Petviashvili equation, referred to as the K-P equation (Matsukawa *et al.*)

$$(u_t+6uu_x+u_{xxx})_x+u_{yy}=0, \tag{1.5.26}$$

rewritten as a bilinear differential equation

$$(D_xD_t+D_y^2+D_x^4)(f,f)=0. \tag{1.5.27}$$

The perturbation modifies (1.5.25) as

$$f=\hat f\,\hat g,\ \hat f = 1+\exp(-2\xi),$$
$$\hat g = 1+g,\ \xi = k(x-Vt). \tag{1.5.28}$$

Substituting (1.5.28) into (1.5.27), we obtain

$$[(D_xD_t+D_y^2+D_x^4)(\hat f,\hat f)]\hat g^2+\hat f^2[(D_xD_t+D_y^2+D_x^4)(\hat g,\hat g)]+$$
$$+6(D_x^2(\hat f,\hat f))(D_x^2(\hat g,\hat g))=0. \tag{1.5.29}$$

By expanding the expression

$$\exp(\varepsilon D_p+\gamma D_q)(ab,ab)=[\exp(\varepsilon D_p+\gamma D_q)(a,b)][\exp(\varepsilon D_p+\gamma D_q)(a,b)], \tag{1.5.30}$$

in a power series of ε and δ, two identities are derived

$$D_pD_q(ab,ab)=(D_pD_q(a,a))b^2+a^2(D_pD_q(b,b)),$$

$$D_p^4(ab,ab)=(D_p^4(a,a))b^2+a^2(D_p^4(b,b))+6(D_p^2(a,a))(D_p^2(b,b)).$$

Inserting (1.5.28) into (1.5.30), the K-P equation (1.5.27) becomes

$$[8(\xi k_t - k^2 x_{0,t})\exp(-2\xi) - 4k_t \exp(-2\xi)(1+\exp(-2\xi))](1+2g)$$
$$+2\hat{f}^2 g_{xt} + 32k^4 \exp(-2\xi)(1+2g) + 2\hat{f}^2 g_{xx} + 96k^2 \exp(-2\xi) g_{xx} - \tag{1.5.31}$$
$$-4(k_{yy}\frac{\xi}{k} - kx_{0,yy})\exp(-2\xi)(1+\exp(-2\xi)) + 2\hat{f}^2 g_{yy} = 0.$$

The zero-th order estimation in (1.5.31) yields

$$x_{0,t} = 4k_0^2\,, \tag{1.5.32}$$

where, we observe that the amplitude of the soliton is constant, as is expected.

From the first order estimation in (1.5.31) we have

$$\begin{aligned} &k^2(4k^2 - x_{0,t})\mathrm{sech}^2\xi + \xi k_t \mathrm{sech}^2\xi - k_t(1-\tanh\xi) + \\ &+g_{xt} + g_{yy} + g_{xx} + 12k^2 g_{xx}\mathrm{sech}^2\xi - (k_{yy}\frac{\xi}{k} - kx_{0,yy})(1-\tanh\xi) = 0, \end{aligned} \tag{1.5.33}$$

where the phase and amplitude of the soliton are dependent on y.

Introducing a new coordinate which is moving at soliton velocity

$$\xi = k(x - 4k^2 t)\,, \tag{1.5.34}$$

we observe that the perturbed solution g depends only on ξ, and then equation (1.5.33) is simplified

$$k^4[g_{\xi\xi\xi\xi} + 4(3\mathrm{sech}^2\xi - 1)g_{\xi\xi} + \xi k_t \mathrm{sech}^2\xi - (k_t + k_{yy}\frac{\xi}{k} - kx_{0,yy})(1-\tanh\xi) = 0. \tag{1.5.35}$$

Multiplying (1.5.35) by $2\mathrm{sech}^2\xi$ and integrating with respect to ξ, it results that $u = k^2\mathrm{sech}^2\xi$, satisfies the KdV equation $u_t + 6uu_x + u_{xxx} = 0$.

From (1.5.35) we have

$$(k_t - kx_{0,yy})\int_{-\infty}^{\infty}(1-\tanh\xi)\mathrm{d}\xi + \frac{k_{yy}}{k}\int_{-\infty}^{\infty}\xi(1-\tanh\xi)\mathrm{d}\xi = 0\,. \tag{1.5.36}$$

Combining (1.5.32) with (1.5.36) the wave equation for the oscillation of the phase is obtained

$$x_{0,tt} = 4k^2 x_{0,yy}\,, \tag{1.5.37}$$

and yields to the conclusion that the soliton is stable against transverse perturbation.

1.6 Linear equivalence method (LEM)

The linear equivalence method (LEM) was introduced by Toma to solve nonlinear differential equations with arbitrary Cauchy data.

Consider the Cauchy problem

$$\frac{\mathrm{d}x}{\mathrm{d}t} = f(x),\ x = \begin{pmatrix} x_1 \\ x_2 \\ \dots \\ x_n \end{pmatrix},\ f(x) = \begin{pmatrix} f_1(x) \\ f_2(x) \\ \dots \\ f_n(x) \end{pmatrix}, \tag{1.6.1}$$

$$x(t_0) = x_0,\quad t_0 \in I \subset \mathrm{R},$$

where $f_j(x) = \sum_{|\nu|\geq 1} f_{j\nu} x^{\nu}$, with the coefficients $f_{j\nu}$ defined on the real interval I. Here $\nu = (\nu_1, \nu_2, \dots \nu_n)$ is a multi-indices vector with n components, and $z^{\nu} = \prod_{j=1}^{n} z_j^{\nu_j}$. LEM consists of applying to (1.6.1) an exponential transform depending on n parameters $\xi \equiv (\xi_1, \xi_2, \dots, \xi_n)$, which maps the equation into a linear first order partial differential equation, with respect to the independent variable

$$v(t,\sigma) = \exp < \sigma, z >,\ \sigma \in \mathrm{R}^n,\ < \sigma, z > = \sum_{j=1}^{n} \sigma_j z_j, \tag{1.6.2}$$

or

$$v(t, \sigma_1, \sigma_2, \dots \sigma_n) = \exp(\sigma_1 z_1 + \sigma_2 z_2 + \dots + \sigma_n z_n),\ \sigma_i \in \mathrm{R}. \tag{1.6.3}$$

Computing $\frac{\partial v}{\partial t}$, and taking account of $\frac{\partial v}{\partial \sigma_i} = v z_i$, $\frac{\partial^2 v}{\partial \sigma_i \sigma_j} = v z_i z_j$, the result is a linear partial differential equation

$$\frac{\partial v}{\partial t} - \sum_{j=1}^{n} \xi_j f_j(D_\xi) v = 0, \tag{1.6.4}$$

where the formal operator from the left-hand side

$$\langle \xi, f(D_\xi) \rangle \equiv \sum_{j=1}^{n} \xi_j f_j(D_\xi), \tag{1.6.5}$$

is obtained by replacing x^{ν} to $\frac{\partial^{\nu}}{\partial \xi^{\nu}}$ in the expression of $f_j(x)$.

The Cauchy conditions $(1.6.1)_2$ become

$$v(t_0, \xi) = \exp\langle x_0, \xi \rangle. \tag{1.6.6}$$

Toma has proved that the solution of the nonlinear Cauchy problem (1.6.1) is equivalent to the analytic solutions in ξ of the linear problem (1.6.4) and (1.6.6). Let us consider v of the form

$$v(t,x) = \sum_{|\nu|\geq 1} v_{\nu}(t) \frac{\xi^{\nu}}{\nu!}. \tag{1.6.7}$$

The coefficients from the right-hand side of (1.6.7) satisfy a linear infinite differential equation with constant coefficients

$$\dot{v}_\nu = \sum_{j=1}^{n} j \sum_{|\gamma|=1}^{\infty} f_{j\gamma} v_{\nu+\gamma-\mathbf{e}_j},\ e_j = \left(\delta_j^k\right)_{k=\overline{1,n}}, \tag{1.6.8}$$

with Cauchy conditions

$$v_\nu\left(t_0\right) = \mathrm{x}^\nu,\ |\nu| \in \mathrm{N}. \tag{1.6.9}$$

The first n components of the solution of (1.6.8) and (1.6.9) coincide with the solution of the initial problem (1.6.1).

For a given vector $g = \left(g_j\left(x\right)\right)_{j=\overline{1,n}}$, the Lie derivative of an analytic function λ is

$$L_g \lambda \equiv \langle g, \operatorname{grad} \lambda \rangle. \tag{1.6.10}$$

So, we can write

$$\langle \xi, f\left(D_\xi\right) \rangle = L_f v. \tag{1.6.11}$$

As a result, the following theorems hold.

THEOREM 1.6.1 (Toma) *The linear equation (1.6.4) is expressed in terms of Lie derivatives as*

$$Lv \equiv \frac{\partial v}{\partial t} - L_{\mathbf{f}} v = 0. \tag{1.6.12}$$

and the analytic solution of (1.6.4) and (1.6.6) have two expansions

$$v(t,\xi) = 1 + \sum_{k=1}^{\infty} L_{\mathbf{f}}^k \left(\exp\langle x_0, \xi \rangle\right) \frac{\left(t-t_0\right)}{k!}, \tag{1.6.13}$$

$$v(t,\xi) = 1 + \sum_{k=1}^{\infty} \langle \xi, f\left(D_\xi\right) \rangle \left(\exp\langle x_0, \xi \rangle\right) \frac{\left(t-t_0\right)}{k!}. \tag{1.6.14}$$

The important point is that the solution of (1.6.1) is written as

$$x(t) = \operatorname{grad}_\xi v(t,\xi)\Big|_{\xi=0}, \tag{1.6.15}$$

where v is replaced by (1.6.13) or (1.6.14).

The linear infinite system (1.6.8) may be written in the matrix form

$$\frac{\mathrm{d}V}{\mathrm{d}t} = AV,\ V = \left(V_j\right)_{j \in IN},\ V_j = \left(v_\gamma\left(t\right)\right)_{|\gamma|=j}, \tag{1.6.16}$$

where the matrix A is defined as

$$A = \begin{bmatrix} A_{11} & A_{12} & A_{13} & \dots & \dots & \dots & \dots \\ 0 & A_{22} & A_{23} & \dots & \dots & \dots & \dots \\ 0 & 0 & A_{33} & \dots & \dots & \dots & \dots \\ \dots & \dots & \dots & \dots & \dots & \dots & \dots \\ 0 & 0 & 0 & \dots & A_{jj} & A_{j,j+1} & \dots \\ \dots & \dots & \dots & \dots & \dots & \dots & \dots \end{bmatrix}. \tag{1.6.17}$$

It is evident that the rectangular matrices $A_{j,j+k-1}$, $k>1, j \in \mathrm{N}$, depend only on those $f_{j\nu}$ for which $|\nu| = k$.

The diagonal cells A_{jj} depend only on the linear part of the differential operator more precisely, A_{11} is the associated Jacobi matrix.

The Cauchy conditions $(1.6.1)_2$ are mapped through (1.6.6) into

$$V(t_0) = \left(x_0^\nu\right)_{|\nu|\in \mathrm{N}}. \tag{1.6.18}$$

The first n components of the solution of (1.6.16) and (1.6.18) coincide to solution of (1.6.1), and this holds even for systems with variable coefficients. The proof of this result is based on the construction of the inverse matrix for nonlinear operators. In the case of constant coefficients, this inverse allows a simplified exponential form and thus, the solution of (1.6.16), (1.6.18) may be written as

$$V(t) = V(t_0)\exp[A(t-t_0)]. \tag{1.6.19}$$

THEOREM 1.6.2 (Toma) *The solution of initial problem (1.6.1) coincides with the first n components of the vector V given by (1.6.19).*

We mention that the block structure of A enables the step-by-step calculus of A^k. A secondary result is that the first n rows of A^k may be expressed by blocks, each of them being computable by a finite number of steps.

Next, denote by P_n the operator that associates to a matrix its first n rows. Then, by theorem 1.6.2 and (1.6.19), it follows that the solution of (1.6.1) coincides with the Taylor series expansion

$$x(t) = x_0 + \sum_{k=1}^{\infty} \frac{(t-t_0)^k}{k!} P_n A^k \left(x_0^\nu\right)_{|\nu|\in \mathrm{N}}. \tag{1.6.20}$$

The above results enable a comparison with the Fliess expansions in the optimal control problems. The Fliess expansion corresponding to solution of (1.6.1) is obtained as a Taylor series

$$x_j(t) = x_{j0} + \sum_{k=1}^{\infty} L_f^k\left(x_{j0}\right)\frac{(t-t_0)}{k!}, \quad j = 1,2\dots,n, \tag{1.6.21}$$

where $L_{\mathbf{f}}^k\left(x_{j0}\right)$ are the k-th order iteratives of the Lie derivative $L_{\mathbf{f}}\left(x_{j0}\right)$. Comparing this with the LEM expansion (1.6.21), we obtain the following result:

THEOREM 1.6.3 (Toma) *The Lie derivative $L_f^k\left(x_{j0}\right)$ of order k with respect to f is calculated as follows*

$$\begin{pmatrix} L_f^k\left(x_{10}\right) \\ L_f^k\left(x_{20}\right) \\ \vdots \\ L_f^k\left(x_{n0}\right) \end{pmatrix} = P_n A^k \left(x_0^{\nu}\right)_{|\nu|\in \mathrm{IN}} . \tag{1.6.22}$$

We have

$$P_n\left(A^m\right) = \left(S_{km}\right)_{k\in \mathrm{IN}} , \tag{1.6.23}$$

where

$$S_{km} = \sum_{|\gamma|=m-k-1} A_{11}^{\gamma_1} A_{12} A_{22}^{\gamma_2} A_{23} \ldots A_{k-1,k} A_{kk}^{\gamma_k} , \tag{1.6.24}$$

$\gamma = \left(\gamma_1, \gamma_2, \ldots, \gamma_k\right)$ is a multi-indices vector. The calculus of S_{km} is easier if the eigenvalues of A_{kk} are specified. These eigenvalues are

$$\langle \iota, \lambda \rangle \equiv \sum_{j=1}^{n} \iota_j \lambda_j , \tag{1.6.25}$$

where $\lambda = \left(\lambda_1, \lambda_2, \ldots, \lambda_n\right)$ are the eigenvalues of A_{11}.

Now we are ready to calculate the *normal LEM representations*, useful for a qualitative study of nonlinear equations.

THEOREM 1.6.4 (Toma) *The solution of (1.6.13) may be expressed as a series with respect to the data*

$$x_j\left(t\right) = x_{j0} + \sum_{|\gamma|=1}^{\infty} u_{\gamma}^j\left(t\right) x_0^{\gamma},\ j = 1, 2, \ldots, n , \tag{1.6.26}$$

where $|\gamma| = k$ and $U_k^j \equiv \left(u_{\gamma}^j\left(t\right)\right)_{|\gamma|=k}$ satisfy the linear finite systems

$$\begin{aligned} \frac{\mathrm{d}U_1}{\mathrm{d}t} &= A_{11}^* U_1, \\ \frac{\mathrm{d}U_2}{\mathrm{d}t} &= A_{22}^* U_2 + A_{12}^* U_1, \\ &\ldots\ldots\ldots\ldots\ldots\ldots \\ \frac{\mathrm{d}U_k}{\mathrm{d}t} &= A_{kk}^* U_k + A_{k,k-1}^* U_{k-1} + \ldots + A_{1k}^* U_1, \end{aligned} \tag{1.6.27}$$

and Cauchy - conditions

$$U_1(t_0) = \left(\delta_m^j\right)_{m=\overline{1,n}},\ U_s(t_0) = 0,\ s = 2,\ldots,k\ . \qquad (1.6.28)$$

Here, the star stands for transpose matrix. This completes the description of the LEM. To illustrate its application, consider two examples.

EXAMPLE 1.6.1 Let the problem be (Toma)

$$y_t = y^2,\ y(0) = y_0\ . \qquad (1.6.29)$$

By applying the exponential transform $v = \exp(\sigma y)$ we obtain the equivalent linear equation

$$\frac{\partial v}{\partial t} = \sigma \frac{\partial^2 v}{\partial \sigma^2}, \qquad (1.6.30)$$

and the condition

$$v(0,\sigma) = \exp(\sigma y_0)\ .$$

Consider

$$v(t,\sigma) = \phi(t)\psi(\sigma)\ ,$$

and then (1.6.30) yields

$$\frac{\mathrm{d}\phi}{\mathrm{d}t} = \lambda\phi\ ,\ \sigma\frac{\mathrm{d}^2\psi}{\mathrm{d}\sigma^2} = \lambda\psi(\sigma)\ ,$$

with λ an arbitrary constant. Next, consider

$$\psi(\sigma) = \sum_{j=0}^{\infty} \varphi_j \frac{\sigma^j}{j!}\ ,$$

to yield

$$j\psi_{j+1} = \lambda\psi_j\ ,\ j \in \mathrm{N}^*\ .$$

We observe that $\psi_0 = 0$. The coefficients are determined from

$$\psi_j = \frac{\lambda^{j-1}}{(j-1)!}\psi_1\ ,\ j \in \mathrm{N}\ .$$

For $v(t,\sigma)$ it results

$$v(t,\sigma) = 1 + \int_{-\infty}^{0} \exp(\lambda t) \sum_{j=1}^{\infty} \frac{\lambda^{j-1}}{(j-1)!}\frac{\sigma^j}{j!}\psi_1 \mathrm{d}\lambda\ , \qquad (1.6.31)$$

when $t \geq 0$. From the initial condition $(1.6.29)_2$ it also results

$$\int_{-\infty}^{0} \frac{\lambda^{j-1}}{(j-1)!}\frac{\sigma^j}{j!}\psi_1 \mathrm{d}\lambda = y_0^j\ ,\ j \in \mathrm{N}\ .$$

So, we have

$$\psi_1 = \exp(-\frac{\lambda}{y_0}),$$

and the solution becomes

$$v(t,\sigma) = 1 + \int_{-\infty}^{0} \exp[\lambda(t - \frac{1}{y_0})] \sum_{j=1}^{\infty} \frac{\lambda^{j-1}}{(j-1)!} \frac{\sigma^j}{j!} d\lambda =$$

$$= 1 + \sum_{j=1}^{\infty} \left(-\frac{1}{t - \frac{1}{y_0}} \right)^j \frac{\sigma^j}{j!} = \exp[-\frac{y_0}{ty_0 - 1}\sigma].$$

Finally, this solution leads to the solution of (1.6.29) given by

$$y(t) = \frac{y_0}{1 - ty_0}.$$

EXAMPLE 1.6.2 The Troesch problem (Toma).
Let us have the problem

$$w_{tt} = n \sinh(nw),\ w(0) = 0,\ w(1) = 1. \qquad (1.6.32)$$

By noting

$$x = nt,\ y(x) = nw, \qquad (1.6.33)$$

we obtain

$$y_{xx} = \sinh y, \qquad (1.6.34)$$

with conditions

$$y(0) = 0,\ y(n) = n,\ y'(0) = \beta. \qquad (1.6.35)$$

The last condition (1.6.35) with a fictitious parameter β, is introduced in order to apply LEM. So, we have

$$v(x,\sigma,\xi) = \exp(\sigma y + \xi y'). \qquad (1.6.36)$$

The linear equivalent equation is given by

$$\frac{\partial v}{\partial x} - \sigma \frac{\partial v}{\partial \xi} - \xi \sum_{k=0}^{\infty} \frac{1}{(2k+1)!} \frac{\partial^{2k+1} v}{\partial \sigma^{2k+1}} = 0, \qquad (1.6.37)$$

and conditions (1.6.35) become

$$v(0,\sigma,\xi) = \exp(\xi\beta). \qquad (1.6.38)$$

The equivalent linear system is

$$\frac{\mathrm{d}V}{\mathrm{d}x} = MV\,, \tag{1.6.39}$$

where V and M are written as

$$V = [V_{2j-1}],\ j \in \mathrm{N}\,,\ V_{2j-1} = [V_{ik}],\ \ i+k = 2j-1\,,$$

$$M = \begin{bmatrix} A_1 & B_{11} & B_{13} & B_{15} & \dots & B_{1,2j-1} & B_{1,2j+1} & \dots \\ 0 & A_3 & B_{33} & B_{35} & \dots & B_{3,2j-1} & B_{3,2j+1} & \dots \\ . & . & . & . & \dots & . & . & \dots \\ 0 & 0 & 0 & 0 & \dots & A_{2j+1} & B_{2j+1,2j+1} & \dots \\ . & . & . & . & \dots & . & . & \dots \end{bmatrix},$$

$$B_{2j-1,2k+2j-1} = [b_{qs}]\,,\ 1 \le q \le j,\ \ 1 \le s \le 2k+2j\,,$$

$$b_{qs} = \frac{(q-1)\delta_s^{q-1}}{(2k+2j-1)!}\,,$$

with δ, the Kronecher symbol.

The three-diagonal matrices A_{2j-1} defined as

$$A_{2j-1} = \begin{bmatrix} 0 & 2j-1 & 0 & 0 & \dots & 0 & 0 & 0 \\ 1 & 0 & 2j-2 & 0 & \dots & 0 & 0 & 0 \\ . & . & . & . & \dots & . & . & . \\ 0 & 0 & 0 & 0 & \dots & 2j-2 & 0 & 1 \\ 0 & 0 & 0 & 0 & \dots & 0 & 2j-1 & 0 \end{bmatrix},$$

have the eigenvalues $\pm(2j-2k-1),\ 0 \le k \le j-1$.

The initial conditions are

$$V(0,\beta) = [\beta^{2k+1}\delta_q^{2k-1}],\ 0 \le q \le 2k-1,\ k \in \mathrm{N}\,. \tag{1.6.40}$$

Then we subtract from (1.6.39) a truncated system of equations given by

$$\frac{\mathrm{d}V^{(m)}}{\mathrm{d}x} = M^{(m)}V^{(m)}\,, \tag{1.6.41}$$

where the finite matrices $M^{(m)}$ are obtained by truncation from M up to the m order, that is, up to A_{2m-1}.

We can show that $V^{(m)}$ satisfy (1.6.40), and admit the representation

$$V^{(m)}(x,\beta) = \exp(A^{(m)}x)V^{(m)}(0,\beta)\,.$$

The equations (1.6.41) and (1.6.40) are solved by the block partitioning method. The solution is obtained as follows

$$y(x) \cong y'(0)\frac{\exp(x)}{4} + 2\ln\frac{1+y'(0)\dfrac{\exp(x)}{\sqrt{32}}}{1-y'(0)\dfrac{\exp(x)}{\sqrt{32}}}, \tag{1.6.42}$$

where $u = y'(0)\exp n$ satisfies

$$\frac{u}{2} = 2\sqrt{2}\ln\frac{1+\dfrac{u}{\sqrt{32}}}{1-\dfrac{u}{\sqrt{32}}}.$$

1.7 Bäcklund transformation

The Bäcklund transformation was developed in 1883, arising from the pseudospherical surfaces construction. The name of Albert Victor Bäcklund (1845–1922) is associated with the transformation of surfaces that bears his name, the extensions of which have a great impact in soliton theory (Coley *et al.*). In 1885 Bianchi shows that the Bäcklund transformation is related to an elegant invariance of the sine-Gordon equation. The geometric origins of the Bäcklund and Darboux transformations and their applications in modern soliton theory represent the subject of the monograph of Rogers and Schief (2002).

One of the simplest Bäcklund transformations are the Cauchy–Riemann relations

$$u_x = v_y, \; u_y = -v_x, \tag{1.7.1}$$

for Laplace equations

$$u_{xx} + u_{yy} = 0, \; v_{xx} + v_{yy} = 0. \tag{1.7.2}$$

If $v(x,y) = xy$ is a solution of the Laplace equation, then another solution of the Laplace equation $u(x,y) = \frac{1}{2}(x^2 - y^2)$, can be determined from

$$u_x = x, \; u_y = -y. \tag{1.7.3}$$

Consider next, the Liouville–Tzitzeica equation

$$u_{xt} = \exp u, \tag{1.7.4}$$

and let us introduce another equation, simple to be solved

$$v_{xt} = 0. \tag{1.7.5}$$

Suppose we have a pair of relations that relate the solutions of (1.7.4) and (1.7.5)

$$u_x + v_x = \sqrt{2}\exp\frac{u-v}{2}, \; u_t - v_t = \sqrt{2}\exp\frac{u+v}{2}. \tag{1.7.6}$$

The important point about (1.7.6) is the possibility to determine a solution of (1.7.4) for a specified solution of (1.7.5). Indeed, we obtain from (1.7.6)

$$u_{xt} + v_{xt} = \frac{1}{\sqrt{2}}(u_t - v_t)\exp\frac{u-v}{2} = \exp u\,,$$

$$u_{tx} - v_{tx} = \frac{1}{\sqrt{2}}(u_x + v_x)\exp\frac{u+v}{2} = \exp u\,,$$

which lead to (1.7.4) and (1.7.5).

Therefore, we have the following definition.

DEFINITION 1.7.1 *Let us have two uncoupled partial differential equations* $P(u(x,t)) = 0$ *and* $Q(v(x,t)) = 0$, *where* P *and* Q *are nonlinear operators. The pair of relations*

$$R_i(u, v, u_x v_x, u_t, v_t; x, t) = 0\,,\ i = 1,2,$$

is a Bäcklund transformation if it is integrable for v *when* $P(u) = 0$ *and if the resulting* v *is a solution of* $Q(v) = 0$, *and vice versa. If* $P = Q$, *so that* v *and* u *satisfy the same equation, then* $R_i = 0$, $i = 1,2$, *is called an auto-Bäcklund transformation.*

The Bäcklund transformation reduces the integration of a nonlinear partial differential equation to the solutions of an ordinary differential equation, in general of low order. The existence of the Bäcklund transformation implies that there is a relation between the solutions of P and Q.

The relations (1.7.6) are therefore the Bäcklund transformation of equations (1.7.4) and (1.7.5).

In the following we will present the original Bäcklund transformation for the sine-Gordon equation (Rogers and Schief 2002).

Let $r = r(x, y, z)$ denote the position vector of a point P on surface Σ in R^3, written under the Monge form

$$r = xe_1 + ye_2 + z(x, y)e_3\,. \tag{1.7.7}$$

The first and second fundamental forms are defined as

$$I = E\mathrm{d}x^2 + 2F\mathrm{d}x\mathrm{d}y + G\mathrm{d}y^2 = (1 + z_x^2)\mathrm{d}x^2 + 2z_x z_y \mathrm{d}x\mathrm{d}y + (1 + z_y^2)\mathrm{d}y^2,$$

$$II = e\mathrm{d}x^2 + 2f\mathrm{d}x\mathrm{d}y + g\mathrm{d}y^2 = \frac{1}{\sqrt{1 + z_x^2 + z_y^2}}(z_{xx}\mathrm{d}x^2 + 2z_{xy}\mathrm{d}x\mathrm{d}y + z_{yy}\mathrm{d}y^2). \tag{1.7.8}$$

The mean and Gaussian (total) curvature of Σ are written as

$$\mathrm{H} = EG - F^2 = \frac{(1 + z_x^2)z_{yy} - 2z_x z_y z_{xy} + (1 + z_y^2)z_{xx}}{2(1 + z_x^2 + z_y^2)^{3/2}}, \tag{1.7.9a}$$

$$K = \frac{eg - f^2}{EG - F^2} = -\frac{z_{xx}z_{yy} - z_{xy}^2}{(1 + z_x^2 + z_y^2)^2}. \tag{1.7.9b}$$

If Σ is a hyperbolic surface, then total curvature is negative and the asymptotic lines on Σ may be taken as parametric curves. Then $e = g = 0$. And the angle ω between the parametric lines is such that

$$\cos\omega = \frac{F}{\sqrt{EG}}, \quad \sin\omega = \frac{H}{\sqrt{EG}}. \tag{1.7.10}$$

In the particular case when $K = -\frac{1}{a^2}$ is a constant, Σ is a pseudospherical surface. If Σ is parametrized by arc length along asymptotic lines corresponding to the transformation

$$\mathrm{d}x \to \mathrm{d}x' = \sqrt{E(x)}\mathrm{d}x, \quad \mathrm{d}y \to \mathrm{d}y' = \sqrt{G(y)}\mathrm{d}y,$$

the fundamental forms become, dropping the prime

$$I = \mathrm{d}x^2 + 2\cos\omega\mathrm{d}x\mathrm{d}y + \mathrm{d}y^2, \tag{1.7.11a}$$

$$II = \frac{2}{a}\sin\omega\mathrm{d}x\mathrm{d}y. \tag{1.7.11b}$$

The Liouville representation of K in terms of E, G and F is given by

$$K = \frac{1}{H}\left[\left(\frac{H\Gamma_{11}^2}{E}\right) - \left(\frac{H\Gamma_{12}^2}{E}\right)\right], \tag{1.7.12}$$

where Γ'_{jk} are the Christoffel symbols. If Σ is a pseudospherical surface, the equation (1.7.12) is reduced to the sine-Gordon equation

$$\omega_{\alpha\beta} = \frac{1}{a^2}\sin\omega. \tag{1.7.13}$$

The original Bäcklund transformation is related to the pseudospherical surfaces construction with the Gaussian curvature

$$K = -\frac{1}{a^2}.$$

If r is the position vector of a pseudospherical surface Σ, then a new pseudospherical surface Σ' with the position vector r' and having the same curvature is written as

$$r' - r = an \times n', \quad n \cdot n' = \cos\zeta, \tag{1.7.14}$$

where ζ is the angle between the unit normals n and n' to Σ and Σ', and it is constant because

$$|r - r'| = L = a\sin\zeta.$$

Mention that $r-r'$ is tangent to both Σ and Σ'. The angle ζ is related to the Bäcklund parameter μ by the relation $\mu=\tan\frac{1}{2}\zeta$. We write (1.7.14) in terms of asymptotic coordinates

$$r'-r=\frac{L}{\sin\omega}[\sin\frac{\omega-\omega'}{2}r_\alpha+\sin\frac{\omega+\omega'}{2}r_\beta], \tag{1.7.15}$$

where ω and ω' are related

$$\left(\frac{\omega'-\omega}{2}\right)_\alpha=\frac{\mu}{a}\sin\frac{\omega'+\omega}{2},$$

$$\left(\frac{\omega'+\omega}{2}\right)_\beta=\frac{1}{\mu a}\sin\frac{\omega'-\omega}{2}. \tag{1.7.16}$$

The relations (1.7.16) represent the standard form of the Bäcklund transformation of the sine-Gordon equation. Here, ω and ω' are solutions of the sine-Gordon equation.

Indeed, by calculating $\left(\frac{\omega'-\omega}{2}\right)_{\alpha\beta}$ and $\left(\frac{\omega'+\omega}{2}\right)_{\beta\alpha}$, we have $\omega_{\alpha\beta}=\frac{1}{a^2}\sin\omega$, and $\omega'_{\alpha\beta}=\frac{1}{a^2}\sin\omega'$. Since both ω and ω' are solutions of the sine-Gordon equations (1.7.16) is an auto-Bäcklund transformation for (1.7.13).

If $\omega=0$, the transformation (1.7.13) becomes

$$\left(\frac{\omega'}{2}\right)_\alpha=\frac{\mu}{a}\sin\frac{\omega'}{2},\quad \left(\frac{\omega'}{2}\right)_\beta=\frac{1}{\mu a}\sin\frac{\omega'}{2}. \tag{1.7.17}$$

Integrating we obtain

$$\frac{\mu}{a}\alpha=\int^{\omega'}\frac{d\omega'}{\sin\frac{\omega'}{2}}=2\log|\tan\frac{1}{4}\omega'|+f(\beta), \tag{1.7.18}$$

and

$$\frac{1}{\mu a}\beta=2\log|\tan\frac{1}{4}\omega'|+g(\alpha), \tag{1.7.19}$$

where f and g are arbitrary functions. From (1.7.18) and (1.7.19) we obtain the *kink solution* of the sine-Gordon equation

$$\tan\frac{1}{4}\omega'=C\exp(\frac{\mu}{a}\alpha+\frac{1}{\mu a}\beta), \tag{1.7.20}$$

or

$$\omega'(\alpha,\beta)=4\arctan\{C\exp(\frac{\mu}{a}\alpha+\frac{1}{\mu a}\beta)\}, \tag{1.7.21}$$

with C a constant. The Bäcklund transformation admits the nonlinear superposition principle, namely, if we refer to (1.7.13)

$$\tan\frac{\omega_{12}-\omega}{4}=\frac{\mu_2+\mu_1}{\mu_2-\mu_1}\tan\frac{\omega_2-\omega_1}{4}. \qquad (1.7.22)$$

If ω_1 and ω_2 are two solutions for the sine-Gordon equation, generated from ω by means of Bäcklund transfomation (1.7.16) of parameters μ_1 and μ_2, then ω_{12} given by (1.7.21) is a new solution for the sine-Gordon equation, that is a two-soliton solution. Writing $\omega\to-\omega$ into (1.7.21) we have

$$\tan\frac{\omega_{12}+\omega}{4}=\frac{\mu_2+\mu_1}{\mu_2-\mu_1}\tan\frac{\omega_2+\omega_1}{4}. \qquad (1.7.23)$$

Relations (1.7.22) and (1.7.23) are known as *permutability theorem of Bianchi*. Any proof of permutability theorem is reduced to equality $\omega_{12}=\omega_{21}$. Luigi Bianchi established in 1892 that the Bäcklund transformation for the sine-Gordon equation admits a commutative property. This property leads to construction of nonlinear superposition formulae for solutions of evolution equations. The Bianchi theorem may be interpreted as an integrable discrete equation, if we assume that ω is a point into a discrete lattice of axes n_1 and n_2

$$\omega=\omega(n_1,n_2)\,, \qquad (1.7.24)$$

and the solutions ω_1, ω_2 and ω_{12} are neighbor points in this lattice (Rogers and Schief 1997)

$$\omega_1=\omega(n_1+1,n_2)\,,$$

$$\omega_2=\omega(n_1,n_2+1)\,, \qquad (1.7.25)$$

$$\omega_{12}=\omega(n_1+1,n_2+1)\,. \qquad (1.7.26)$$

Therefore, the Bäcklund transformation (1.7.16) can generate discrete pseudospherical surfaces Σ, associated to the sine-Gordon equation.

Let us illustrate the application of the Bäcklund transformation with some examples (Lamb, Drazin).

EXAMPLE 1.7.1 (Lamb) A general class of soluble nonlinear evolution equations are derived from linear equations. Let us consider the linear equations

$$v_{1x}+\mathrm{i}\zeta v_1=qv_2\,,\ v_{2x}+\mathrm{i}\zeta v_2=-qv_1\,, \qquad (1.7.27)$$

where q must be determined, and introduce a set of additional equations

$$w_{1x}+\mathrm{i}\zeta w_1=\varphi w_2\,,\ w_{2x}+\mathrm{i}\zeta w_2=-\varphi w_1\,, \qquad (1.7.28)$$

where φ must be determined. Suppose that v and w are related by

$$\begin{pmatrix} w_1 \\ w_2 \end{pmatrix} = \begin{pmatrix} v_1 \\ v_2 \end{pmatrix} + \wp \begin{pmatrix} v_1 \\ v_2 \end{pmatrix}, \tag{1.7.29}$$

where $\wp$ is a 2×2 matrix of elements a_{ij}. Taking account of (1.7.27), we rewrite (1.7.29) under the form

$$w_1 = Av_1 + Bv_2 , \; w_2 = Cv_1 + Dv_2 , \tag{1.7.30}$$

where

$$A = a_{11} - \mathrm{i}\zeta , \; B = a_{12} + q , \; C = a_{21} - q , \; D = a_{22} + \mathrm{i}\zeta .$$

By choosing $a_{21} = -a_{12}$, so that $C = -B$, and substituting (1.7.30) into (1.7.28) we obtain some relations between q and φ

$$A_x = (q - \varphi)B , \; B_x = -2\mathrm{i}\zeta B - qA + \varphi D ,$$

$$B_x = 2\mathrm{i}\zeta B - \varphi B - qD , \; D_x = (q - \varphi)B . \tag{1.7.31}$$

These equations admit first integrals $AD + B^2 = f^2(t)$ and $A - D = 2g(t)$, that imply

$$B^2 = f^2 + g^2 - (A - g)^2 .$$

So, the relation $(A + D)_x = 2B(q - \varphi)$ can be written as

$$\frac{A_x}{\sqrt{h^2 - (A-g)^2}} = q - \varphi . \tag{1.7.32}$$

Denoting $q = z_x$, $\varphi = z'_x$ and integrating, we obtain

$$A = g + h\cos(z - z') , \; B = h\sin(z - z') , \; D = -g + h\cos(z - z') . \tag{1.7.33}$$

The integration constant has been chosen so that B vanishes when z and z' vanish. Next, we set for simplicity $g = h = \dfrac{1}{2}$, and obtain from (1.7.31)

$$(z + z')_x = -2\mathrm{i}\zeta \sin(z - z') . \tag{1.7.34}$$

The equation (1.7.34) relates z and z', and is a Bäcklund transformation. The second Bäcklund transformation, which relates z_t and z'_t, depends upon the particular evolution equation being considered. According to the permutability theorem, by introducing $z \to -z$ $z' \to -z'$ into (1.7.34), we obtain

$$\tan(\frac{z_3 - z_0}{z}) = \pm \frac{a_1 + a_2}{a_1 - a_2} \tan(\frac{z_2 - z_1}{2}) , \tag{1.7.35}$$

where $a_i = -2\mathrm{i}\zeta_i$. Applying (1.7.35) to all evolution equations for which (1.7.27) are the appropriate linear equations, we can calculate another solution $q = z_x$.

EXAMPLE 1.7.2 Let us consider the KdV equation

$$P(u) = u_t - 6uu_x + u_{xxx} = 0 .$$

We introduce a new dependent variable w, $w_x = u$, and define the operator

$$Q(w) = w_t - 3w_x^2 + w_{xxx} ,$$

so that

$$(Q(w))_x = P(u) .$$

The Bäcklund transformation is given by

$$w_x + w'_x = 2\lambda + \frac{1}{2}(w - w')^2 , \tag{1.7.36}$$

$$w_t + w'_t = -(w - w')(w_{xx} - w'_{xx}) + 2(u^2 + uu' + u'^2) . \tag{1.7.37}$$

Suppose that w_1 and w_2 are two solutions that verify (1.7.36) and (1.7.37), and w_0 is a solution of $Q(w) = 0$, that is

$$w_{1x} + w_{0x} = -2\lambda_1 + \frac{1}{2}(w_1 - w_0)^2 , \tag{1.7.38a}$$

$$w_{2x} + w_{0x} = -2\lambda_2 + \frac{1}{2}(w_2 - w_0)^2 . \tag{1.7.38b}$$

Similarly, we can construct a solution w_{12} from w_1 and λ_2, and another solution w_{21} from w_2 and λ_1

$$w_{12x} + w_{1x} = -2\lambda_2 + \frac{1}{2}(w_{12} - w_1)^2 , \tag{1.7.38c}$$

$$w_{21x} + w_{2x} = -2\lambda_1 + \frac{1}{2}(w_{21} - w_2)^2 . \tag{1.7.38d}$$

From the Bianchi permutability theorem it results

$$w_{12} = w_{21} .$$

Then, subtract equations (1.7.38a) and (1.7.38b), and similar equations (1.7.38c) and (1.7.38d), and add the results to obtain

$$0 = -4(\lambda_1 - \lambda_2) + \frac{1}{2}(w_1 - w_0)^2 - \frac{1}{2}(w_2 - w_0)^2 - \frac{1}{2}(w_{12} - w_1)^2 + \frac{1}{2}(w_{12} - w_2)^2 ,$$

or

$$w_{12} = w_0 + \frac{4(\lambda_1 - \lambda_2)}{w_1 - w_2} .$$

Therefore, this method is remarkable for generation of new solutions. In particular, for

$$w_0 = 0\,,\ \lambda_1 = -1\,,\ w_1 = -2\tanh(x-4t)\,,\ \lambda_2 = -4\,,$$

and

$$w_2 = -4\coth(2x-32t)\,,$$

we have

$$w_{12}(x,t) = -\frac{12}{[2\coth(2x-32t) - \tanh(x-4t)]}\,,$$

that yields to a two-soliton solution

$$u_{12}(x,t) = -12\frac{3+4\cosh(2x-8t)+\cosh(4x-64t)}{[3\cosh(x-28t)+\cosh(3x-36t)]^2}\,.$$

1.8 Painlevé analysis

Before discussing the Painlevé property of an ordinary differential equation, we give a short review of singular points and their classification (Gromak).

Let us consider a nonlinear differential equation in the complex plane

$$w' = f(z,w)\,, \tag{1.8.1}$$

where $f : D \subset C^{n+1} \to C^n$ is an analytic function of complex variable. A solution of (1.8.1) is an analytic function, which is determined by its singular points. Here $w' = \frac{\mathrm{d}w}{\mathrm{d}z}$.

DEFINITION 1.8.1 *A point at which $w(z)$ fails to be analytic is called a singular point or singularity of $w(z)$. Singular points can belong to the following classes:*

a) *Isolated singular points.*
b) *Nonisolated singular points.*
c) *Single-valued points, for which the function does not change its value as z goes around a given initial point z_0.*
d) *Multi-valued or branch, or critical points.*
e) *The points for which the function has a limit, whether finite or infinite, as $z \to z_0$.*
f) *The points for which the function $w(z)$ has a limit as $z \to z_0$, namely z_0 is an essential singular point.*

We recall that a critical point is a singular point at which the solution is not analytic, which is not a pole.

PROPOSITION 1.8.1. *Consider the system*

$$w_j' = f_j(z, w_1, w_2, ..., w_n),\ w_j(z_0) = w_j^0,\ j = 1, 2, ..., n, \tag{1.8.2}$$

and let f *be analytic in the domain* $|z - z_0| \leq a$, $|w_j - w_j^0| \leq b$ *and* M *be the upper bound of the set* f *in this domain. Then the system admits a unique solution* $w(z)$, *which is analytic within the circle* $|z - z_0| \leq \rho$ *and which reduces to* w_0, *when* $z \to z_0$, $\rho = a[1 - \exp\frac{b}{M(n+1)}]$. *In the linear case* $\rho = a$.

A singular point may have more than one of the above properties. If (z_0, w_0) is a singularity of f then $w(z)$ can have a singularity at z_0. For example, the equation

$$zw' + w = 0,$$

admits the general solution

$$w(z) = \frac{c}{z},$$

where c is an arbitrary constant. The solution has a simple pole $z = 0$ for $c \neq 0$. The nonlinear equations may have movable singularities, whose position does depend on the arbitrary constants of integration. The equation

$$w' + w^2 = 0,$$

admits the general solution

$$w(z) = \frac{1}{z - z_0},$$

where z_0 is an arbitrary complex constant, which is a simple movable pole.

The equation

$$\frac{dw}{dz} = \exp(-w),$$

has the general solution

$$w(z) = \log(z - z_0),$$

where z_0 is an arbitrary complex constant. This solution has a logarithmic branch point (critical point) at the movable point $z = z_0$.

DEFINITION 1.8.2 *The movable singularities of the solution are the singularities whose location depends on the constant of integration. Fixed singularities occur at points where the coefficients of equation are singular.*

DEFINITION 1.8.3 *An ordinary differential equation is said to possess the Painlevé property when all movable singularities are single-valued (poles), that is when solutions are free from movable critical points but can have fixed multivalued singularities.*

Consider in the following those equations which do not contain movable singular points. In 1884, Fuchs shows that if the first order equation (Drazin and Johnson)

$$w' = F(z,w)\,,$$

where F is a rational function in w, and analytic in z, does not contain any movable crtical points, then

$$w' = F(z,w) = a(z) + b(z)w + c(z)w^2\,, \tag{1.8.3}$$

with a, b, c analytic functions. The above equation is the Riccati equation.

Painlevé and Gambier (see Gromak) extended these ideas to equations of the second order

$$w' = F(z,w,w')\,,$$

and showed that there are only 50 equations which have the property of having no movable critical points. They showed that 44 equations were integrable in terms of known functions, such as elliptic functions and solutions of linear equations, or were reductible to one of six new nonlinear differential equations, namely (Gromak)

$$w'' = 6w^2 + z\,,$$

$$w'' = zw + 2w^3 + \alpha\,,$$

$$w'' = \frac{1}{w}w'^2 - \frac{1}{z}w' + \frac{1}{z}(\alpha w^2 + \beta) + \gamma w^3 + \frac{\delta}{w}\,,$$

$$w'' = \frac{1}{2w}w'^2 + \frac{3}{2}w^3 + 4zw^2 + 2(z^2 - \alpha)w + \frac{\beta}{w}\,,$$

$$w'' = \frac{3w-1}{2w(w-1)}w'^2 - \frac{1}{z}w' + \frac{1}{z^2}(w-1)^2(\alpha w + \frac{\beta}{w}) + \frac{\gamma}{z} + \frac{\delta w(w+1)}{w-1}\,,$$

$$w'' = \frac{1}{2}(\frac{1}{w} + \frac{1}{w-1} + \frac{1}{w-z})w'^2 - (\frac{1}{z} + \frac{1}{w-1} + \frac{1}{w-z})w' + $$
$$+ \frac{w(w-1)(w-z)}{z^2(z-1)^2}[\alpha + \frac{\beta z}{(w-1)^2} + \frac{\gamma(z-1)}{(w-1)^2} + \frac{\delta z(z-1)}{(w-z)^2}],$$

with α, β, γ and δ arbitrary constants. These equations are known as the Painlevé equations. The first three equations were discovered by Painlevé, and the next three by Gambier, Ablowitz, Ramani and Segur expressed in 1980, the sufficient and necessary conditions of integrability for a nonlinear partial differential equation; every ordinary differential equation derived from it by exact reduction, must satisfy the Painlevé property.

Each of the above equations can be analyzed in terms of meromorphic functions theory. In particular, let us consider the second Painlevé equation

$$w'' = zw + 2w^3 + \alpha\,, \tag{1.8.4}$$

which is a meromorphic function of z. In the neighborhood of each pole z_0, the solution has the expansion (Gromak)

$$w = \frac{\varepsilon}{z - z_0} - \frac{\varepsilon}{6} z_0 (z - z_0) - \frac{\alpha + \varepsilon}{4}(z - z_0)^2 + h(z - z_0)^3 - \ldots.. ,\ \varepsilon^2 = 1\,, \tag{1.8.5}$$

where h is an arbitrary constant. The theorem of meromorphic functions implies that each solution of (1.8.4) can be expressed as the ratio of two entire functions, that is

$$w(z) = \frac{v(z)}{u(z)}\,. \tag{1.8.6}$$

The expansion (1.8.5) leads to

$$w^2 = \frac{1}{(z - z_0)^2} - \frac{z_0}{3} + O(z - z_0)\,, \tag{1.8.7}$$

and

$$\int_{z_1}^{z} w^2 \mathrm{d}z = -\frac{1}{z - z_0} + O(z - z_0)\,. \tag{1.8.8}$$

Therefore the function

$$u(z) = \exp\left(-\int_{z_1}^{z} \mathrm{d}z \int_{z_1}^{z} w^2 \mathrm{d}z \right), \tag{1.8.9}$$

has a simple zero at z_0, and it is entire. The function $v(z)$ is also entire. From (1.8.6) it follows the equations for $u(z)$ and $v(z)$

$$uu'' = u'^2 - v^2\,, \tag{1.8.10}$$

$$v''u^2 + vu'^2 - 2u'v'u = v^3 + zvu^2 + \alpha u^3\,. \tag{1.8.11}$$

When v is expressed as a polynomial in u, the equation (1.8.10) reduces to a Weierstrass equation (1.4.14). It can be proved that all the solutions of (1.8.9) are entire, and, if (u,v), $u \neq 0$ is an arbitrary solution of (1.8.9), then (1.8.6) is a solution of (1.8.4).

The equation (1.8.3) has a one-parameter family of solutions, which is defined by the general solution of Riccati equation (1.8.3). Indeed from this equation we obtain

$$w'' = a'w^2 + b'w + c' + (2aw + b)(aw^2 + bw + c)\,,$$

and the coefficients a,b,c are obtained, $a^2 = 1$, $b = 0$, $c = \dfrac{z}{2a}$, $\alpha = \dfrac{a}{2}$.

By employing the Painlevé analysis, the explicit solutions for several nonlinear equations can be obtained. To illustrate this, we present further some examples of Painlevé analysis application.

EXAMPLE 1.8.1 (Satsuma) Exact solutions of a nonlinear reaction–diffusion equation of the type

$$u_t = (u^2)_{xx} + F(u). \tag{1.8.12}$$

In (1.8.12) $F(u)$ is a polynomial. In population dynamics, this equation is a model for the spatial diffusion of biological populations. Let us consider the case

$$F(u) = u(u-1)(\alpha - u). \tag{1.8.13}$$

Substituting $u = u(z) = u(x-ct)$ into (1.8.12), we have

$$(u^2)_{zz} + cu_z + u(u-1)(\alpha - u) = 0. \tag{1.8.14}$$

We suppose that the solution is a meromorphic function which can be written as a Laurent series, that is

$$u(z) = \sum_{n=-m}^{\infty} a_n (z-z_0)^n. \tag{1.8.15}$$

For simplicity we take $z_0 = 0$. Substitution of (1.8.15) into (1.8.14) and equating the terms with the same power in z, yield to $m = 2$, $a_0 = \frac{5}{12}(1+\alpha)$, $a_1 = -\frac{1}{28}c$, $a_{-1} = 0$, $a_{-2} = 20$, and so on. It is easy to verify that the determination of a_4 leads to $c = 0$. For this particular value of c, the solution (1.8.15) becomes

$$u = 20z^{-2} + \frac{5}{12}(1+\alpha) + \frac{1}{60}\{\frac{5}{16}(1+\alpha)^2 - \alpha\}z^2 + \ldots . \tag{1.8.16}$$

Analyzing the nature of the singularities in (1.8.16), we choose for u the form

$$u = \frac{f}{g},$$

$$f = A + B\exp(kz) + C\exp(2kz), \tag{1.8.17}$$

$$g = [1+\exp(kz)]^2,$$

where K, A, B and C are constants. Substituting (1.8.17) into (1.8.14) we find two nontrivial cases:

Case 1. $A = 1$, $B = -2$, $C = 1$, $k = \frac{1}{\sqrt{5}}$, $\alpha = \frac{3}{5}$. It follows that

$$u = \begin{cases} \tanh^2 \frac{1}{2\sqrt{5}} z, & z < 0, \\ 0, & z \ge 0. \end{cases} \tag{1.8.18}$$

Case 2. $A = 0$, $B = 5$, $C = 0$, $k = \frac{1}{2}\mathrm{i}$, $\alpha = 0$. It results

$$u = \frac{5}{4}\sec^2\frac{1}{4}z. \quad (1.8.19)$$

Solution (1.8.18) corresponds to an equilibrium solution. The equilibrium solution has the property

$$\lim_{z\to -0} u_z = 0, \ \lim_{z\to -0} u_{zz} = \text{const}.$$

By setting $z \to \mathrm{i}z$, the solution (1.8.18) can be written as a soliton

$$u = \frac{5}{4}\mathrm{sech}^2\frac{1}{4}z. \quad (1.8.20)$$

EXAMPLE 1.8.2 (Satsuma) Let us consider the equation

$$u_t = (u^2)_{xx} + 16u(u - \frac{3}{4}). \quad (1.8.21)$$

By expressing the solution u into a Laurent series, the same procedure as above, yields to an equilibrium solution

$$u = \begin{cases} \cos^2 x, & |x| < \frac{\pi}{2}, \\ 0, & |x| \geq 0. \end{cases} \quad (1.8.22)$$

Suppose that $u = f(t)\cos^2 x$, and substituting it into (1.8.21) we obtain

$$u = \begin{cases} \frac{1}{1 - \exp(12(t - t_0))}\cos^2 x, & |x| < \frac{\pi}{2}, \\ 0, & |x| \geq \frac{\pi}{2}, \end{cases} \quad (1.8.23)$$

for $u(x = 0) > 1$, and

$$u = \begin{cases} \frac{1}{1 + \exp(12(t - t_0))}\cos^2 x, & |x| < \frac{\pi}{2}, \\ 0, & |x| \geq \frac{\pi}{2}, \end{cases} \quad (1.8.24)$$

for $u(x = 0) < 1$. The solution (1.8.23) blows up at $t = t_0$, and the solution (1.8.24) tends to zero as $t \to \infty$. This result implies that the equilibrium solution (1.8.22) is unstable.

EXAMPLE 1.8.3 (Satsuma) For the equation

$$u_t = (u^2)_{xx} - 6u^2(u^2 - 1), \quad (1.8.25)$$

the Painlevé analysis leads to a traveling wave solution

$$u = \begin{cases} \tanh(2t - x), & x < 2t, \\ 0, & x \geq 2t. \end{cases} \tag{1.8.26}$$

This wave is a *kink*, which propagates in the positive direction of the axis x. From the symmetry of the equation, the solution

$$u = \begin{cases} \tanh(2tx), & x > -2t, \\ 0, & x \leq -2t, \end{cases} \tag{1.8.27}$$

is a traveling wave which propagates in the negative direction.

EXAMPLE 1.8.4 (Satsuma) For equation

$$u_t = (u^2)_{xx} + \alpha u, \tag{1.8.28}$$

the Painlevé analysis leads to

$$u = \begin{cases} \dfrac{\alpha \exp(\alpha t)}{12(1 - \exp(\alpha t))}[x^2 - c(1 - \exp(\alpha t))^{2/3}], & |x| < \sqrt{c}(1 - \exp(\alpha t))^{1/3}, \\ 0, & |x| \geq \sqrt{c}(1 - \exp(\alpha t))^{1/3}, \end{cases} \tag{1.8.29}$$

for $\alpha < 0$, with c a positive constant. The solution is meaningful for $t > 0$. The amplitude monotonically decreases to zero, and the width increases to $2\sqrt{c}$. For $\alpha > 0$, two types of solutions are derived

$$u = \begin{cases} -\dfrac{\alpha \exp(\alpha t)}{12(1 + \exp(\alpha t))}[x^2 - c(1 + \exp(\alpha t))^{2/3}], & |x| < \sqrt{c}(1 + \exp(\alpha t))^{1/3}, \\ 0, & |x| \geq \sqrt{c}(1 + \exp(\alpha t))^{1/3}, \end{cases} \tag{1.8.30}$$

$$u = \begin{cases} -\dfrac{\alpha \exp(\alpha t)}{12(\exp(\alpha t) - 1)}[x^2 - c(\exp(\alpha t) - 1)^{2/3}], & |x| < \sqrt{c}(\exp(\alpha t) - 1)^{1/3}, \\ 0, & |x| \geq \sqrt{c}(\exp(\alpha t) - 1)^{1/3}, \end{cases} \tag{1.8.31}$$

where c are c' are constants.

The amplitude of the solution (1.8.30) monotonically increases to infinity, and the amplitude of the solution (1.8.31) admits a minimum for $t = \dfrac{\ln 3/2}{\alpha}$, and then increases to infinity.

Chapter 2

SOME PROPERTIES OF NONLINEAR EQUATIONS

2.1 Scope of the chapter

The stability of solitons is explained by the existence of infinitely many conservation laws. The conserved geometric features of solitons are related also to the symmetries. A symmetry group of an equation consists of variable transformations that leave the equation invariant.

In this chapter we summarize some of the elementary principles of linear and nonlinear evolution equations, including the symmetries and conservation laws. In the classical Sophus Lie theory, the symmetry groups consist of geometric symmetries, which are transformations of independent and dependent variables. For example, the KdV equation has four such linear independent symmetries, namely arbitrary translations in the space and time coordinates, the Galilean boost and the scaling.

The theorem of Emmy Noether (1918) gives a one-by-one correspondence between symmetry groups and conservation laws for Euler–Lagrange equations. The generalized symmetries introduced by Noether, are groups whose infinitesimal generators depend not only on the independent and dependent variables, but also the derivatives of the dependent variables.

The generalized symmetries are able to explain the existence of infinitely many conservation laws for a given nonlinear evolution equation.

Many authors have studied the properties of nonlinear equations with solitonic behavior. We refer to the monographs of Dodd *et al.* (1982), Teodorescu and Nicorovici-Porumbaru (1985) and Engelbrecht (1991) and to works of Wang (1998) and Bâlă (1999).

2.2 General properties of the linear waves

The soliton, described by the hyperbolic secant shape, is a localized disturbance with non-oscillatory motion, having the velocity dependent of its amplitude. This contrasts strongly with the linear waves for which the velocity is independent of the amplitude. Let us consider the one-dimensional string motion equation

$$u_{tt} - c^2 u_{xx} = 0, \tag{2.2.1}$$

where c is a real positive number. Let x range from $-\infty$ to ∞. For the transverse vibrations of a string $c^2 = \frac{T}{\lambda}$, where T is the constant tension and λ, the mass per unit length at the position x. For the compressional vibrations of an isotropic elastic solid in which the density and elastic constants are functions of x only (laminated medium), we have $c^2 = \frac{\lambda + 2\mu}{\rho}$. For the transverse vibrations of such laminated solid it follows that $c^2 = \frac{\mu}{\rho}$. The characteristics are given by $\frac{\mathrm{d}x}{\mathrm{d}t} = \pm c$, namely the straight lines inclined to the axis at $c = \tan\varphi$.

D'Alembert solution of (2.2.1) is written as

$$u(x,t) = f(x-ct) + g(x+ct). \tag{2.2.2}$$

Functions $f, g : \mathrm{R} \to \mathrm{R}$ are determined from the initial conditions attached to (2.2.1)

$$u(x,0) = \Phi(x), \quad u_t(x,0) = \Psi(x). \tag{2.2.3}$$

Thus, we have

$$f(x) = \frac{1}{2}\Phi(x) + \frac{1}{2c}\int_0^x \Psi(\alpha)\mathrm{d}\alpha + \frac{a}{2}, \tag{2.2.4}$$

$$g(x) = \frac{1}{2}\Phi(x) - \frac{1}{2c}\int_0^x \Psi(\alpha)\mathrm{d}\alpha + \frac{a}{2}, \quad a \in \mathrm{R}. \tag{2.2.5}$$

The solution (2.2.2) describes two waves $f(x-ct)$, and respectively $g(x+ct)$. Geometrically, the function $u(x,t)$ can be represented as a surface in the space (u, x, t). A section through this surface in the plane $t = t_0$, is $u = u(x, t_0)$, and represents the profile of the vibrating string (a wave) at the time $t = t_0$. A section through the surface in the plane $x = x_0$, is $u = u(x_0, t)$, and represents the motion phenomenon of the point x_0. The modified profiles of $f(x-ct)$ can be determined in the following way. Consider one observer with a system of coordinates (x', t') so that, at the time $t_0 = 0$, the observer occupies the position $x = 0$, and at the time t, the position ct, since it has a rectilinear motion with the velocity c.

In the new coordinate system (x', t'), attached to the observer $(t' = t, x' = x - ct)$, the function $f(x-ct)$ is specified, at any time t', by $f(x')$.

The observer sees, at any moment of time the unchanged profile $f(x)$ at the initial time $t_0 = t'_0 = 0$. This is the way the function $f(x-ct)$ represents a right traveling wave or a forward-going wave with the velocity c. For a similar reason, $g(x+ct)$ represents a left traveling wave or a backward-going wave with the velocity c. As a consequence, both waves are not interacting between them and do not change their shape during the propagation. These waves can be called *solitary waves*, for the reason they are not changing their shape during propagation process, and do not interact one with the other These waves can be superposed by a simple sum, because of the linearity of (2.2.1). The most elementary linear wave is a harmonic wave

$$v(x,t) = A\exp i(kx - \omega t)\,, \tag{2.2.6}$$

and represents the solution of (2.2.1).

The real number k is the wave number, related to the wavelength by $k = \dfrac{2\pi}{\lambda} = \dfrac{\omega}{c}$, ω is the angular frequency related to the frequency by $\omega = 2\pi f$, and the real number A is the amplitude.

The phase velocity is the speed of the phase $\varphi = \omega t - kx$, and represents the velocity of propagation of a surface with constant phase

$$c_p = \frac{\omega}{k}\,. \tag{2.2.7}$$

The group velocity is the speed of the bulk of the wave

$$c_g = \frac{d\omega}{dk}\,. \tag{2.2.8}$$

Introducing (2.2.6) into (2.2.1) we obtain the dispersion relation, that is a relation between k and ω

$$F(\omega, k) = 0\,. \tag{2.2.9}$$

In particular, for homogeneous and isotropic media the classical one-dimensional longitudinal wave propagation equation is

$$(\lambda + 2\mu)\phi_{xx} = \rho\phi_{tt}\,, \tag{2.2.10}$$

where ϕ is displacement, λ, μ the Lamé elastic constants, and ρ density. The dispersion relation is written as

$$k^2(\lambda + 2\mu) = \omega^2\rho\,. \tag{2.2.11}$$

Here, $\dfrac{\omega}{k} = \text{const.}$, and we say that the waves are not dispersive. If ω is real, we say we have *dispersion of waves*, if the phase velocity depends on the wavenumber. For example, in the case of the linearised KdV equation

$$u_t - u_x + u_{xxx} = 0\,, \tag{2.2.12}$$

the dispersion relation is

$$\omega + k + k^3 = 0\,. \tag{2.2.13}$$

For $\text{Re}\dfrac{\omega}{k} = \text{const.}$, the waves are nondispersive, and for $\text{Im}\dfrac{\omega}{k} = 0$, the waves are nondissipative. The phase velocity $c_p = -1 - k^2$, depends on the wavenumber and then we have the dispersion phenomenon. The group velocity $c_g = -1 - 3k^2$ differs by the phase velocity for $k \neq 0$, and in consequence, the components of waves scatter and disperse in the propagation process.

The linear Klein–Gordon equation

$$\phi_{xx} - \phi_{tt} = m^2\phi \,, \tag{2.2.14}$$

yields dispersion relation

$$\omega^2 = m^2 + k^2 \,. \tag{2.2.15}$$

If ω is complex $\omega = \mathrm{Re}\,\omega + \mathrm{i}\,\mathrm{Im}\,\omega$, $\mathrm{Im}\,\omega < 0$, we say we have *dissipation of waves.* The solution in this case is

$$v(x,t) = A\exp(t\,\mathrm{Im}\,\omega)\exp\mathrm{i}[kx - \mathrm{Re}(\omega t)] \,, \tag{2.2.16}$$

and the amplitude is exponential decreasing at $t \to \infty$.

For the linearised Burgers equation

$$u_t + u_x + u_{xx} = 0 \,, \tag{2.2.17}$$

the dispersion relation is

$$\omega = k - \mathrm{i}k^2 \,. \tag{2.2.18}$$

The phase velocity of the harmonic waves $v(x,t) = A\exp(-tk^2)\exp\mathrm{i}k(x-t)$ is $c_p = 1$, and the group velocity, $c_g = 1 - 2\mathrm{i}k$.

The dissipation appears because $\mathrm{Im}\,\omega = -k^2$ is negative for any real k. In conclusion, the KdV equation is dispersive due to the term u_{xxx}, and Burgers equation is dissipative due to u_{xx}.

Consider next the linearised Schrödinger equation

$$\phi_t = \delta\phi_{xx} \,. \tag{2.2.19}$$

If $\delta = -\mathrm{i}$, (2.2.19) becomes

$$\mathrm{i}\phi_t = \phi_{xx} \,. \tag{2.2.20}$$

It follows that $\omega = -k^2$, $c_p = -k$, $c_g - 2k$, and, as we mentioned before, this is a purely dispersive equation. If δ is real and positive, we obtain $\omega = -\mathrm{i}\delta k^2$, and the real part of ω vanishes. We do not have dispersion, and the waves decay as $\exp(-\delta k^2 t)$ when $t \to \infty$. In this case (2.2.19) is the heat equation, and is purely dissipative.

Coming back to (2.2.19), let us consider the initial conditions $\phi(x,0) = f(x)$, with a nonharmonic function $f(x)$. In this case we represent $\phi(x,0)$ by a Fourier integral

$$\phi(x,0) = \frac{1}{\sqrt{2\pi}}\int_{-\infty}^{\infty} A(k)\exp(\mathrm{i}kx)\,\mathrm{d}k \,, \tag{2.2.21}$$

where

$$A(k) = \frac{1}{\sqrt{2\pi}} \int_{-\infty}^{\infty} \phi(x,0) \exp(-ikx)\,\mathrm{d}x\,. \tag{2.2.22}$$

A solution for (2.2.19) is given by

$$\phi(x,t) = \frac{1}{\sqrt{2\pi}} \int_{-\infty}^{\infty} A(k) \exp\{\mathrm{i}[kx - \omega(k)t]\}\,\mathrm{d}k\,, \tag{2.2.23}$$

with the amplitude $A(k)$ determined from the initial conditions $\phi(x,0)$.

The solution (2.2.23) can be written under the form (Dodd *et al.*).

$$\phi(x,t) = \frac{1}{\sqrt{2\pi}} \int_{-\infty}^{\infty} \mathrm{d}k \left[\int_{-\infty}^{\infty} \phi(\xi,0) \exp(-\mathrm{i}k\xi)\mathrm{d}\xi \right] \exp[\mathrm{i}(kx + \mathrm{i}\delta k^2 t)]\,, \tag{2.2.24}$$

or

$$\phi(x,t) = \frac{1}{2\pi} \int_{-\infty}^{\infty} \left[\int_{-\infty}^{\infty} \exp\{\mathrm{i}[(x-\xi)k - \delta k^2 t]\}\mathrm{d}k \right] \phi(\xi,0)\mathrm{d}\xi\,. \tag{2.2.25}$$

Noting $\eta = x - \xi$, (2.2.25) reduces to the evaluation of two integrals

$$I_1 = \int_{-\infty}^{\infty} \cos k\eta \exp(-\delta k^2 t)\,\mathrm{d}k\,,\quad I_2 = \int_{-\infty}^{\infty} \sin k\eta \exp(-\delta k^2 t)\,\mathrm{d}k\,. \tag{2.2.26}$$

The integrant of I_2 is odd, and then $I_2 = 0$.

We define $\lambda^2 = \delta k^2 t$ and $a = \frac{\eta}{\sqrt{\delta t}}$, so that

$$I_1(a) = \frac{1}{\sqrt{\delta t}} \int_{-\infty}^{\infty} \cos a\lambda \exp(-\lambda^2)\,\mathrm{d}\lambda\,. \tag{2.2.27}$$

Differentiating (2.2.27) with respect to a, and integrating by parts with respect to λ, yield

$$\frac{\mathrm{d}I_1}{\mathrm{d}a} = -\frac{1}{2} a I_1,\quad I_1(0) = \sqrt{\frac{\pi}{\delta t}}\,, \tag{2.2.28}$$

$$I_1 = \sqrt{\frac{\pi}{\delta t}} \exp(-\frac{1}{4}a^2)\,, \tag{2.2.29}$$

where $\int_{-\infty}^{\infty} \exp(-\lambda^2)\,\mathrm{d}\lambda = \sqrt{\pi}$. Finally, the solution is

$$\phi(x,t) = \frac{1}{\sqrt{4\pi\delta t}} \int_{-\infty}^{\infty} \phi(\xi,0) \exp[-\frac{(x-\xi)^2}{4\delta t}]\,. \tag{2.2.30}$$

The integral (2.2.30) can be evaluated in a particular case, namely $\phi(x,0)$ is a Gaussian function, $\phi(x,0) = A\exp(-\alpha x^2)$. The result is

$$\phi(x,t) = \frac{A}{\sqrt{1+4\pi\alpha\delta t}} \exp[\frac{-\alpha x^2}{1+4\pi\delta t}]. \tag{2.2.31}$$

If δ is real and positive, then $\phi \to 0$ as $t \to \infty$, the equation being dissipative. If $\delta = -i$, then ϕ is an oscillatory function, and the initial wave is dispersed.

Finally, we may say that the cnoidal waves, expressed by Jacobi elliptic functions, form a special class of nonlinear waves that bridge the linear waves to solitons.

Such a class of waves is represented by solutions of the KdV equation

$$u_t - 6uu_x + u_{xxx} = 0. \tag{2.2.32}$$

Introducing the new variable $\theta = x - c\tau$, equation KdV becomes

$$u'' = -6u^2 + (c-1)u. \tag{2.2.33}$$

Multiplying (2.2.33) by $2u'$ and integrating we obtain a Weierstrass equation (1.4.8)

$$u'^2 = -4u^3 + (c-1)u^2 + C, \tag{2.2.34}$$

where C is a constant. For $g_2^3 - 27g_3^2 > 0$ and $C = 0$, we obtain the solution

$$\wp(u) = e_3 + (e_1 - e_3)\mathrm{cn}^2(u^*, m), \tag{2.2.35}$$

where $m = \frac{e_2 - e_3}{e_1 - e_3}$ and $u^* = \sqrt{e_1 - e_3}u$, with $e_1 \geq e_2 > e_3$, the roots of cubic polynomials from the right-hand side of (2.2.34). Coming back to the KdV equation, the solution becomes

$$u(\theta) = -\frac{1}{2}c\,\mathrm{cn}^2\{\frac{1}{2}\sqrt{c}(\theta - \theta_0)\}. \tag{2.2.36}$$

These constants c and θ_0 are determined from the initial conditions $u(0) = u_0$, $u'(0) = c_0$. The solution (2.2.36) is called a cnoidal wave. In the limit, when u and its derivatives tend to zero at infinity, we have

$$\mathrm{cn}^2\{\frac{1}{2}\sqrt{c}(\theta-\theta_0)\} \to \mathrm{sech}^2\{\frac{1}{2}\sqrt{c}(\theta-\theta_0)\}, \quad m \to 1, \tag{2.2.37}$$

and the cnoidal wave transforms into a soliton.

On the other hand, for small amplitude waves, when the linearised version of the KdV equation is appropriate, we have in the limit $m \to 0$

$$\mathrm{cn}^2\{\frac{1}{2}\sqrt{c}(\theta-\theta_0)\} \to \cos^2\{\frac{1}{2}\sqrt{c}(\theta-\theta_0)\}, \quad m \to 0, \tag{2.2.38}$$

and the cnoidal wave transforms into a cosine oscillation.

2.3 Some properties of nonlinear equations

In nature, each effect admits an opposite, that means a contrary effect. The opposite of dispersion is the energy concentration due to nonlinearities of the medium, which manifests as a focusing of waves (Munteanu and Donescu, Munteanu and Boştină, Whitham 1974).

The opposite of dissipation is amplification. The amplification of waves arises from an influx of energy in active media, where the energy is pumping from a source to wave motion, or due to some interactions between waves and medium (Chiroiu *et al.* 2001a).

The mechanism, called the dilaton mechanism, is recently proposed for explaining the possible amplification of nonlinear seismic waves (Engelbrecht and Khamidullin, Engelbrecht).

Zhurkov and Petrov introduce the dilaton concept to explain the fracture of solids. The dilaton is a fluctuation of internal energy of a medium with loosened bonds between its structural elements. The dilaton is able to absorb energy from the surrounding medium, and when the accumulated energy in it has reached its critical value, the dilaton breaks up releasing the stored energy, causing the amplification of waves. This mechanism is controlled by the intensity of the propagating wave. Low-intensity waves give a part of their energy away to the dilatons, and high-intensity waves cause the dilatons to break up.

The dilatons may be distributed in a medium according to a certain order, may be absent in some regions, or may be randomly distributed. Sadovski and Nikolaev have shown in 1982 that the phase transitions in a multiphased medium can activate the dilatonic mechanism of energy pumping from a dilaton to wave motion. We can say also, that the tectonic stress concentrations in solids may grow until the stress field components reach their critical values at the points of their maximum concentrations, causing a seismically active event with the releasing of energy by the dilatons broken (Kozák and Šileny).

Here is an example based on the description of long seismic waves in an elastic layer of thickness h resting on an elastic halfspace. The motion equation of a continuous medium is (Eringen 1970)

$$(\sigma_{KL}x_{k,L})_{,K} + \rho_0(f_k - A_k) = 0\,, \tag{2.3.1}$$

where ρ_0 is the density in underformed case, σ_{KL} is the Kirchhoff stress tensor, f_k the body force, A_k the acceleration, and x_k the Eulerian coordinates.

The constitutive laws are given by

$$\sigma_{KL} = \sigma_{KL}(E_{KL}, \frac{\partial E_{KL}}{\partial t})\,, \tag{2.3.2}$$

$$f_k = f_k(E_{KL})\,, \tag{2.3.3}$$

where E_{KL} is the Green deformation tensor. The constitutive law (2.3.3) is referred to the body forces that result from long-range effects, and depend on E_{KL}.

Noting with U the transverse displacement, the motion equation of waves in the positive direction of the axis X is obtained from (2.3.2) (Engelbrecht)

$$\frac{\partial^2 U}{\partial t^2} = c_{s2}^2 \frac{\partial^2 U}{\partial X^2} + c_{s2}^2 m (\frac{\partial U}{\partial X})^2 \frac{\partial^2 U}{\partial X^2} + c_{s2}^2 l_0^2 \frac{\partial^4 U}{\partial X^4} + f \,, \tag{2.3.4}$$

where X is directed along the layer, m is a nonlinear material parameter expressed in terms of second, third and fourth order elastic moduli, ν_i, $i = 1,2$ are Poisson ratio in the layer ($i = 1$) and the halfspace ($i = 2$), and c_{si}, $i = 1,2$, are the velocity of transverse waves in layer and in the halfspace

$$\omega^2 = c_{s2}^2 (k^2 - l_0^2 k^4) \,, \; l_0^2 = h^2 (\frac{c_{s2}^2}{c_{s1}^2} - 1)^2 \frac{\nu_1^2}{\nu_2^2} > 0 \,. \tag{2.3.5}$$

In (2.3.5) ω is the angular frequency and k the wave number. The equation (2.3.4) may be written under the form

$$\frac{\partial u}{\partial \tau} - \frac{m}{2\varepsilon^2 c_{s2}^2} u^2 \frac{\partial u}{\partial \xi} + \frac{1}{2} \frac{l_0}{\varepsilon} \frac{\partial^3 u}{\partial \xi^3} + \frac{1}{2\varepsilon \, c_{s2}} f = 0 \,, \tag{2.3.6}$$

where $\xi = c_{s2} t - X$, $\tau = \varepsilon^2 X$, $u = \frac{\partial U}{\partial t}$. The small parameter ε is related to the absolute value of the deformation $\frac{\partial U}{\partial X} << 1$.

The constitutive law (2.3.3) is written under the form

$$f = -(b_1 U_X - b_2 U_X^2 + b_3 U_X^3) \,, \tag{2.3.7}$$

with b_i, $i = 1,2,3$, positive constants.

Since the motion equation (2.3.6) is written in terms of $u = \frac{\partial U}{\partial t}$, we express also (2.3.7) under the form

$$f = B_1 u - B_2 u^2 + B_3 u^3 \,, \tag{2.3.8}$$

where

$$B_1 = \frac{b_1}{c_{s2}} \,, \; B_2 = \frac{b_2}{c_{s2}^2} \,, \; B_3 = \frac{b_3}{c_{s2}^3} \,.$$

Using the dimensionless variables

$$v = \frac{u}{u_0} \,, \; \sigma = \frac{a_1 u_0^2 \tau}{\tau_0} \,, \; \zeta = \frac{\xi}{\tau_0} \,, \; a_1 = \frac{1}{2} |m| \varepsilon^{-2} c_{s2}^{-2} \,,$$

where u_0 is the initial amplitude of the velocity, and τ_0 the wavelength. The equation (2.3.6) becomes

$$\frac{\partial v}{\partial \sigma} - \text{sign}(m) v^2 \frac{\partial v}{\partial \zeta} + \mu \frac{\partial^3 v}{\partial \zeta^3} + f(v) = 0 \,, \tag{2.3.9}$$

where $\mu = \dfrac{\varepsilon l_0^2 c_{s2}^2}{\tau_0 u_0^2 \,|\, m \,|}$, and

$$f(v) = \beta_1 v - \beta_2 v^2 + \beta_3 v^3 , \tag{2.3.10a}$$

$$\beta_1 = Q\frac{B_1}{u_0^2} > 0 ,\ \beta_2 = Q\frac{B_2}{u_0} > 0 ,\ \beta_3 = QB_3 > 0 ,\ \ Q = \frac{\tau_0 \varepsilon c_{s2}}{|\, m \,|} . \tag{2.3.10b}$$

The equation (2.3.9) governs the motion of long transversal waves in a layer in the case of a driving force f. Due to the cubic nonlinearity of f (2.3.10a), the equation (2.3.9) can be reduced to modified KdV equation that contains the term $v^2 \dfrac{\partial v}{\partial \zeta}$.

Introducing the initial condition $v(0) = A_0 \mathrm{sech}^2 \sigma$, the equation (2.3.9) admits a dilatonic behavior in some circumstances. For $A_0 / A_{cr} = 1$, and $m = 1$, the initial soliton transforms into an asymmetric soliton, which becomes unstable at the finite time $\tau = \tau_c = (\frac{8}{45}\frac{a_2}{\mu})^{-1/2}$, $a_2 = \frac{8}{5} b_2 > 0$ (Figure 2.3.1).

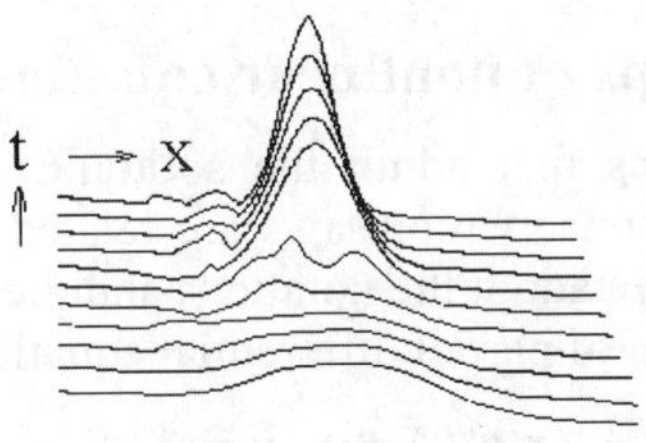

Figure 2.3.1 The dilatonic behavior of an initial soliton of amplitude $A_0 / A_{cr} = 1$ due to a driving force $f\ (m = 1)$.

The dilatonic behavior depends on the properties of the medium and on critical amplitude of the initial soliton $A_{cr} = \dfrac{a_2}{15\mu \,|\, u_0 \,|}$. For $A_0 < A_{cr}$ the amplitude of the soliton decreases, and for $A_0 \geq A_{cr}$ the amplitude of the soliton dramatically increases.

When the nonlinear and dispersive effects are equilibrated, the perturbed soliton by the initial data remains unchanged and propagates without changing its identity (velocity, amplitude and shape).

In Figure 2.3.2 it is represented as the explosive amplification of the dimensionless amplitude v (solution of (2.3.9)) with respect to time t [sec] and coordinate X_1 [m], for three values of the ratio $r_0 = A_0 / A_{cr}$, namely $r = 0.98$, $r = 1$ and $r = 1.2$, for the case $m = -1$.

As we said, the soliton is a perfect balance between nonlinear and dispersive effects, exhibiting a remarkable survivability under conditions where a wave might normally be destroyed.

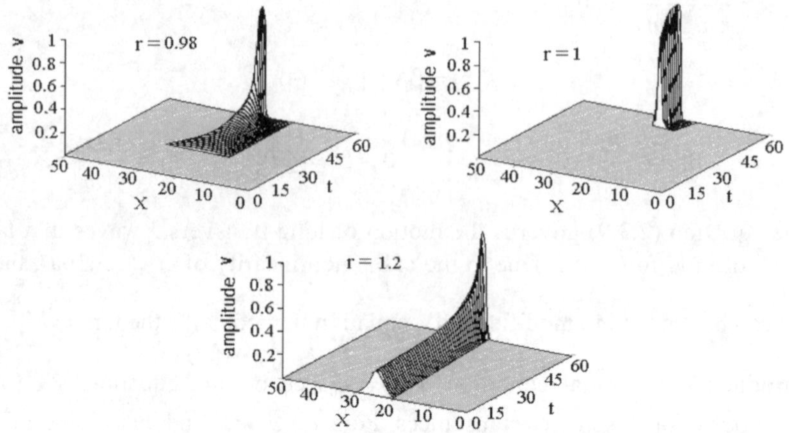

Figure 2.3.2 Amplitude of v with respect to time t and X, for $r = 0.98$, $r = 1$ and $r = 1.2$ ($m = -1$).

2.4 Symmetry groups of nonlinear equations

The evolution equations that admit the solitonic solutions have infinitely many conservation laws. These conservation laws are related to the symmetry groups of the equations. In this section, we adopt the notations and the point of view of Bâlă.

Let us consider a system of partial differential equations

$$\Delta_\nu(x,u^{(n)}) = 0\,,\ \ \nu = 1,2,...,l\,, \tag{2.4.1}$$

where

$$x = (x^1,x^2,...,x^p)\,,\ \ u = (u^1,u^2,...,u^p)\,,$$

and

$$\Delta(x,u^{(n)}) = (\Delta_1(x,u^{(n)}),...,\Delta_l(x,u^{(n)}))\,,$$

is a differentiable function. All the derivatives of u are denoted by $u^{(n)}$. Any function $u = h(x)$, $h : D \subset R^p \to U \subset R^q$, $h = (h^1,h^2,...,h^q)$, induces the function $u^{(n)} = pr^{(n)}h$ called the n-th prolongation of h, which is defined by $u_j^\alpha = \partial_j h^0$, $pr^{(n)}h : D \to U^{(n)}$, and for each $x \in D$, $pr^{(n)}h$ is a vector whose $qp^{(n)} = C_{p+n}^n$ entries represent the values of h and all its derivatives up to order n at the point x. The space $D \times U^{(n)}$, whose coordinates represent the independent variables, the dependent variables and the derivatives of the dependent variables up to the order n, is called the n-th order jet space of the underlying space $D \times U$. Therefore, Δ is a map from the jet space $D \times U^{(n)}$ to R^l. The system of equations (2.4.1) determine the subvariety

$$S = \{(x,u^{(n)}),\ \Delta(x,u^{(n)}) = 0\}\,,$$

of the total jet space $D\times U^{(n)}$. We can identify the system of equations (2.4.1) with its corresponding subvariety S.

Let $M\subset D\times U^{(n)}$ be an open set. A symmetry group of (2.4.1) is a local group of transformations G acting on M with the property that wherever $u=f(x)$ is a solution of (2.4.1) and whenever $g\cdot f$ is defined for $g\in G$, then $u=g\cdot f(x)$ is also a solution of the system.

The system (2.4.1) is called invariant with respect to G. If X is a vector on M with corresponding 1-parameter group $\exp(\varepsilon X)$ that is the infinitesimal generator of the symmetry group of (2.4.1). The infinitesimal generator of the corresponding prolonged 1-parameter group $pr^{(n)}[\exp(\varepsilon X)]$

$$pr^{(n)}X\,|_{(x,u^{(n)})}=\frac{\mathrm{d}}{\mathrm{d}\varepsilon}pr^{(n)}[\exp(\varepsilon X)](x,u^{(n)})\,|_{\varepsilon=0},$$

for any $(x,u^{(n)})\in M^{(n)}$, is a vector field on the n-jet space $M^{(n)}$ called the n-prolongation of X and denoted by $pr^{(n)}X$. The system (2.4.1) is called to be of maximal rank if the Jacobi matrix

$$J_\Delta(x,u^{(n)})=\left(\frac{\partial\Delta_\nu}{\partial x^i},\frac{\partial\Delta_\nu}{\partial u_j^0}\right),$$

of Δ, with respect to all the variables $(x,u^{(n)})$ is of rank l whenever $\Delta(x,u^{(n)})=0$.

THEOREM 2.4.1 (Bâlă) *Let*

$$X=\sum_{i=1}^{p}\zeta^i(x,u)\frac{\partial}{\partial x^i}+\sum_{\alpha=1}^{q}\phi_\alpha(x,u)\frac{\partial}{\partial u^\alpha},$$

be a vector field on $M\subset D\times U$. *The* n-*th prolongation of* X *is the vector field*

$$pr^{(n)}X=X+\sum_{\alpha=1}^{q}\sum_{J}(x,u^{(n)})\frac{\partial}{\partial u_j^\alpha}\phi_\alpha^J,$$

defined on the corresponding jet space $M^{(n)}\subset D\times U^{(n)}$, *and* J *is the multi-indices* $J=(j_1,\dots,j_k)$, $1\le j_k\le p$, $1\le k\le n$. *The coefficient functions* ϕ_α^J *are given by*

$$\phi_\alpha^J(x,u^{(n)})=D_J(\phi_\alpha-\sum_{i=1}^{p}\zeta^i u_i^\alpha)+\sum_{i=1}^{p}\zeta^i u_{J,i}^\alpha.$$

THEOREM 2.4.2 (criterion of infinitesimal invariance). *Consider the system of equations (2.4.12) of maximal rank defined over* $M\subset D\times U$. *If* G *is a local group of transformations acting on* M *and*

$$pr^{(n)}X[\Delta_\nu(x,u^{(n)})]=0,\ \nu=1,2,\dots,l, \tag{2.4.2}$$

whenever $\Delta_v(x,u^{(n)})=0$, *for every infinitesimal generator* X *of* G, *then* G *is a symmetry group of (2.4.1).*

To find the symmetry group G of the system (2.4.1) we consider the vector field X on M and write the infinitesimal invariance condition (2.4.2). Then eliminate any dependence between partial derivatives of u^α and write the condition (2.4.2) like polynomials in the partial derivatives of u^α. Equate with zero the coefficients of the partial derivatives of u^α in (2.4.2) and obtain a partial differential equations system with respect to the unknown functions ζ^i, ϕ_α, and this system defines the symmetry group G of (2.4.1).

Now let us study the symmetry group of the system of equations arising from the Tzitzeica equations of the surface (Bâlă)

$$\theta_{uu}=a\theta_u+b\theta_v,$$

$$\theta_{uv}=h\theta, \tag{2.4.3}$$

$$\theta_{vv}=a''\theta_u+b''\theta_v,$$

where the independent variables are u and v, and the dependent variable is θ. We associate to (2.4.3) the integrability conditions (see section 10.2)

$$ah=h_u,$$

$$a_v=ba''+h,$$

$$b_v+bb''=0, \tag{2.4.4}$$

$$b''h=h_v,$$

$$a''+aa''=0,$$

$$b''_u+a''b=h,$$

where a, b, a'', h..., are invariant functions of u and v. The solutions $x=x(u,v)$, $y=y(u,v)$, $z=z(u,v)$ of (2.4.3) and (2.4.4) define a Tzitzeica surface. Denote $x^1=u$, $x^2=v$ and $u^1=\theta$.

Let $D\times\bar{U}^2$ be the second jet space attached to (2.4.3) and $\bar{M}\subset D\times\bar{U}^2$ an open set. An infinitesimal generator of the symmetry group G of (2.4.3) is given by

$$\bar{X}=\zeta\frac{\partial}{\partial u}+\eta\frac{\partial}{\partial v}+\alpha\frac{\partial}{\partial\theta}, \tag{2.4.5}$$

where ζ, η and α are functions of u, v and θ. The theorem 2.4.1 gives the first and the second prolongations of the vector $\bar{X}$, namely

$$pr^{(1)}\bar{X}=\bar{X}+\alpha^u\frac{\partial}{\partial\theta_u}+\alpha^v\frac{\partial}{\partial\theta_v},$$

$$pr^{(2)}\overline{X} = pr^{(1)}\overline{X} + \alpha^{uu}\frac{\partial}{\partial\theta_{uu}} + \alpha^{uv}\frac{\partial}{\partial\theta_{uv}} + \alpha^{vv}\frac{\partial}{\partial\theta_{vv}}, \tag{2.4.6}$$

where

$$\begin{aligned}
\alpha^{u} &= D_u(\alpha - \zeta\theta_u - \eta\theta_v) + \zeta\theta_{uu} + \eta\theta_{uv},\\
\alpha^{v} &= D_v(\alpha - \zeta\theta_u - \eta\theta_v) + \zeta\theta_{uv} + \eta\theta_{vv},\\
\alpha^{uu} &= D_{uu}(\alpha - \zeta\theta_u - \eta\theta_v) + \zeta\theta_{uuu} + \eta\theta_{uuv},\\
\alpha^{uv} &= D_{uv}(\alpha - \zeta\theta_u - \eta\theta_v) + \zeta\theta_{uuv} + \eta\theta_{uvv},\\
\alpha^{vv} &= D_{vv}(\alpha - \zeta\theta_u - \eta\theta_v) + \zeta\theta_{uvv} + \eta\theta_{vvv}.
\end{aligned} \tag{2.4.7}$$

The existence of a subgroup $\overline{G}_1$ of the symmetry group $\overline{G}$, which acts on the space of the dependent variable θ, and the existence of a subgroup $\overline{G}_2$ of the symmetry group $\overline{G}$, which acts on the space of the independent variables u and v, are proved by the following theorems (Bâlă):

THEOREM 2.4.3 *The Lie algebra of the infinitesimal symmetries associated to the subgroup $\overline{G}_1$ of the full symmetry group $\overline{G}$ of (2.4.3) is generated by the vector field*

$$Y_1 = \theta\frac{\partial}{\partial\theta}. \tag{2.4.8}$$

THEOREM 2.4.4 *The general vector field of the algebra of the infinitesimal symmetries associated to the subgroup $\overline{G}_2$ of the full symmetry group $\overline{G}$ of (2.4.3) is*

$$\overline{Z} = \zeta(u)\frac{\partial}{\partial u} + \eta(v)\frac{\partial}{\partial v}, \tag{2.4.9}$$

where ζ and η satisfy

$$\begin{aligned}
&\zeta a_u + \eta a_v + a\zeta_u + \zeta_{uu} = 0,\\
&\zeta b_u + \eta b_v - b\eta_u + 2b\zeta_u = 0,\\
&\zeta h_u + \eta h_v + h(\zeta_u + \eta_v) = 0,\\
&\zeta a''_u + \eta a''_v - a''\zeta_u + 2a''\eta_v = 0,\\
&\zeta b''_u + \eta b''_v + b''\eta_v + \eta_{vv} = 0.
\end{aligned} \tag{2.4.10}$$

2.5 Noether theorem

In 1918, Emmy Noether proved the remarkable theorem giving a one-by-one correspondence between symmetry groups and conservation laws for the Euler–Lagrange equations (Teodorescu and Nicorovici-Porumbaru). To understand the Noether theorem, let us consider a mechanical system with n degrees of freedom, whose states are determined by the independent variable, time t, and the dependent variables (state functions or generalized coordinates) q_i, $i=1,2,...,n$.

Suppose that the equations of motion derive from the functional

$$\mathbb{F} = \int_{t_0}^{t_1} L(q_1,q_2,...,q_n,\dot{q}_1,\dot{q}_2,...,\dot{q}_n,t)\mathrm{d}t\ , \tag{2.5.1}$$

where L is the Lagrangian function, $\dot{q}_i$, $i=1,2,...,n$, the generalized velocities, and $t\in[t_0,t_1]$. To obtain the motion equation, we introduce the infinitesimal transformation of time

$$t' = t + \delta t\ , \tag{2.5.2}$$

with δt an arbitrary infinitesimal quantity.

In terms of $q_i' = q_i + \delta q_i$, and $\dot{q}_i' = \dot{q}_i + \delta\dot{q}_i$, $i=1,2,...,n$, the variation of the functional (2.5.1) is

$$\delta\mathbb{F} = \int_{t_0'}^{t_1'} L(q_i',\dot{q}_i',t')\mathrm{d}t' - \int_{t_0}^{t_1} L(q_i,\dot{q}_i,t)\mathrm{d}t\ ,\ i=1,2,...,n\ . \tag{2.5.3}$$

When the Lagrangian L does not explicitly depend on time, then $\delta\dot{q}_i = \frac{\mathrm{d}}{\mathrm{d}t}\delta q_i$. Writing the Taylor series expansion, neglecting the second derivatives

$$L(q_i',\dot{q}_i',t') = L(q_i,\dot{q}_i,t) + \frac{\partial L}{\partial q_i}\delta q_i + \frac{\partial L}{\partial \dot{q}_i}\delta\dot{q}_i + \frac{\partial L}{\partial t}\delta t\ ,$$

where the Einstein's summation convention for dummy indices is used, the variation of the integral becomes

$$\delta\mathbb{F} = \int_{t_0}^{t_1}\left[\frac{\mathrm{d}}{\mathrm{d}t}(L\delta t - \frac{\partial L}{\partial\dot{q}_i}\dot{q}_i\delta t + \frac{\partial L}{\partial\dot{q}_i}\delta q_i) + \left(\frac{\partial L}{\partial q_i} - \frac{\mathrm{d}}{\mathrm{d}t}\left(\frac{\partial L}{\partial\dot{q}_i}\right)\right)(\delta q_i - \dot{q}_i\delta t)\right]\mathrm{d}t\ , \tag{2.5.4}$$

where $i=1,2,...,n$.

The motion equations are obtained from (2.5.4) for assuming that the variation δt and the variation of generalized coordinates are identically zero at $t=t_0$ and $t=t_1$.

Thus, the first term in (2.5.4) vanishes, and the Lagrange equations of motion are derived

$$\frac{\partial L}{\partial q_i} - \frac{\mathrm{d}}{\mathrm{d}t}\left(\frac{\partial L}{\partial\dot{q}_i}\right) = 0\ . \tag{2.5.5}$$

The Lagrange equations are invariant if we replace L by αL, where α is an arbitrary non-zero constant. This transformation is called the scale transformation.

The Lagrange equations are also invariant if we replace L by $L+\frac{\mathrm{d}f}{\mathrm{d}t}$, where $f(t,q_i)$, is an arbitrary function. This transformation is called the gauge transformation.

The transformation of independent variable under which the form of the equations of motion (2.5.5) remains invariant is a symmetry transformation.

A general form of a transformation of independent variable is

$$t' = \varphi(t)\,, \tag{2.5.6}$$

and the changes of generalized coordinates are

$$q_i'(t') = Q_i(t,q_i)\,. \tag{2.5.7}$$

The functional (2.5.1) is invariant with respect to (2.5.6) if

$$L'(q_i',\dot{q}_i',t')\mathrm{d}t' = L(q_i,\dot{q}_i,t)\mathrm{d}t\,. \tag{2.5.8}$$

The Lagrange equations are invariant if

$$L'(q_i',\dot{q}_i',t') = L(q_i',\dot{q}_i',t') + \frac{\mathrm{d}f}{\mathrm{d}t'}\,. \tag{2.5.9}$$

We can say that (2.5.6) is a symmetry transformation for a mechanical system if and only if the conditions (2.5.8) and (2.5.9) are satisfied. For the infinitesimal transformation (2.5.2) the conditions (2.5.6) and (2.5.7) yield to

$$\left(\delta t\frac{\partial}{\partial t_i} + \delta q_i\frac{\partial}{\partial q_i} + \delta\dot{q}_i\frac{\partial}{\partial t}\right)L = -\frac{\mathrm{d}}{\mathrm{d}t}\delta f(t,q_i)\,. \tag{2.5.10}$$

The transformation (2.5.2) is a symmetry transformation if, for a given L, there exists a function $f(t,q_i)$, so that the equation (2.5.10) is satisfied. For a given L, the symmetry transformation forms a group, which represents the symmetry group of the system.

Consequently, when the equations of motion (2.5.5) are satisfied, it results an equation of conservation of the form

$$\frac{\mathrm{d}}{\mathrm{d}t}\left(L\delta t - \frac{\partial L}{\partial \dot{q}_i}\dot{q}_i\delta t + \frac{\partial L}{\partial \dot{q}_i}\delta\dot{q}_i + \delta f\right) = 0\,. \tag{2.5.11}$$

On this basis, a symmetry transformation of a mechanical system is associated with an equation of conservation.

This result is proved by the Noether theorem.

THEOREM 2.5.1. (Noether) *If the Lagrangian of a mechanical system is invariant with respect to a continuous group of transformation with* p *parameters, then exist* p *quantities which are conserved during the evolution of the system.*

In the case when the Lagrangian $L(q,\dot{q})$ does not depend explicitly on time, it relates to the Hamiltonian $H(q,p)$ by

$$\dot{q}_i = \frac{\partial H}{\partial p_i}, \; i = 1,2,\ldots,n, \tag{2.5.12}$$

where p_i, $i = 1,2,\ldots,n$, are the generalized momenta in the Hamilton formulation.

Let $C(q,p)$ be an integral of motion in the Hamilton formulation. In the Lagrange formulation, this integral becomes $F(q,\dot{q}) = C(q,p)$. The integral of motion along trajectories of the system satisfies the condition

$$\frac{\mathrm{d}}{\mathrm{d}t} F(q,\dot{q}) = 0. \tag{2.5.13}$$

Along the trajectories system, the variation of the Lagrangian due to the transformations $q_i' = q_i + \delta q_i$, $i = 1,2,\ldots,n$, is given by

$$\delta L = \frac{\mathrm{d}}{\mathrm{d}t}\left(\frac{\partial L}{\partial \dot{q}_i}\delta q_i\right). \tag{2.5.14}$$

From (2.5.14) it follows certain conserved quantities (Teodorescu and Nicorovici-Porumbaru).

Case 1. If $\delta L = 0$, the integral of motion is given by

$$\frac{\partial L}{\partial \dot{q}_i} = p_i \delta q_i. \tag{2.5.15}$$

Case 2. If there exists a function $f(q_i)$ such that $\delta L = \frac{\mathrm{d}}{\mathrm{d}t}\delta f(q_i)$, then the integral of motion is

$$\left(\frac{\partial L}{\partial \dot{q}_i} - \frac{\partial f}{\partial q_i}\right)\delta q_i = \left(p_i - \frac{\partial f}{\partial q_i}\right)\delta q_i. \tag{2.5.16}$$

Case 3. If there exists a function $g(q_i, \dot{q}_i)$ such that

$$\delta L = \frac{\mathrm{d}}{\mathrm{d}t} g(q_i, \dot{q}_i), \tag{2.5.17}$$

the integral of motion becomes

$$\frac{\partial L}{\partial \dot{q}_i}\delta q_i - g(q_i, \dot{q}_i) = p_i \delta q_i - g(q_i, \dot{q}_i). \tag{2.5.18}$$

We see that (2.5.15) and (2.5.16) are particular cases of the third case.

THEOREM 2.5.2 (Noether) *To every infinitesimal transformation of the form (2.5.10), due to a variation of the Lagrangian of the form (2.5.17), it corresponds to a conserved quantity defined by (2.5.18).*

2.6 Inverse Lagrange problem

The study of the inverse problem for a given evolution equation consists of the calculus of variations to determine if this equation is identical to a Lagrange equation (Santilli 1978, 1983).

Zamarreno deduced in 1992 the known Helmholtz conditions (1887) for the existence of a matrix of multipliers allowing an indirect Lagrangian representation of the Newtonian system $\ddot{q}_i = F_i(\dot{q}, q, t)$. He considers the case when F_i is time independent, and gives the functional relation between the Lagrangian and a first integral of motion equations. This section presents the principal results obtained by Zamarreno.

The inverse Lagrange problem consists in determining the Lagrangian from a given evolution equation.

DEFINITION 2.6.1. *Given a system of second order differential equations*

$$\ddot{q}_i = F_i(q_1, q_2, ..., q_n, \dot{q}_1, \dot{q}_2, ..., \dot{q}_n, t)\,,\ i = 1, 2, ..., n\,, \tag{2.6.1}$$

which describes the motion of a mechanical system with n *degrees of freedom, the inverse problem consists of determining a Lagrangian* $L(q, \dot{q}, t)$ *such that*

$$\ddot{q}_i = F_i(q_i, \dot{q}_i, t) \Leftrightarrow \frac{\mathrm{d}}{\mathrm{d}t}\left(\frac{\partial L}{\partial \dot{q}_i}\right) - \frac{\partial L}{\partial q_i} = 0\,,\ \forall i \in \{1, 2, ..., n\}\,. \tag{2.6.2}$$

We assume that the dynamic functions $F_i \in C^2$ are defined for a time interval $(t_1 < t < t_2)$ for which the generalized coordinates and velocities vary in $[q', q_i'']$ and $[\dot{q}', \ddot{q}_i'']$ intervals of the phase space. The operator $\mathrm{d}/\mathrm{d}t$ is applied over the integral trajectories of (2.6.1) and is defined as

$$\frac{\mathrm{d}}{\mathrm{d}t} = F_i \frac{\partial}{\partial \dot{q}_i} + q_i \frac{\partial}{\partial q_i} + \frac{\partial}{\partial t}\,. \tag{2.6.3}$$

Lagrange equations (2.6.2) can be written under the form

$$F_k \frac{\partial^2 L}{\partial \dot{q}_i \partial \dot{q}_k} + \dot{q}_k \frac{\partial^2 L}{\partial \dot{q}_i \partial q_k} + \frac{\partial^2 L}{\partial \dot{q}_i \partial t} - \frac{\partial L}{\partial q_i} = 0\,,\ i = 1, 2, ..., n\,, \tag{2.6.4}$$

and represent an over determined system of n second order partial differential equations in $L(q, \dot{q}, t)$. The results are presented in the following propositions (Zamarreno).

PROPOSITION 2.6.1 *If the system (2.6.1) is equivalent to the Euler–Lagrange equations corresponding to a variational principle whose Lagrangian is* $L(q, \dot{q}, t)$, *then the partial derivatives of generalized momenta with respect to the generalized velocities*

$$\alpha_{ij} = \frac{\partial p_i}{\partial \dot{q}_j} = \frac{\partial^2 L}{\partial \dot{q}_i \partial \dot{q}_j} = \frac{\partial p_j}{\partial \dot{q}_i} = \alpha_{ji}\,, \tag{2.6.5}$$

satisfy identically the system of equations

$$\frac{\mathrm{d}\alpha_{ij}}{\mathrm{d}t}+\frac{1}{2}\frac{\partial F_k}{\partial \dot{q}_j}\alpha_{ik}+\frac{1}{2}\frac{\partial F_k}{\partial \dot{q}_i}\alpha_{jk}=0\,. \tag{2.6.6}$$

PROPOSITION 2.6.2 *In the same conditions as before, the equations*

$$\frac{\partial(\alpha_{ik}F_k)}{\partial q_j}-\frac{\partial(\alpha_{jk}F_k)}{\partial q_i}+\frac{1}{2}\left(\frac{\partial}{\partial t}+\dot{q}_k\frac{\partial}{\partial q_k}\right)\left(\frac{\partial F_h}{\partial \dot{q}_j}\alpha_{ih}-\frac{\partial F_h}{\partial \dot{q}_i}\alpha_{jh}\right)=0\,, \tag{2.6.7}$$

are identically verified.

Let us denote by $\overline{\alpha}_{ij}$ the multipliers that allow an indirect Lagrangian representation of (2.6.1)

$$\overline{\alpha}_{ij}(q,\dot{q},t)(\ddot{q}_j-F_j)\equiv\frac{\mathrm{d}}{\mathrm{d}t}\left(\frac{\partial L}{\partial \dot{q}_i}\right)-\frac{\partial L}{\partial q_i},\ i,j=1,2,...,n\,. \tag{2.6.8}$$

Expanding the time derivatives on the right-hand side, we obtain

$$\overline{\alpha}_{ij}=\alpha_{ij}\,. \tag{2.6.9}$$

In this way the equations (2.6.6) and (2.6.7) coincide according to well-known Helmholtz conditions. We also have

$$\frac{\partial\overline{\alpha}_{ij}}{\partial \dot{q}_k}=\frac{\partial\overline{\alpha}_{ik}}{\partial \dot{q}_j}\,, \tag{2.6.10}$$

because

$$\frac{\partial\alpha_{ij}}{\partial \dot{q}_k}=\frac{\partial^3 L}{\partial \dot{q}_k\partial \dot{q}_i\partial \dot{q}_j}=\frac{\partial^3 L}{\partial \dot{q}_j\partial \dot{q}_i\partial \dot{q}_k}=\frac{\partial\alpha_{ik}}{\partial \dot{q}_j}\,. \tag{2.6.11}$$

Consider only regular systems for which

$$\det(\alpha_{ij})\neq 0\,. \tag{2.6.12}$$

PROPOSITION 2.6.3 *If α_{ij} is a symmetric solution of (2.6.6), then its determinant Δ satisfies the equation*

$$\frac{\mathrm{d}\Delta}{\mathrm{d}t}+\Delta\sum_{i=1}^{n}\frac{\partial F_i}{\partial \dot{q}_i}=0\,. \tag{2.6.13}$$

COROLLARY 2.6.1 *If α_{ij} and α'_{ij} are two nonsingular symmetrical matrix solutions of (2.6.5), the determinant $\Delta^*=|(\alpha_{ij})(\alpha'_{ij})^{-1}|$ is a constant of evolution of the system.*

PROPOSITION 2.6.4 *If α_{ij} is a symmetric nonsingular matrix, solution of (2.6.6), then Δ is a Jacobi multiplier for the dynamical system written in the equivalent form*

$$\mathrm{d}t = \frac{\mathrm{d}q_1}{\dot{q}_1} = \frac{\mathrm{d}q_2}{\dot{q}_2} = \ldots = \frac{\mathrm{d}q_n}{\dot{q}_n} = \frac{\mathrm{d}\dot{q}_1}{F_1} = \frac{\mathrm{d}\dot{q}_2}{F_2} = \ldots = \frac{\mathrm{d}\dot{q}_n}{F_n}. \tag{2.6.14}$$

PROPOSITION 2.6.5 *If α_{ij} is a symmetric matrix, solution of (2.6.5) and the matrix F of elements $\frac{\partial F_i}{\partial \dot{q}_j}$ is skew symmetric, the trace of any integer power of α_{ij} are constants of motion for the dynamical system.*

We apply the variation of constants method to integrate (2.6.6), for the uncoupled case

$$\frac{\mathrm{d}\alpha_{ij}}{\mathrm{d}t} + \frac{1}{2}\frac{\partial F_i}{\partial \dot{q}_j}\alpha_{ij} + \frac{1}{2}\frac{\partial F_j}{\partial \dot{q}_i}\alpha_{ji} = 0. \tag{2.6.15}$$

The solution of (2.6.15) is expressed as

$$\alpha_{ij}^{*} = \exp\left(-\frac{1}{2}\int\left(\frac{\partial F_j}{\partial \dot{q}_j} + \frac{\partial F_i}{\partial \dot{q}_i}\right)\mathrm{d}t\right). \tag{2.6.16}$$

Here, the repeated indices do not imply summation. Writing $\alpha_{ij} = \mu_{ij}\alpha_{ij}^{*}$, we obtain

$$\frac{\mathrm{d}\alpha_{ij}^{*}}{\mathrm{d}t}\mu_{ij} + \alpha_{ij}^{*}\frac{\mathrm{d}\mu_{ij}}{\mathrm{d}t} + \frac{1}{2}\frac{\partial F_k}{\partial \dot{q}_j}\alpha_{ik}^{*}\mu_{ik} + \frac{1}{2}\frac{\partial F_k}{\partial \dot{q}_i}\alpha_{jk}^{*}\mu_{jk} = 0. \tag{2.6.17}$$

From $\mu_{ij} = \mu_{ji}$, and taking into account that α_{ij}^{*} are solutions of (2.6.15), it follows

$$\frac{\mathrm{d}\alpha_{ij}^{*}}{\mathrm{d}t}\mu_{ij} + \frac{1}{2}\frac{\partial F_k}{\partial \dot{q}_j}\alpha_{ik}^{*}\mu_{ik} + \frac{1}{2}\frac{\partial F_h}{\partial \dot{q}_i}\alpha_{jh}^{*}\mu_{jh} = 0,\ k \neq j,\ h \neq i. \tag{2.6.18}$$

Replacing α_{ij}^{*} from (2.6.16) and multiplication by α_{ij}^{*-1}, yield

$$\begin{aligned}&\frac{\mathrm{d}\mu_{ij}}{\mathrm{d}t} + \frac{1}{2}\frac{\partial F_k}{\partial \dot{q}_j}\mu_{ik}\exp\left(\frac{1}{2}\int\left(\frac{\partial F_j}{\partial \dot{q}_j} - \frac{\partial F_k}{\partial \dot{q}_k}\right)\mathrm{d}t\right) + \\ &+ \frac{1}{2}\frac{\partial F_h}{\partial \dot{q}_i}\alpha_{jh}^{*}\exp\left(\frac{1}{2}\int\left(\frac{\partial F_i}{\partial \dot{q}_i} - \frac{\partial F_h}{\partial \dot{q}_h}\right)\mathrm{d}t\right) = 0,\ j \neq k,\ i \neq h.\end{aligned} \tag{2.6.19}$$

Denoting by a_{ij} the expressions

$$a_{ij} = \frac{\partial F_i}{\partial \dot{q}_j}\exp\left(\frac{1}{2}\int\left(\frac{\partial F_j}{\partial \dot{q}_j} - \frac{\partial F_i}{\partial \dot{q}_i}\right)\mathrm{d}t\right),$$

the system (2.6.19) becomes

$$\frac{\mathrm{d}\mu_{ij}}{\mathrm{d}t}+\frac{1}{2}(\mu_{ik}a_{kj}+\mu_{jh}a_{hi})=0,\quad k\neq j,\ h\neq i. \tag{2.6.20}$$

PROPOSITION 2.6.6. *The determinant* Γ *of elements* μ_{ij}, *is a constant of evolution for the dynamical system.*

PROPOSITION 2.6.7. *Sufficient conditions for the matrix* (α_{ij}) *to be diagonal are* $a_{ij}a_{ji}^{-1}=k_{ij}=\text{const.}$

If F_i does not depend explicitly on time, the following results hold:

PROPOSITION 2.6.8 *If the multipliers* α_{ij} *satisfy the Helmholtz conditions and the dynamical system is autonomous, there exists a first integral* $I(q,\dot{q})$ *for the dynamical system given by*

$$\alpha_{ik}\dot{q}_k=\frac{\partial I}{\partial \dot{q}_i},$$

$$\alpha_{ik}F_k+\left(\frac{\partial^2 L}{\partial \dot{q}_i\partial q_k}-\frac{\partial^2 L}{\partial q_i\partial \dot{q}_k}\right)\dot{q}_k=-\frac{\partial I}{\partial q_i}, \tag{2.6.21}$$

for which $\mathrm{d}I=0$, *along the system trajectories.*

We can define $I(q,\dot{q})$ as

$$\alpha_{ik}\dot{q}_k=\frac{\partial I}{\partial \dot{q}_i},$$

$$\alpha_{ik}F_k+\frac{\dot{q}_k}{2}\left(\frac{\partial F_k}{\partial \dot{q}_k}\alpha_{ih}-\frac{\partial F_h}{\partial \dot{q}_i}\alpha_{kh}\right)=-\frac{\partial I}{\partial q_i}. \tag{2.6.22}$$

PROPOSITION 2.6.9 *The Lagrangian that allows an indirect representation for the autonomous system*

$$\ddot{q}_i=F_i(q_i,\dot{q}_i), \tag{2.6.23}$$

by using α_{ij}, *must verify*

$$I=\frac{\partial L}{\partial \dot{q}_i}\dot{q}_i-L, \tag{2.6.24}$$

where $I(q,\dot{q})$ *is an integral for (2.6.22).*

As an example, consider the simple pendulum equation

$$\ddot{\theta}+\frac{g}{l}\sin\theta=0.$$

The Helmholtz conditions (2.6.6) and (2.6.7) are satisfied for $\alpha = ml^2$. A prime integral I is obtained by integrating (2.6.22)

$$I = \frac{1}{2} ml^2 \dot{\theta}^2 + mgl(1-\cos\theta),$$

and the Lagrange function is obtained from (2.6.24)

$$L = \frac{1}{2} ml^2 \dot{\theta}^2 - mgl(1-\cos\theta).$$

Another example is given by (Santilli 1978)

$$\ddot{q}_1 + \frac{1}{2}\gamma(\dot{q}_1^2 + 2\dot{q}_1\dot{q}_3) = 0,$$

$$\ddot{q}_2 + \frac{1}{2}\gamma(\dot{q}_2^2 + 2\dot{q}_2\dot{q}_3) = 0,$$

$$\ddot{q}_3 + \frac{1}{2}\gamma(\dot{q}_3^2 - \dot{q}_1^2 \exp(\gamma q_1) - \dot{q}_2^2 \exp(\gamma q_2)) = 0.$$

This system of equations satisfies the conditions of *Proposition 2.6.7*, and therefore (2.6.6) admits a diagonal symmetric matrix as a solution

$$\alpha_{11} = \exp\{\gamma(q_1 + q_3)\},$$

$$\alpha_{22} = \exp\{\gamma(q_2 + q_3)\},$$

$$\alpha_{33} = \exp(\gamma q_3),$$

which satisfy (2.6.7) The first integral is obtained from (2.6.22)

$$I = \frac{c}{2}\dot{q}_3^2 \exp(\gamma q_3) + \frac{c}{2}\dot{q}_1^2 \exp\{\gamma(q_1 + q_3)\} + \frac{c}{2}\dot{q}_2^2 \exp\{\gamma(q_2 + q_3)\}.$$

The Lagrange function is determined then from (2.6.24)

$$L = \frac{c}{2}\dot{q}_3^2 \exp(\gamma q_3) + \frac{c}{2}\dot{q}_1^2 \exp\{\gamma(q_1 + q_3)\} + \frac{c}{2}\dot{q}_2^2 \exp\{\gamma(q_2 + q_3)\}.$$

Among other works dedicated to inverse problems in mechanics we mention Chiroiu and Chiroiu (2003a), Frederiksen, Tanaka and Nakamura.

2.7 Recursion operators

The existence of infinitely conserved densities for evolution equations is explained by generalized symmetries that are groups, whose infinitesimal generators depend not only on the independent and dependent variables of the system, but also the derivatives of the dependent variables.

The existence of infinitely many symmetries for evolution equations of the form

$$u_t = u_k + f(u, u_1, ..., u_{k-1}), \tag{2.7.1}$$

where $u_1 = u_x$, $u_2 = u_{xx}$, and so on, and the right-hand sides are homogeneous with respect to the scaling symmetry $T = xu_x + \lambda u$, with $\lambda \geq 0$, it was studied by Wang.

The equations of the type (2.7.1) are called λ-homogeneous equations. If an equation (2.7.1) has a nonlinear symmetry, it has infinitely many and these can be found using recursion operators.

Integrable evolution equations in one space variable, like the KdV equation, admit a recursion operator, which is an operator invariant under the flow of the equation, carrying symmetries of the equation into new symmetries.

In 1977, Olver provided a method for the construction of infinitely many symmetries of evolution equations, originally due to Lenard. This recursion operator maps a symmetry to a new symmetry. For the KdV equation, for example, a recursion operator defined by $\mathrm{Im}(D_x)$ is given by

$$D_x^2 + \frac{2}{3}u + \frac{1}{3}u_x D_x^{-1}, \tag{2.7.2}$$

where D_x^{-1} is the left inverse of D_x. Magri studied in 1978 the connections between conservation laws and symmetries from the geometric point of view. He observed that the gradients of the conserved densities are related to the theory of symmetries.

This problem required the introduction to Hamiltonian operators. Magri found that some systems admit a pair of Hamiltonians. For example, the KdV equation can be written in two forms corresponding to a pair of Hamiltonians

$$u_t = D_x(u_{xx} + \frac{1}{2}u^2) = (D_x^3 + \frac{2}{3}uD_x + \frac{1}{3}u_x)u.$$

The Nijenhuis operator (hereditary operator) is a special kind of recursion operator found as an integrability condition. For any vector field A_0 leaving the Nijenhuis operator $\mathfrak{A}$ invariant, the $A_j = \mathfrak{A}^j(A_0)$, $j = 0,1,...$, leave $\mathfrak{A}$ invariant and commute in pairs. Gelfand and Dorfman gave the relation between Hamiltonian pairs and Nijenhuis operators. For the KdV equation the operator

$$\mathfrak{A} = D_x^2 + \frac{2}{3}u + \frac{1}{3}u_x D_x^{-1},$$

is a Nijenhuis recursion operator and it produces the higher KdV equations, the KdV hierarchy, which shares infinitely many commuting symmetries produced by the same recursion operator

$$u_t = \mathfrak{A}^j(u_x), \; j = 0,1,....$$

The recursion method can be applied also to $\lambda > 0$, and $\lambda \leq 0$. In the particular case $\lambda = 1$, we have

- Potential Korteweg–de Vries equation $f = u_{xxx} + u_x^2$,
- Modified Korteweg–de Vries equation $f = u_{xxx} + u^2 u_x$,
- Burgers equation $f = u_{xx} + uu_x$.

When $\lambda = 2$, we have

– Korteweg–de Vries equation $f = u_{xxx} + uu_x$.

A selected list of integrable evolution equations, scaling symmetry transformation T, and the recursion operators are given below (Wang)

– Korteweg–de Vries equation

$$u_t = u_{xxx} + uu_x,$$

$$T = xu_x + 2u,$$

$$\mathfrak{A} = D_x^2 + \frac{2}{3}u + \frac{1}{3}u_x D_x^{-1}.$$

– Burgers equation

$$u_t = u_{xx} + uu_x,$$

$$T = -2t\partial_t + (xu_x + u)\partial_u,$$

$$\mathfrak{A}_1 = D_x + u_x, \ \mathfrak{A}_2 = tD_x + tu_x + \frac{1}{2}x.$$

– Potential Burgers equation

$$u_t = u_{xx} + u_x^2,$$

$$T = xu_x,$$

$$\mathfrak{A}_1 - D_x + u_x, \ \mathfrak{A}_2 - tD_x + tu_x + \frac{1}{2}x.$$

– Diffusion equation

$$u_t = u^2 u_{xx},$$

$$T = axu_x + bu, \ a, b \in C,$$

$$\mathfrak{A} = uD_x + u_{xx}D_x^{-1}.$$

– Nonlinear diffusion equation

$$u_t = \mathrm{D}_x(u_x / u^2),$$

$$T = axu_x + bu, \ a, b \in C,$$

$$\mathfrak{A} = \frac{1}{u}D_x - \frac{2u_x}{u^2} - u_t D_x^{-1}.$$

– Potential KdV equation

$$u_t = u_{xxx} + 3u_x^2,$$

$$T = xu_x + u,$$

$$\mathfrak{A} = D_x^2 + 4u_x - 2D_x^{-1}u_{xx}.$$

– Modified KdV equation

$$u_t = u_{xxx} + u^2 u_x,$$

$$T = xu_x + u,$$

$$\mathfrak{A} = D_x^2 + \frac{2}{3}u^2 + \frac{2}{3}u_x D_x^{-1} u.$$

– Sine-Gordon equation

$$u_{xt} = \sin u,$$

$$T = axu_x + bu, \ a, b \in C,$$

$$\mathfrak{A} = D_x^2 + u_x^2 - u_x D_x^{-1} u_{xx}.$$

– Liouville–Tzitzeica equation

$$u_{xt} = \exp u,$$

$$T = axu_x + bu, \ a, b \in C,$$

$$\mathfrak{A} = D_x^2 - u_x^2 + u_x D_x^{-1} u_{xx}.$$

– Tzitzeica equation

$$u_{xt} = \alpha \exp(-2u) + \beta \exp(u),$$

$$T = xu_x,$$

$$\begin{aligned}
\mathfrak{A} = {} & D_x^6 + 6(u_{xx} - u_x^2)D_x^4 + 9(u_{xxx} - 2u_x u_{xx})D_x^3 + \\
& + (5u_{xxxx} - 22u_x u_{xxx} - 13u_{xx}^2 - 6u_x^2 u_{xx} + 9u_x^4)D_x^2 + \\
& + (u_{xxxxx} - 8u_x u_{xxxx} - 15u_{xx}u_{xxx} - 3u_x^2 u_{xxx} - 6u_x u_x^2 + 18u_x^3 u_{xx})D_x - \\
& - 4u_x u_{xxxxx} + 20u_x^3 u_{xxx} - 20u_x u_{xx} u_{xxx} + 20u_x^2 u_{xx}^2 - 4u_x^6 + \\
& + 2u_x D_x^{-1}(u_{xxxxxx} + 5u_{xx}u_{xxxx} + 5u_{xxx}^2 - 5u_x^2 u_{xxxx} - 20u_x u_{xx} u_{xxx} - \\
& - 5u_{xx}^3 + 5u_x^4 u_{xx} + + 2(u_{xxxxx} + 5u_{xx}u_{xxx} - 5u_x^2 u_{xxx} - \\
& - 5u_x u_{xx}^2 + u_x^5)D_x^{-1}u_{xx}.
\end{aligned}$$

– Dispersiveless long wave system

$$u_t = u_x v + u v_x, \quad v_t = u_x + v v_x,$$

$$T = \begin{pmatrix} x u_x \\ x v_x \end{pmatrix} + a \begin{pmatrix} 2u \\ v \end{pmatrix}, \quad a \in C,$$

$$\mathfrak{A} = \begin{pmatrix} v & 2u + u_x D_x^{-1} \\ 2 & v + v_x D_x^{-1} \end{pmatrix}.$$

– Diffusion system

$$u_t = u_{xx} + v^2, \quad v_t = v_{xx},$$

$$T = \begin{pmatrix} x u_x + 2u \\ x v_x + 2v \end{pmatrix} + a \begin{pmatrix} 2u \\ v \end{pmatrix}, \quad a \in C,$$

$$\mathfrak{A} = \begin{pmatrix} D_x & v D_x^{-1} \\ 0 & D_x \end{pmatrix}.$$

– Nonlinear Schrödinger equation

$$v_t = -u_{xx} \pm u(u^2 + v^2), \quad u_t = v_{xx} \mp v(u^2 + v^2),$$

$$T = \begin{pmatrix} x u_x + u \\ x v_x + v \end{pmatrix},$$

$$\mathfrak{A} = \begin{pmatrix} \mp 2 v D_x^{-1} u & D_x \mp 2 v D_x^{-1} v \\ -D_x \pm 2 u D_x^{-1} u & \pm 2 u D_x^{-1} v \end{pmatrix}.$$

– Boussinesq system

$$u_t = v_x, \quad v_t = \frac{1}{3} u_{xxx} + \frac{8}{3} u u_x,$$

$$T = \begin{pmatrix} x u_x + 2u \\ x v_x + 3v \end{pmatrix},$$

$$\mathfrak{A} = \begin{pmatrix} 3v + 2 v_x D_x^{-1} & D_x^2 + 2u + u_x D_x^{-1} \\ \mathfrak{A}_{21} & 3v + v_x D_x^{-1} \end{pmatrix},$$

$$\mathfrak{A}_{21} = \frac{1}{3} D_x^4 + \frac{10}{3} u D_x^2 + 5 u_x D_x + 3 u_{xx} + \frac{16}{3} u^2 + 2 v_t D_x^{-1}.$$

Chapter 3

SOLITONS AND NONLINEAR EQUATIONS

3.1 Scope of the chapter

Solitons or solitary waves are localized waves that travel without change in shape. In the mathematics literature the word *soliton* refers to solitary traveling waves which preserve their identities after a pair-wise collision. Solitons were first discovered in shallow water by the Scottish engineer John Scott Russell in 1844, but they exist everywhere, in many kind of systems. The study of solitons is an exciting branch in the science of nonlinear physics and their importance as nonlinear waves is well-recognized in the past two decades. The basic theory of solitons is simple and relies on mathematical methods well-known to any applied mathematician.

In the last years of the nineteenth century, mathematicians as Korteweg, De Vries, Bianchi, Boussinesq, Darboux and Bäcklund have analyzed some remarkable nonlinear partial differential equations which possess the cnoidal or solitonic solutions.

Engelbrecht is right when he affirms that the story of solitons is the story of three equations, namely Korteweg and de Vries, sine-Gordon and nonlinear Schrödinger equations. In this chapter the first five evolution equations are briefly discussed and the solitonic solutions outlined. The stability and particle-like behavior of the solitons can be explained by the existence of an infinite number of conservation laws. So, these equations can be interpreted as completely integrable Hamiltonian systems in the same sense as finite dimensional integrable Hamiltonian systems, where we find for every degree of freedom a conserved quantity.

The reader is referred to the remarkable monographs of Ablowitz and Segur (1981), Dodd *et al.* (1982), Lamb (1980), Drazin (1983), Drazin and Johnson (1989) and Engelbrecht (1991). We mention also the monographs of Carroll (1991), Nettel (1992) and Newell (1985) for enlarged topics in soliton theory.

3.2 Korteweg and de Vries equation (KdV)

To explain the nature of solitons, we will consider the behavior of water waves on shallow water. The scenario could be set in one of the canals, which was the 19th century's analogue to highways nowadays. Indeed, it was in such a location that the Scottish engineer John Scott Russell first noticed a soliton in 1844.

The solitary wave, or great wave of translation, was first observed on the Edinburgh to Glasgow canal in 1834 by Russell. Russell reported his discovery to the British Association in 1844 as follows (Drazin):

"*I believe I shall best introduce this phenomenon by describing the circumstances of my own first acquaintance with it. I was observing the motion of a boat which was rapidly grown along a narrow channel by a pair of horses, when the boat suddenly stopped – not so the mass of water in the channel which it had put in motion; it accumulated round the prow of the vessel in a state of violent agitation, then suddenly leaving it behind, rolled forward with great velocity, assuming the form of a large solitary elevation, a rounded, smooth and well-defined heap of water, which continued its course along the channel apparently without change of form or diminution of speed. I followed it on horseback, and overtook it still rolling on at a rate of some eight or nine miles an hour, preserving its original figure some thirty feet long and a foot to a foot and a half in height. Its height gradually diminished and after a chase of one or two miles I lost it in the windings of the channel. Such in the month of August 1834 was my first chance interview with that singular and beautiful phenomenon which I have called the Wave of Translation…*".

Russell performed some experiments in the laboratory in a small-scale wave tank in order to study the phenomenon more carefully. This is shown in Figure 3.2.1, which is one of Russell's original diagrams (Dodd *et al.*). Figure 3.2.1 is represented as a raised area of fluid behind an obstacle. When this obstacle is removed, a long heap-shaped wave propagates down the channel. Figure 3.2.1b is identical except that the initial volume of trapped fluid is larger.

In this case two solitary waves are found. If a wave is somehow initiated in such a canal, we expect that the wave rolls along the canal while it spreads out and soon ends its life as small wiggles on the surface. But, in certain conditions a soliton can be excited, and the wave will continue to roll along the canal without changing shape. It turns out that a soliton is very robust against perturbations.

The bottom of the canal may be uneven and bumpy, but the soliton will gently pass all obstacles.

Let us suppose the canal has depth h and note by $h+\eta$ (η small) the elevation of the surface above the bottom. Korteweg and de Vries derived in 1895 the partial differential equation, which governs the wave motion (Dodd *et al.*)

$$\frac{\partial \eta}{\partial t} = \frac{3}{2}\sqrt{\frac{g}{h}}\frac{\partial}{\partial x}\left(\frac{2}{3}\alpha\eta + \frac{1}{2}\eta^2 + \frac{1}{3}\sigma\frac{\partial^2 \eta}{\partial x^2}\right), \tag{3.2.1}$$

where $\sigma = \frac{h^3}{3} - \frac{Th}{\rho g}$, α an arbitrary constant, T the surface tensions, ρ the density of the fluid and g gravity acceleration.

This equation bears the Korteweg and de Vries names, usually shortened to KdV, and becomes one of the most celebrated equations, which is related to solitons.

Introducing

$$\eta = 8\alpha u, \quad \xi = \sqrt{\frac{2\alpha}{\sigma}}x, \quad \tau = \sqrt{\frac{2\alpha^3 g}{\sigma l}}t,$$

the equation (3.2.1) becomes

$$u_\tau + u_\xi + 12uu_\xi + u_{\xi\xi\xi} = 0, \qquad (3.2.2)$$

where the subscripts denote partial differentiation $u_\tau = \partial u/\partial\tau$, $u_\xi = \partial u/\partial\xi$, $u_{\xi\xi\xi} = \partial^3 u/\partial\xi^3$, etc. The factor of 12 in (3.2.2) is a matter of choice and can be rescaled by the transformation $u \to \beta u + \gamma$

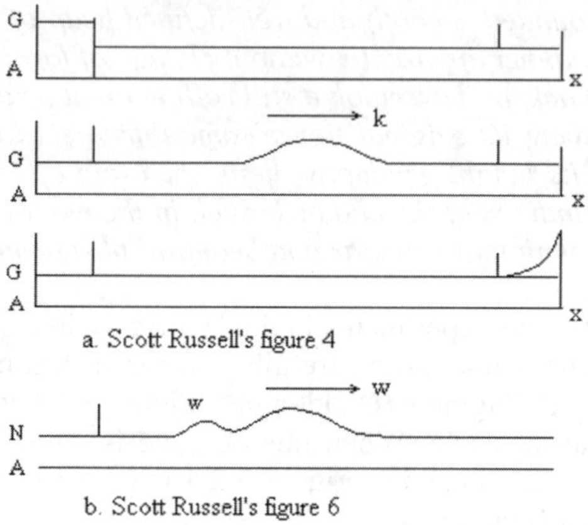

Figure 3.2.1 From Scott Russell's *Report on Waves* 1844 (Dodd *et al.*).

For $\beta = -\dfrac{1}{2}$ and $\gamma = -\dfrac{1}{12}$ the equation (3.1.2) becomes

$$u_\tau - 6uu_\xi + u_{\xi\xi\xi} = 0. \qquad (3.2.3)$$

The equation (3.2.3) is known as the standard KdV equation. By applying a Lie transformation

$$u = k_1\eta + k_0, \quad \xi = k_3 x + k_2, \quad \tau = k_4\eta + k_5, \qquad (3.2.4)$$

we may write the KdV equation in the forms

$$u_\tau \pm 6uu_\xi + u_{\xi\xi\xi} = 0, \; u_\tau \pm (1+u)u_\xi \pm u_{\xi\xi\xi} = 0, \qquad (3.2.5)$$

by choosing in an adequate way the constants k_i, $i = 0,1,2,3,4,5$.

Returning to the equation (3.2.3) we look for a solution with the permanent

$$u = u(\theta), \quad \theta = \xi - c\tau, \qquad (3.2.6)$$

where c is the velocity of wave.

Substitution of (3.2.6) into (3.2.5) gives

$$-cu' - 6uu' + u''' = 0,$$

where prime means the differentiation with respect to the new variable θ.

The above equation becomes

$$-cu' - 3(u^2)' + u''' = 0 .$$

Integration with respect to θ yields

$$u'' = 3u^2 + cu + A ,$$

where A is an integration constant. Multiplying this equation with u' and integrating with respect to θ we have

$$\frac{1}{2}u'^2 = u^3 + \frac{1}{2}cu^2 + Au + B , \tag{3.2.7}$$

with B an integration constant. The equation (3.2.7) can be reduced, in certain conditions, to the Weierstrass equation (1.4.8). For $g_2^3 - 27g_3^2 > 0$, the solution is expressed in terms of the elliptic Jacobi function.

At the limit $m = 1$, the solution of the KdV equation is described in terms of hyperbolic function

$$\wp(z) = e_3 + (e_1 - e_3)\,\mathrm{sech}^2(u^*, m) .$$

The boundary conditions $u, u', u'' \to 0$ as $\theta \to \infty$, give $A = B = 0$, and

$$\frac{1}{2}u'^2 = u^2(u + \frac{1}{2}c) . \tag{3.2.8}$$

From $du = u'd\theta$ we have

$$\theta = \int\frac{du}{u'} = \int\frac{du}{u\sqrt{2u+c}} ,$$

or

$$\theta = \frac{2}{\sqrt{c}}a\tanh(\frac{1}{\sqrt{c}}\sqrt{2u+c}) + \theta_0 ,$$

with θ_0 an integration constant.

Substituting $u = -\frac{1}{2}c\,\mathrm{sech}^2\phi$ into the above equation we have

$$\theta = \frac{2}{\sqrt{c}}a\tanh(\frac{1}{\sqrt{c}}\sqrt{c(1-\mathrm{sech}^2\phi)}) + \theta_0 ,$$

or

$$\theta = \frac{2}{\sqrt{c}}a\tanh(\frac{1}{\sqrt{c}}\sqrt{c\tanh^2\phi}) + \theta_0 ,$$

with $\phi = \frac{1}{2}\sqrt{c}(\theta - \theta_0)$. Therefore, we obtain the solution

$$u(\theta) = -\frac{1}{2}c\,\mathrm{sech}^2\{\frac{1}{2}\sqrt{c}(\theta-\theta_0)\}, \tag{3.2.9}$$

for any $c \geq 0$ and θ_0. These constants are determined from the initial conditions $u(0) = u_0$, $u'(0) = c_0$.

We see from (3.2.9) that the amplitude of the wave depends on the velocity. Thus, the wave having higher amplitude travels faster than the wave having lower amplitude.

Similarly the solution of (3.2.2) was found looking at solutions of the form $u = u(\theta)$, with $\theta = a\xi - \omega t + \delta$ and δ the phase constant. Requiring that u and its derivatives to be zero as $|\theta| \to \infty$ the equation (3.2.2) may be integrated to give the solution

$$u = \frac{1}{4}a^2\mathrm{sech}^2\frac{1}{2}\left[a\xi - \left(a + a^3\right)\tau + \delta\right]. \tag{3.2.10}$$

The constants a and δ are arbitrary, δ playing the role of a phase.

The solution (3.2.10) represents a solitary wave, which has the shape as Russell observed. This wave propagates at a constant velocity without change of shape. Its velocity is $\omega / a = 1 + a^2$, and depends on amplitude and vice versa the amplitude depends on the velocity. Figure 3.2.2 represents the solution (3.2.10) for three values of a. For smaller values of a, the wave is low and broad, and becomes narrow and sharper as a increases.

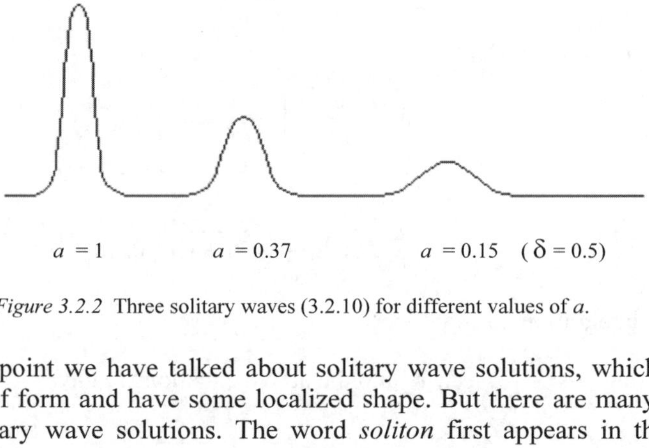

Figure 3.2.2 Three solitary waves (3.2.10) for different values of *a*.

Up until this point we have talked about solitary wave solutions, which propagate without change of form and have some localized shape. But there are many equations, which have solitary wave solutions. The word *soliton* first appears in the paper of Zabusky and Kruskal. They have studied the problem of Fermi, Pasta and Ulam (1955) of motion of a line of identical particles of unit mass, with fixed end points, and put into evidence a remarkable property of the solitary solutions (Kawahara and Takaoka). Their interaction with one another as they passed through the cycles of evolution forced by the periodic boundary conditions shows no change in the form, but only a small change in the phase. Zabusky and Kruskal called these solitary waves *solitons* where the ending 'on' is Greek for particle. This particle-like behavior does not depend on periodic boundary conditions. Zabusky and Kruskal numerically integrated the KdV equation on boundary conditions $u \to 0$ as $\xi \to \infty$ and considered two well-separated solitary wave solutions of the KdV equation as initial data with different velocities $v_1 > v_2$. The taller and therefore faster wave v_1 is situated on the left. As time increases throughout the numerical integration, the tall, fast soliton overtakes the broad, slow one, as shown in Figure 3.2.3. A larger soliton travels with a high velocity so that it overtakes the smaller one and after the collision they reappear without changing their form identity.

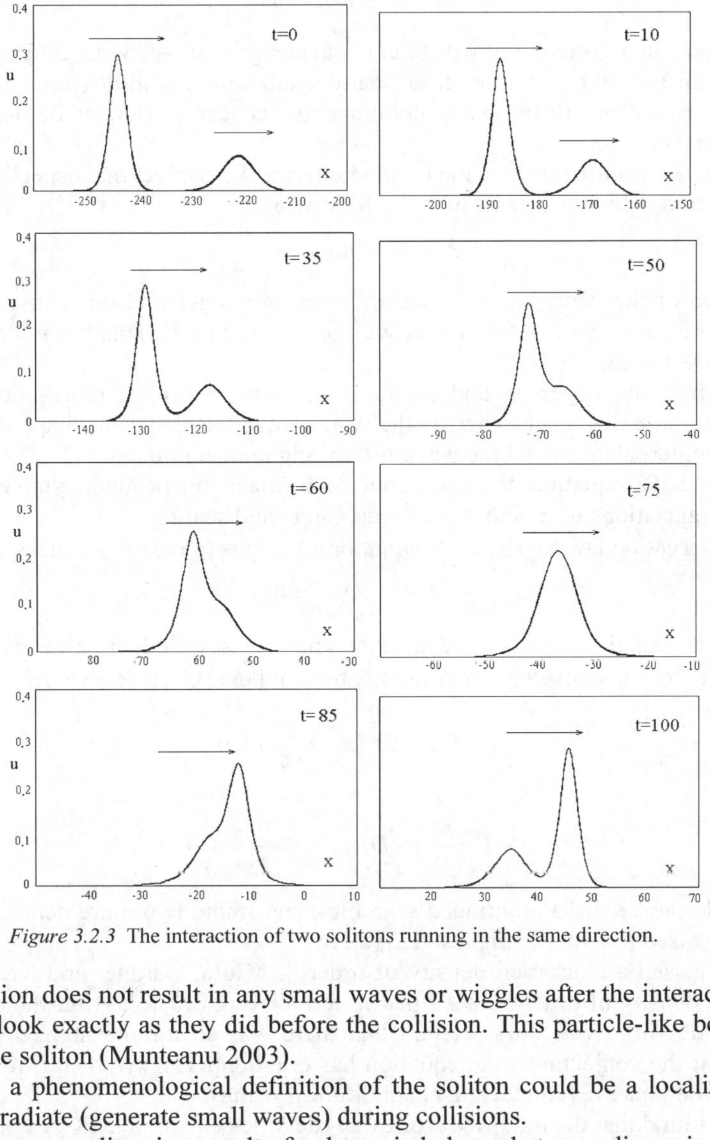

Figure 3.2.3 The interaction of two solitons running in the same direction.

The collision does not result in any small waves or wiggles after the interaction. The two solitons look exactly as they did before the collision. This particle-like behavior is generic for the soliton (Munteanu 2003).

Therefore a phenomenological definition of the soliton could be a localized wave that does not radiate (generate small waves) during collisions.

The water wave soliton is a result of a dynamic balance between dispersion, namely the wave's tendency to spread out, and nonlinear effects. In order to substantiate this statement, we have to pass to some mathematics.

The dynamics of water waves in shallow water is described mathematically by the Korteveg–de Vries (KdV) equation (3.2.3). The second and the third term in the equation are the nonlinear and the dispersive term, respectively.

Let us first investigate the effect of the dispersive term. Thus, we neglect the nonlinear term in the KdV equation. This leaves us with the following:

$$u_\tau + u_{\xi\xi\xi} = 0 .$$

We know that waves with different wavelengths travel with different velocities. Since the initial wave is composed of many small waves with different wavelengths, it will soon spread out in the many components and can no longer be described as an entity or object.

Now let us see the effect of the nonlinear term. We neglect the dispersive term in the KdV equation, which leaves us with the following

$$u_\tau - 6uu_\xi = 0 .$$

The top of the wave moves faster than the low sides and this causes the wave to shock in the same way as the waves we see on the beach. This behavior is known as *breaking the waves*.

When both the dispersive and the nonlinear term are present in the equation the two effects can neutralize each other. If the water wave has a special shape the effects are exactly counterbalanced and the wave rolls along undistorted.

For the KdV equation, there are four such linear independent symmetries, namely arbitrary translations in x and t, Galilean gauge and scaling.

A conservation law for the KdV equation (3.2.3) is (Kruskal, Zabusky and Kruskal)

$$D_\tau U + D_\xi F = 0 , \tag{3.2.11}$$

where U is called the conserved density and F is called the conserved flux. The expressions for the conservation of momentum and energy are known as

$$D_\tau u - D_x(u_{\xi\xi} + \frac{u^2}{2}) = 0 ,$$

$$D_\tau(\frac{u^2}{2}) - D_x(uu_{\xi\xi} - \frac{u_\xi^2}{2} + \frac{u^3}{3}) . \tag{3.2.12}$$

Zabusky and Kruskal continued searching and found two more densities or order 2 and 3 expressed in terms of highest derivative.

They missed a conserved density of order 4. Miura, Gardner and Kruskal found a conserved density of order 5, and filled in the missing order 4. After the order 6 and 7 were found, Miura was fairly certain that there was an infinite number. Later it was proved that the conjecture if an equation has one nontrivial symmetry, it has infinitely many, is true under certain technical conditions (Wang).

Miura found that the modified Korteweg–de Vries equation (MKdV)

$$v_\tau = v_{\xi\xi\xi} + v^2 v_\xi , \tag{3.2.13}$$

has an infinite number of conserved densities. He introduced the transformation

$$u = v^2 + \sqrt{-6}v_\xi , \tag{3.2.14}$$

which now bears his name. Applying the Miura transformation to (3.2.14) we have (Wang)

$$u_\tau - (u_{\xi\xi\xi} + uu_\xi) = (2v + \sqrt{-6}\mathrm{D}_\xi (v_\tau - (v_{\xi\xi\xi} + v^2 v_\xi)) , \tag{3.2.15}$$

from which we see that, if $v(\xi,\tau)$ is a solution of (3.2.13), then $u(\xi,\tau)$ is a solution of (3.2.3). From this observation, the famous inverse scattering method was developed and the Lax pair was found (Lax, Lax and Phillips).

Gardner also observed that the KdV equation might be written in a Hamiltonian way. Notice that the Hamiltonian of KdV is $H = \dfrac{u^2}{2}$.

The integrability of the KdV equation can be understood in the same sense as finite dimensional integrable Hamiltonian systems, where we can find for every degree of freedom a conserved quantity named the action.

Let us return to the linearised KdV equation

$$u_t + u_x + u_{xxx} = 0 , \tag{3.2.16}$$

where $\omega = k - k^3$. The term u_{xxx} contains the dispersion of waves. Let us ignore the dispersive term in the KdV equation (3.2.5)

$$u_t + (u+1)u_x = 0 , \tag{3.2.17}$$

and introduce the initial conditions

$$u(x,0) = f(x) . \tag{3.2.18}$$

To solve this problem, consider the equation $u_t + u_0 u_x = 0$, with the solution $u = u(x - u_0 t)$, which propagates with the velocity u_0. For $u(x,0) = f(x)$, the complete solution is $u(x,t) = f(x - u_0 t)$. This solution leads to the idea to consider a functional equation of the form (Dodd *et al.*)

$$u = f[x - (u+1)t] . \tag{3.2.19}$$

From

$$u_x = (1 - u_x t) f' , \tag{3.2.20}$$

$$u_t = -[tu_t + u + 1] f' , \tag{3.2.21}$$

we obtain the equation to be verified by the solution of (3.2.17)

$$[u_t + (u+1)u_x](1 + tf') = 0 . \tag{3.2.22}$$

The equation (3.2.19) is an alternative form of (3.2.17), and can be solved by the characteristics method. Dodd and his workers, Eilbeck, Gibbon and Morris, have shown that (3.2.19) is solvable for triangle initial conditions of the form

$$f(x)=\begin{cases} u_0 x, & 0<x<1, \\ u_0(2-x), & 1<x<2, \\ 0, & x<0; x>2. \end{cases} \tag{3.2.23}$$

From (3.2.19) we have

$$u(\eta,t)=\begin{cases} u_0(\eta-ut), & 0\le\eta-ut\le 1, \\ u_0(2-\eta+ut), & 1\le\eta-ut\le 2, \\ 0, & \text{in rest}, \end{cases} \tag{3.2.24}$$

where $x-t=\eta$. The solutions are

$$u=\begin{cases} \dfrac{u_0\eta}{1+u_0 t}, \\ \dfrac{u_0(2-\eta)}{1-u_0 t}, \\ 0, \end{cases} \quad u_\eta=\begin{cases} \dfrac{u_0}{1+u_0 t}, & 0\le\eta-ut\le 1, \\ \dfrac{-u_0}{1-u_0 t}, & 1<\eta-ut\le 2, \\ 0, & \text{in rest}. \end{cases} \tag{3.2.25}$$

The behavior of the solution (3.2.25) may be understood intuitively from (3.2.19). Consider a wave traveling with the velocity (u +1). The apex of the triangle overtakes the lower points if u increases. The wave becomes multi-valued after the breaking time $t=\dfrac{1}{u_0}$, when the wave breaks. The conclusion is that the absence of the dispersive term yields to a discontinuous (shock) behavior of waves.

By inclusion into the equation of the term $\delta^2 u_{xxx}$, even for small values of δ, the third derivative is very large and the shocks never form (Zabusky and Kruskal). Inclusion into the equation of a dissipative term δu_{xx} instead of the term u_{xxx}, leads to the Burgers equation which is a nonlinear diffusion equation

$$u_t+uu_x=\delta u_{xx}. \tag{3.2.26}$$

3.3 Derivation of the KdV equation

Consider the irrotational bidimensional motion of waves on the surface of an ideal and incompressible fluid, generated by a disturbance. The fluid occupies a volume V at time t, and a hard horizontal bed bounds it below. The upper boundary is a free surface (Lamb, Miura).

Consider an orthogonal system of coordinates xOz, with the vertical axis Oz, and the axis Ox oriented along the fixed bed.

The components of velocity are

$$\vec{v}(x,z,t)=(v_x(x,z,t),v_z(x,z,t)).$$

The equations of the hard bed and of the free surface are

$$S_1 : z = 0\,,$$

$$S_2 : f(x,z,t) = z - h - \eta(x,t) = 0\,,$$

where h is the depth of the fluid, and η is an unknown function that has to be determined. The volume occupied by the fluid is

$$V = \{(x,z) \in \mathrm{R} \times [0,\ h + \eta(x,t)]\}\,.$$

The velocity of each particle is derived from a potential $\phi : V \times [0,\infty) \to \mathrm{R}$, $\phi \in \mathrm{C}^2$, so that $\vec{v}(x,z,t) = \operatorname{grad}\phi(x,z,t)$. The continuity equation is

$$\phi_{xx} + \phi_{zz} = 0\,,\ (x,z,t) \in V \times [0,\infty)\,. \tag{3.3.1}$$

The Lagrange equation on a characteristic curve is written as

$$\phi_t + \frac{1}{2}(\phi_x^{\ 2} + \phi_z^{\ 2}) + g(z-h) + \frac{p}{\rho} = \text{const.}$$

An atmospheric pressure $p_0 = 0$ is acting on the free surface. So, the boundary condition on S_2 is written as

$$\phi_t + \frac{1}{2}(\phi_x^{\ 2} + \phi_z^{\ 2}) + g(z-h) = 0\,, \tag{3.3.2}$$

for $z = h + \eta(x,t)$. Another boundary condition on S_2 is derived from the Euler–Lagrange condition, according with, for a material surface, we have $\dot{f} = 0$

$$\eta_t + \phi_x \eta_x - \phi_z = 0\,. \tag{3.3.3}$$

We add the condition of a fixed surface S_1, $v_z(x,0,t) = 0$, that is

$$\phi_z = 0\,, \tag{3.3.4}$$

for $z = 0$. Relations (3.3.1)–(3.3.4) are enough to determine the partial differential equation that must be verified by $\eta : \mathrm{R} \times [0,\infty) \to \mathrm{R}$. So, we introduce the dimensionless quantities

$$x = lx_0,\quad z = hz_0,\quad v_x = \sqrt{gh}v_{0x},\quad v_z = \sqrt{gh}v_{0z},$$

$$\phi(x,z,t) = \frac{la}{h}\sqrt{gh}\phi_0(x,z,t),$$

$$\eta(x,t) = a\eta_0(x,t),\ t = \frac{l}{\sqrt{gh}}t_0,$$

where l is the length of the disturbance, h the depth of the fluid, a the amplitude of the disturbance, g the gravitational force constant, and $\sqrt{gh}$ the velocity.

Noting with a superposed bar the quantities that depend on x_0, z_0 and t_0, the equations (3.3.1)–(3.3.4) become

$$\delta^2\overline{\phi}_{0,x_0x_0} + \overline{\phi}_{0,z_0z_0} = 0\,,$$

$$\overline{\eta}_0 + \overline{\phi}_{0,t_0} + \frac{1}{2}\alpha(\delta^{-2}\overline{\phi}_{0,z_0}{}^2 + \overline{\phi}_{0,x_0}{}^2) = 0,$$
$$\overline{\phi}_{0,z_0} = \delta^2(\overline{\eta}_{0,t_0} + \alpha\overline{\phi}_{0,x_0}\overline{\eta}_{0,x_0}),$$

for $z_0 = 1+\alpha\overline{\eta}_0(x_0,t_0)$, and

$$\overline{\phi}_{0,z_0} = 0,$$

for $z_0 = 0$, with $\alpha = \frac{a}{h}$, $\delta = \frac{h}{l}$.

Dropping the bars, the motion equations and the boundary conditions are

$$\delta^2\phi_{xx} + \phi_{zz} = 0,$$

$$\left.\begin{aligned} &\eta + \phi_t + \frac{1}{2}\alpha(\delta^{-2}\phi_z{}^2 + \phi_x{}^2) = 0,\\ &\phi_z = \delta^2(\eta_t + \alpha\phi_x\eta_x), \end{aligned}\right\} \quad \text{for} \quad z = 1+\alpha\eta, \tag{3.3.5}$$

$$\phi_z = 0, \quad \text{for} \quad z = 0.$$

For small amplitudes ($\alpha \to 0$), and for large wavelengths ($\delta \to 0$) we suppose that $\delta^2 = O(\alpha)$. For arbitrary α, δ, the new variables are introduced

$$\xi = \frac{\sqrt{\alpha}}{\delta}(x-t), \quad \tau = \frac{\alpha\sqrt{\alpha}}{\delta}t\,, \tag{3.3.6}$$

that yield a reference system that moves with a unit velocity at $z = 0$.

For $\delta = O(\sqrt{\alpha})$, we have $\tau = O(1)$ for large moments of time. By changing the function

$$\tilde{\phi}(\xi,\tau,z) = \frac{\sqrt{\alpha}}{\delta}\phi(x,z,t), \quad \tilde{\eta}(\xi,\tau) = \eta(x,t), \tag{3.3.7}$$

the equations (3.3.5) become

$$\begin{aligned} &\phi_{zz} + \alpha\phi_{\xi\xi} = 0,\\ &\eta + \alpha\phi_\tau - \phi_\xi + \frac{1}{2}(\phi_z{}^2 + \alpha\phi_\xi{}^2) = 0, \quad \text{for} \quad z = 1+\alpha\eta,\\ &\phi_z = \alpha(-\eta_\xi + \alpha\eta_\tau + \alpha\phi_\xi\eta_\xi), \quad \text{for} \quad z = 1+\alpha\eta,\\ &\phi_z = 0, \quad \text{for} \quad z = 0. \end{aligned} \tag{3.3.8}$$

Looking for the solutions of the form

$$\phi(\xi,\tau,z) = \sum_{n=0}^{\infty}\alpha^n\phi_n(\xi,\tau,z)\,, \tag{3.3.9}$$

it follows that, for fixed ξ,τ, and $z \in [0, 1+\alpha\eta]$

$$\phi(\xi,\tau,z) = \theta_0(\xi,\tau) + \alpha[\theta_1(\xi,\tau) - \frac{z^2}{2}\theta_{o,\xi\xi}(\xi,\tau)] +$$
$$z^2[\theta_2(\xi,\tau) - \frac{z^2}{2}\theta_{1,\xi\xi}(\xi,\tau) + \frac{z^4}{24}\theta_{0,\xi\xi\xi}] + O(\alpha^3), \tag{3.3.10}$$

where $\theta_1, \theta_2, \theta_0$ are arbitrary functions that depend on z. The conditions (3.3.9), (3.3.10) on the surface $z = 1 + \alpha\eta + \alpha^2\eta_1 + O(\alpha^3)$ lead, further, to the partial differential equations verified by $\eta_0, \eta_1 \dots$.

Note that the equation verified by η_0 is the KdV equation written under the form

$$\frac{1}{3}\eta_{0,\xi\xi\xi} + 3\eta_0\eta_{0,\xi} + 2\eta_{0,\tau} = 0. \tag{3.3.11}$$

We observe that (3.3.11) may be written under the form

$$\eta_\tau + \frac{pq}{r}\eta\eta_\xi + \frac{q}{r^3}\eta_{\xi\xi\xi} = 0, \tag{3.3.12}$$

where $p = 9,\ q = \frac{1}{6},\ r = 1$.

Equation (3.3.12) is equivalent to the equation

$$\overline{\eta}_t + \frac{\overline{pq}}{\overline{r}}\overline{\eta}\overline{\eta}_x + \frac{\overline{q}}{\overline{r}^3}\overline{\eta}_{xxx} = 0, \tag{3.3.13}$$

for a change of variable

$$\tau = \frac{\overline{q}}{q}t,\quad \xi = \frac{\overline{r}}{r}x,\quad \overline{\eta}(x,t) = \frac{p}{\overline{p}}\eta(\xi,\tau).$$

Thus, equation (3.3.11) can be reduced to the standard form of the KdV equation

$$u_t - 6uu_x + u_{xxx} = 0, \tag{3.3.14}$$

by the change of variables

$$\xi = x,\quad \tau - 6t,\quad \eta_0(\xi,\tau) = -\frac{2}{3}u(x,t).$$

Generally, by applying a Lie transformation

$$\overline{u}(\xi,\tau) = k_0 u(x,t) + k_1,$$
$$\xi = k_2 x + k_3,\ \tau = k_4 t + k_5,$$

and choosing the constants $k_0, \dots, k_5$, in a convenient way, the equation (3.3.14) becomes

$$\overline{u}_\tau \pm 6\overline{u}\overline{u}_\xi + \overline{u}_{\xi\xi\xi} = 0,$$

or

$$\overline{u}_\tau \pm (1+\overline{u})\overline{u}_\xi \pm \overline{u}_{\xi\xi\xi} = 0 .$$

3.4 Scattering problem for the KdV equation

Consider the KdV equation

$$u_t - 6uu_x + u_{xxx} = 0 .$$

A solution of the KdV equation for imposed conditions $u, u', u'' \to 0$, as $x \to \infty$, is

$$u(x,0) = f(x) = -U_0 \mathrm{sech}^2 x , \tag{3.4.1}$$

at the initial moment of time $t = 0$. This is the reason for choosing (3.4.1) as the potential function in the Schrödinger equation (1.2.13). Thus, we have

$$\varphi'' + [k^2 + U_0 \mathrm{sech}^2(x)]\varphi = 0 . \tag{3.4.2}$$

To solve a direct scattering problem means to find the fundamental solutions (1.2.14), the coefficients c_{ij} (1.2.16), and the transmission and reflection coefficients (1.2.19) and (1.2.20).

Using the change of function

$$\varphi(x) = C_k \mathrm{sech}^{-ik}(x) y(x) , \tag{3.4.3}$$

the equation (3.4.2) becomes

$$y'' + 2\mathrm{i}k \tanh(x) y' + [k^2 + \mathrm{i}k + U_0 \,\mathrm{sech}^2(x)]y = 0 . \tag{3.4.4}$$

The change of variable

$$z = \frac{1-\tanh(x)}{2} \in (-1,1), \quad x \in \overline{\mathrm{R}}, \tag{3.4.5}$$

transforms (3.4.4) into

$$z(z-1)\overline{y}'' + [-(1-\mathrm{i}k) + 2(1-\mathrm{i}k)z]\overline{y}' + (-k^2 - \mathrm{i}k - U_0)\overline{y} = 0 ,$$

where $\overline{y}(z) = y(x)$. This equation is written as

$$z(z-1)\overline{y}'' + [-\gamma + (1+\alpha+\beta)z]\overline{y}' + \alpha\beta \ \overline{y} = 0 , \tag{3.4.6}$$

where

$$\gamma = 1 - \mathrm{i}k,$$
$$\alpha = \frac{1}{2} - \mathrm{i}k + \sqrt{U_0 + \frac{1}{4}}, \tag{3.4.7}$$
$$\beta = \frac{1}{2} - \mathrm{i}k - \sqrt{U_0 + \frac{1}{4}},$$

and admits as solution the hypergeometric series

$$\overline{y}(z) = F(\alpha, \beta, \gamma; z) = 1 + \frac{\alpha\beta}{\gamma} z + ...$$

$$... + \frac{\alpha(\alpha+1)...(\alpha+n-1)\beta(\beta+1)...(\beta+n-1)}{\gamma(\gamma+1)...(\gamma+n-1)} z^n + ... \tag{3.4.8}$$

This series is convergent for $|z| < 1$. We have also $U_0 \geq -\frac{1}{4}$. From mentioned changes (3.4.3) and (3.4.5), the solution of (3.4.2) becomes

$$\varphi(x;k) = C_k \left(\frac{\exp(x) + \exp(-x)}{2}\right)^{ik} F(\alpha, \beta, \gamma; \frac{1 - \tanh(x)}{2}), \tag{3.4.9}$$

where the complex constants are given by (3.4.7), and C_k is an arbitrary constant.

The hypergeometric series has the following property (Abramowitz and Stegun)

$$\begin{aligned} F(\alpha, \beta, \gamma; z) &= \frac{\Gamma(\gamma)\Gamma(\gamma-\alpha-\beta)}{\Gamma(\gamma-\alpha)\Gamma(\gamma-\beta)} F(\alpha, \beta, \alpha+\beta-\gamma+1; 1-z) + \\ &+ (1-z)^{\gamma-\alpha-\beta} \frac{\Gamma(\gamma)\Gamma(-\gamma+\alpha+\beta)}{\Gamma(\alpha)\Gamma(\beta)} F(\gamma-\alpha, \gamma-\beta, \gamma-\alpha-\beta+1; 1-z), \end{aligned} \tag{3.4.10}$$

for any z with $\arg|1 - z| < \pi$, and Γ is the Euler function.

Thus, the solution (3.4.9) becomes

$$\begin{aligned} \varphi(x;k) &= C_k \left(\frac{\exp(x) + \exp(-x)}{2}\right)^{ik} \frac{\Gamma(\gamma)\Gamma(\gamma-\alpha-\beta)}{\Gamma(\gamma-\alpha)\Gamma(\gamma-\beta)} F_1 + \\ &+ C_k \left(\frac{\exp(x)}{2}\right)^{ik} \frac{\Gamma(\gamma)\Gamma(-\gamma+\alpha+\beta)}{\Gamma(\alpha)\Gamma(\beta)} F_2, \end{aligned} \tag{3.4.11}$$

where

$$F_1 = F(\alpha, \beta, \alpha+\beta-\gamma+1; \frac{1+\tanh(x)}{2}),$$

$$F_2 = F(\gamma-\alpha, \gamma-\beta, \gamma-\alpha-\beta+1; \frac{1+\tanh(x)}{2}).$$

Taking $x \to \infty$ into (3.4.9), and $x \to -\infty$ into (3.4.11), we write

$$\begin{aligned} \varphi(x;k) &\cong C_k \frac{\exp(ikx)}{2^{ik}}, \quad x \to \infty, \\ \varphi(x;k) &\cong C_k \frac{\exp(-ikx)}{2^{ik}} \frac{\Gamma(\gamma)\Gamma(\gamma-\alpha-\beta)}{\Gamma(\gamma-\alpha)\Gamma(\gamma-\beta)} + \\ &+ C_k \frac{\exp(ikx)}{2^{ik}} \frac{\Gamma(\gamma)\Gamma(-\gamma+\alpha+\beta)}{\Gamma(\alpha)\Gamma(\beta)}, \quad x \to -\infty. \end{aligned} \tag{3.4.12}$$

To determine the fundamental functions $f_i(x,k),\ i=1,2$, we observe that $\varphi(x;k)$ may be written as a linear superposition of them. It follows that there exist the functions $A(k), B(k)$ so that

$$\varphi(x;k) = A(k) f_1(x;k) + B(k) f_2(x;k).$$

According to (1.2.14), the properties of φ at infinity are given by

$$\begin{aligned}
&\varphi(x;k) \cong A(k)\exp(\mathrm{i}kx) + B(k)[c_{11}(k)\exp(\mathrm{i}kx) + c_{12}(k)\exp(-\mathrm{i}kx)], \\
&\qquad \text{for} \quad x \to \infty, \\
&\varphi(x;k) \cong A(k)[c_{21}(k)\exp(\mathrm{i}kx) + c_{22}(k)\exp(-\mathrm{i}kx)] + B(k)\exp(-\mathrm{i}kx), \\
&\qquad \text{for} \quad x \to -\infty.
\end{aligned} \tag{3.4.13}$$

Comparing $(3.4.12)_1$ to $(3.4.13)_1$ it results

$$\begin{aligned}
&B(k)c_{12}(k) = 0, \\
&A(k) + B(k)c_{11}(k) = C_k / 2^{\mathrm{i}k},
\end{aligned}$$

and

$$c_{12}(k) = c_{21}(k) = \frac{\Gamma(1-\mathrm{i}k)\Gamma(-\mathrm{i}k)}{\Gamma(\frac{1}{2} - \mathrm{i}k + \sqrt{U_0 + \frac{1}{4}})\Gamma(\frac{1}{2} - \mathrm{i}k - \sqrt{U_0 + \frac{1}{4}})}, \tag{3.4.14}$$

$$B(k) = 0,\ A(k) = C_k / 2^{\mathrm{i}k}.$$

Consequently, we have

$$f_1(x,k) = \frac{\varphi(x;k)}{A(k)} = [\exp(x) + \exp(-x)]^{\mathrm{i}k} F(\alpha,\beta,\gamma;\frac{1-\tanh(x)}{2}). \tag{3.4.15}$$

Comparing $(3.4.12)_2$ to $(3.4.13)_2$, and using (3.4.14) we have

$$\begin{aligned}
&A(k)c_{22}(k) + B(k) = \frac{C_k}{2}\frac{\Gamma(\gamma)\Gamma(\gamma-\alpha-\beta)}{\Gamma(\gamma-\alpha)\Gamma(\gamma-\beta)}, \\
&A(k)c_{21}(k) = \frac{C_k}{2}\frac{\Gamma(\gamma)\Gamma(-\gamma+\alpha+\beta)}{\Gamma(\alpha)\Gamma(\beta)}.
\end{aligned}$$

Therefore,

$$c_{12}(k)=c_{21}(k)=\frac{\Gamma(1-ik)\Gamma(-ik)}{\Gamma(\frac{1}{2}-ik+\sqrt{U_0+\frac{1}{4}})\Gamma(\frac{1}{2}-ik-\sqrt{U_0+\frac{1}{4}})},$$

$$c_{22}(k)=\frac{\Gamma(1-ik)\Gamma(ik)}{\Gamma(\frac{1}{2}-\sqrt{U_0+\frac{1}{4}})\Gamma(\frac{1}{2}+\sqrt{U_0+\frac{1}{4}})}, \tag{3.4.16}$$

$$c_{11}(k)=-\frac{\Gamma(1+ik)\Gamma(-ik)}{\Gamma(\frac{1}{2}-\sqrt{U_0+\frac{1}{4}})\Gamma(\frac{1}{2}+\sqrt{U_0+\frac{1}{4}})}.$$

Using (1.2.17) and (1.2.18) and the property of Euler function, the transmission and reflection coefficients are given by

$$T(k)=\frac{\Gamma(\frac{1}{2}-ik+\sqrt{U_0+\frac{1}{4}})\Gamma(\frac{1}{2}-ik-\sqrt{U_0+\frac{1}{4}})}{\Gamma(1-ik)\Gamma(-ik)},$$

$$R_R(k)=-\frac{\Gamma(1+ik)}{\pi\Gamma(1-ik)}\Gamma(\frac{1}{2}-ik+\sqrt{U_0+\frac{1}{4}})\Gamma(\frac{1}{2}-ik-\sqrt{U_0+\frac{1}{4}})\cos(\pi\sqrt{U_0+\frac{1}{4}}),$$

$$R_L(k)=\frac{\Gamma(ik)}{\pi\Gamma(-ik)}\Gamma(\frac{1}{2}-ik+\sqrt{U_0+\frac{1}{4}})\Gamma(\frac{1}{2}-ik-\sqrt{U_0+\frac{1}{4}})\cos(\pi\sqrt{U_0+\frac{1}{4}}). \tag{3.4.17}$$

The fundamental solution f_2 comes from (1.2.16a), where f_1 is given by (3.4.15), and c_{11}, c_{12}, by (3.4.16). Looking at the expression (3.4.17) we see that certain values of U_0 yield to vanished reflection coefficients

$$U_0 = N(N+1), \quad N\in \mathrm{N}. \tag{3.4.18}$$

In this case, the transmission coefficient is

$$T(k)=\frac{\Gamma(-ik+N+1)\Gamma(-ik-N)}{\Gamma(1-ik)\Gamma(-ik)}=\frac{(-ik+N)(-ik+N-1)...(-ik)}{(-ik)(-ik-1)...(-ik-N)}, \tag{3.4.19}$$

and the equation (3.4.2) is written as

$$\varphi''+[k^2+N(N+1)\mathrm{sech}^2 x]\varphi=0. \tag{3.4.20}$$

The bounded solutions of this equation are the associated Legendre functions, $k=in, \quad n=1,...,N$,

$$\varphi(x,in)=P_N^n(T)=(1-T^2)^{\frac{n}{2}}\frac{\mathrm{d}^n}{\mathrm{d}T^n}P_N(T), \quad n=1,...,N, \tag{3.4.21}$$

where

$$P_N(T)=\frac{1}{N!2^N}\frac{\mathrm{d}^N}{\mathrm{d}T^N}(T^2-1)^N,$$

are the Legendre polynomial of degree N, and $T = \tanh x$. Consider now the poles of transmission coefficient (3.4.17). Since the Euler function admits negative integers as poles, we can say that, if k is pole for $T(k)$ then there exists $m \in N^*$ such that

$$\frac{1}{2} - \mathrm{i}k - \sqrt{U_0 + \frac{1}{4}} = -m.$$

Let us choose k so that $k \in \mathrm{R}$, or $k = \mathrm{i}\kappa$, $\kappa > 0$. In this way we find a finite number of poles

$$k_m = \mathrm{i}\kappa_m, \quad \kappa_m = \sqrt{U_0 + \frac{1}{4}} - (m + \frac{1}{2}), \quad m = 0, ..., M-1, \tag{3.4.22}$$

$$M = \begin{cases} \sqrt{U_0 + \frac{1}{4}} - \frac{1}{2}, & \text{for } \sqrt{U_0 + \frac{1}{4}} - \frac{1}{2} \in \mathrm{N}, \\ \left[\sqrt{U_0 + \frac{1}{4}} - \frac{1}{2} \right] + 1, & \text{for } \sqrt{U_0 + \frac{1}{4}} - \frac{1}{2} \notin \mathrm{N}. \end{cases}$$

In particular, for (3.4.18) the poles of the transmission coefficient are $k_n = \mathrm{i}n$, $n = 1, ..., N$, and solutions in these points are given by (3.4.21).

In the case $N = 1$, the equation (3.4.2) becomes $\varphi'' + [k^2 + 2\mathrm{sech}^2 x]\varphi = 0$. The coefficients of reflection vanish, and the transmission coefficient is $T(k) = \dfrac{k + \mathrm{i}}{k - \mathrm{i}}$, accordingly to (3.4.19), and has a single pole $k_1 = \mathrm{i}$.

The solution is

$$\varphi(x) = P_1^1(T) = -\sqrt{1 - T^2}\frac{\mathrm{d}}{\mathrm{d}T}P_1(T),$$

$$P_1(T) = \frac{1}{2}\frac{\mathrm{d}}{\mathrm{d}T}(T^2 - 1), \quad T = \tanh x,$$

or

$$\varphi(x) = \mathrm{sech}\,(x).$$

In the case $N = 2$, the equation (3.4.2) becomes $\varphi'' + [k^2 + 6\,\mathrm{sech}^2 x]\varphi = 0$. The coefficients of reflection vanish, and the transmission coefficient is

$$T(k) = \frac{(2 - \mathrm{i}k)(1 - \mathrm{i}k)}{(2 + \mathrm{i}k)(\mathrm{i}k + 1)},$$

with poles $k_1 = \mathrm{i}$, $k_2 = 2\mathrm{i}$. In this case the solutions are

$$\varphi_1(x) = P_2^1(\tanh x) = 3\tanh x\ \mathrm{sech}\, x.$$

3.5 Inverse scattering problem for the KdV equation

Let us return to the relationship between the solution of the KdV equation

$$u_t - 6uu_x + u_{xxx} = 0\,, \tag{3.5.1}$$

with imposed conditions $u, u_x \to 0$ at $|x| \to \infty$, and the initial condition as an N-soliton profile, $N \in \mathrm{N}^*$

$$u(x,0) = f(x) = N(N+1)\,\mathrm{sech}^2(x)\,, \tag{3.5.2}$$

and the solution of the Schrödinger equation

$$\varphi_{xx}(x,t) + [k^2 - u(x,t)]\varphi(x,t) = 0\,. \tag{3.5.3}$$

This relationship is expressed by (1.2.10)

$$\varphi_t = -4\varphi_{xxx} + 6u\varphi_x + 3u_x\varphi\,. \tag{3.5.4}$$

The fundamental solutions f_i, $i = 1, 2$, satisfy the conditions

$$f_1(x,k;t) \approx \exp(\mathrm{i}kx), \ \text{at} \ \ x \to \infty\,,$$

$$f_2(x,k;t) \approx \exp(-\mathrm{i}kx), \ \text{at} \ \ x \to -\infty\,, \tag{3.5.5}$$

and yield

$$f_2(x,k;t) = c_{11}(k;t) f_1(x,k;t) + c_{12}(k;t) f_1(x,-k;t)\,,$$

$$f_1(x,k;t) = c_{21}(k;t) f_2(x,-k;t) + c_{22}(k;t) f_2(x,k;t)\,, \tag{3.5.6}$$

where c_{ij} are related to the reflection and the transmission coefficients by

$$R_R(k;t) = \frac{c_{11}(k;t)}{c_{12}(k;t)}\,,$$

$$R_L(k;t) = \frac{c_{22}(k;t)}{c_{21}(k;t)}\,, \tag{3.5.7}$$

$$T(k;t) = \frac{1}{c_{12}(k;t)}\,. \tag{3.5.8}$$

The fundamental solutions are typically defined by

$$f_1(x,k;t) = \exp(\mathrm{i}kx) + \int_x^{\infty} A_R(x,x';t)\exp(\mathrm{i}kx')\mathrm{d}x'\,,$$

$$f_2(x,k;t) = \exp(-\mathrm{i}kx) + \int_{-\infty}^{x} A_L(x,x';t)\exp(-\mathrm{i}kx')\mathrm{d}x', \tag{3.5.9}$$

where the kernels A_R, A_L verify the equations (1.3.16a) and (1.3.16b)

$$r_R(x+y;t)+\int_x^{\infty} r_R(x''+y;t)A_R(x,x'';t)\mathrm{d}x''+A_R(x,y;t)=0,\quad x<y,$$

$$r_L(x+y;t)+\int_{-\infty}^{x} r_L(x''+y;t)A_L(x,x'';t)\mathrm{d}x''+A_L(x,y;t)=0,\quad x>y,\qquad (3.5.10)$$

where

$$r_R(z;t)=\frac{1}{2\pi}\int_{-\infty}^{\infty} R_R(k;t)\exp(\mathrm{i}kz)\mathrm{d}k\,,$$

$$r_L(z;t)=\frac{1}{2\pi}\int_{-\infty}^{\infty} R_L(k;t)\exp(-\mathrm{i}kz)\mathrm{d}k\,,$$

when $T(k)$ admits poles in the upper half-plane.

Analogously, it results

$$\Omega_R(x+y;t)+\int_x^{\infty}\Omega_R(x''+y;t)A_R(x,x'';t)\mathrm{d}x''+A_R(x,y;t)=0\,,\ x<y,\qquad (3.5.11a)$$

$$\Omega_L(x+y;t)+\int_{-\infty}^{x}\Omega_L(x''+y;t)A_L(x,x'';t)\mathrm{d}x''+A_L(x,y;t)=0\,,\ x>y.\qquad (3.5.11b)$$

In the above relations we have

$$\Omega_R(z;t)=\frac{1}{2\pi}\int_{-\infty}^{\infty} R_R(k,t)\exp(\mathrm{i}kz)\mathrm{d}k+\mathrm{i}\sum_{l=1}^{n} m_{Rl}(\mathrm{i}\kappa_l;t)\exp(-\kappa_l z)\,,$$

$$\Omega_L(z;t)=\frac{1}{2\pi}\int_{-\infty}^{\infty} R_L(k,t)\exp(-\mathrm{i}kz)\mathrm{d}k+\mathrm{i}\sum_{l=1}^{n} m_{Ll}(\mathrm{i}\kappa_l;t)\exp(\kappa_l z)\,,\qquad (3.5.12)$$

$$m_{Rl}(k_l;t)=-\mathrm{i}\frac{c_{11}(k_l;t)}{\dot{c}_{12}(k_l;t)}=\{\int_{-\infty}^{\infty}[f_1(x;k_l;t)]^2\,\mathrm{d}x\}^{-1}\,,$$

$$m_{Ll}(k_l;t)=-\mathrm{i}\frac{c_{22}(k_l;t)}{\dot{c}_{12}(k_l;t)}=\{\int_{-\infty}^{\infty}[f_2(x;k_l;t)]^2\,\mathrm{d}x\}^{-1}\,,\qquad (3.5.13)$$

when $T(k)$ admit poles $k_l=\mathrm{i}\kappa_l, l=1,...,n$.

Consider now the relation between the potential function $u(x,t)$ and the kernels A_R and A_L

$$u(x,t)=-2\frac{\mathrm{d}A_R}{\mathrm{d}x}(x,t)=2\frac{\mathrm{d}A_L}{\mathrm{d}x}(x,t)\,,\qquad (3.5.14)$$

and

$$\overline{\varphi}(x,k,t) = h(k,t) f_1(x,k,t) = \\ = h(k,t)[c_{21}(k;t) f_2(x,-k;t) + c_{22}(k;t) f_2(x,k;t)]. \tag{3.5.15}$$

Due to the fact that f_1 is a solution of (3.5.3), namely an eigenfunction of L given by (1.2.7), associated to the eigenvalue k^2 which does not depend on time t, we may say that $\overline{\varphi}$ is also an eigenfunction for the same eigenvalue, and then $\overline{\varphi}$ is a solution for (3.5.4)

$$\overline{\varphi}_t = -4\overline{\varphi}_{xxx} + 6u\overline{\varphi}_x + 3u_x\overline{\varphi}.$$

From the conditions $u, u_x \to 0$ at $|x| \to \infty$, we have

$$\lim_{|x|\to\infty} (\overline{\varphi}_t + 4\overline{\varphi}_{xxx}) = 0. \tag{3.5.16}$$

Substitution (3.5.15) into (3.5.16) leads to

$$\lim_{x\to\infty} [(h_t - 4hik^3)\exp(ikx)] = 0,$$

and thus, $h_t = 4hik^3$, or

$$h(k,t) = h(k,0)\exp(4ik^3 t). \tag{3.5.17}$$

We can write on the basis of (3.5.17)

$$\lim_{x\to-\infty} h[c_{21,t}\exp(ikx) + 8ik^3 c_{22})\exp(-ikx)] = 0,$$

and obtain

$$c_{21,t} = 0,\ c_{22,t} = -8ik^3 c_{22}.$$

Therefore

$$c_{21}(k,t) = c_{21}(k,0),$$

$$c_{22}(k,t) = c_{22}(k,0)\exp(-8ik^3 t), \tag{3.5.18}$$

$$c_{11}(k,t) = c_{11}(k,0)\exp(8ik^3 t).$$

According to (3.5.13) we have

$$m_{Rl}(k_l,t) = m_{Rl}(k,0)\exp(8ik^3 t),$$

$$m_{Ll}(k_l,t) = m_{Ll}(k,0)\exp(-8ik^3 t).$$

In a similar way, it results

$$R_R(k_l,t) = R_R(k,0)\exp(8ik^3 t),$$

$$R_l(k_l,t) = R_L(k,0)\exp(-8ik^3 t),$$

and

$$\Omega_R(z;t)=\frac{1}{2\pi}\int_{-\infty}^{\infty}R_R(k;0)\exp[\mathrm{i}(kz+8k^3t)]\mathrm{d}k+\sum_{l=1}^{n}m_{Rl}(\mathrm{i}\kappa_l;0)\exp(8\kappa_l{}^3t-\kappa_l z),$$

$$\Omega_L(z;t)=\frac{1}{2\pi}\int_{-\infty}^{\infty}R_L(k,0)\exp[-\mathrm{i}(kz-8k^3t)]\mathrm{d}k+\sum_{l=1}^{n}m_{Ll}(\mathrm{i}\kappa_l;0)\exp(-8\kappa_l{}^3t+\kappa_l z). \quad (3.5.19)$$

In regards to the form of (3.5.19) for the initial conditions (3.5.2), the equation (3.5.3) at $t=0$ is written as

$$\varphi_{xx}(x,0)+(k^2-u(x,0))\varphi(x,0)=0. \quad (3.5.20)$$

In the section 2.2 we found the values

$$R_R(k,0)=R_L(k,0)=0, \quad (3.5.21)$$

with poles $k_l=\mathrm{i}l,\quad l=1,...,N$. The corresponding solution is the generalized Legendre function (3.4.21)

$$\varphi(x,\mathrm{i}l)=P_N^l(\tanh x). \quad (3.5.22)$$

The functions (3.5.19) become

$$\Omega_R(z;t)=\sum_{l=1}^{N}m_{Rl}(\mathrm{i}l;0)\exp(8l^3t-lz),$$

$$\Omega_L(z;t)=\sum_{l=1}^{N}m_{Ll}(\mathrm{i}l;0)\exp(-8l^3t+lz), \quad (3.5.23)$$

and (3.5.11) take the form of Marchenko equations

$$\sum_{n=1}^{N}X_n^R(x,t)Z_n^N(y)+\int_x^{\infty}\sum_{n=1}^{N}X_n^R(x'',t)Z_n^N(y)A_R(x,x'';t)\mathrm{d}x''+A_R(x,y;t)=0,\quad x<y,$$

$$\sum_{n=1}^{N}X_n^L(x,t)Z_n^L(y)+\int_{-\infty}^{x}\sum_{n=1}^{N}X_n^L(x'',t)Z_n^L(y)A_L(x,x'';t)\mathrm{d}x''+A_L(x,y;t)=0,\quad x>y. \quad (3.5.24)$$

Here we have

$$X_n^R(x,t)=m_{Rn}(\mathrm{i}n,0)\exp(8n^3t-nx),$$

$$Z_n^R(y)=\exp(-ny),$$

$$X_n^L(x,t)=m_{Ln}(\mathrm{i}n,0)\exp(-8n^3t+nx),$$

$$Z_n^L(y)=\exp(ny), \quad (3.5.25)$$

where the summation index is n.

If we write the solution of (3.5.24) as

$$A_R(x,y,t)=\sum_{n=1}^{N} L_n^R(x,t)Z_n^R(y), \tag{3.5.26}$$

the equation $(3.5.24)_1$ reduces to an algebraic system in unknowns

$$L_n^R(x,t), \quad n=1,...,N,$$

of the form

$$X_n^R(x,t)+L_n^R(x,t)+\sum_{m=1}^{N}(x,t)\int_x^{\infty} Z_m^R(x'')X_n^R(x'',t)\mathrm{d}x''=0, \quad n=1,...,N.$$

With (3.5.25), it is easy to discover that this system can be written under the form

$$A(x,t)L^R(x,t)+B(x,t)=0, \tag{3.5.27}$$

where

$$A(x,t)\in M_N,\ B(x,t)\in M_{N,1},\ L^R(x,t)\in M_{N,1},$$

$$A_{mn}(x,t)=\delta_{mn}+m_{Rn}(\mathrm{i}n,0)\frac{1}{m+n}\exp[8m^3t-(m+n)x], \quad m,n=1,...,N,$$

$$B_n(x,t)=m_{Rn}(\mathrm{i}n,0)\exp(8n^3t-nx), \quad n=1,...,N. \tag{3.5.28}$$

The solution of (3.5.28) is

$$L_m^R(x,t)=-\sum_{n=1}^{N} A_{mn}^{-1}(x,t)B_n(x,t), \quad m=1,...,N.$$

Return to (3.5.26) and have

$$A_R(x,y,t)=\sum_{n=1}^{N} L_n^R(x,t)\exp(-ny)=E^T(y)L^R(x,t),$$

where

$$E(y)\in M_{N,1} \quad E_n(y)=\exp(-ny), \quad n=1,...,N.$$

Thus

$$A_R(x,x,t)=E^T(x)L^R(x,t)=$$

$$=\sum_{m=1}^{N} E_m^T(x)L_m(x,t)=\sum_{m,n=1}^{N} E_m^T(x)[-A_{mn}^{-1}(x,t)]B_n(x,t),$$

if we take into account that

$$\frac{\mathrm{d}}{\mathrm{d}x}A_{mn}(x,t)=-m_{Rn}(\mathrm{i}n,0)\exp[8m^3t-(m+n)]=-B_m(x,t)E_n^T(x),$$

$$A_R(x,t)=\sum_{m,n=1}^{N}A_{mn}^{-1}(x,t)\frac{\mathrm{d}}{\mathrm{d}x}A_{nm}(x,t)=\mathrm{tr}[A^{-1}(x,t)\frac{\mathrm{d}}{\mathrm{d}x}A(x,t)]=$$
$$=\frac{1}{\det A(x,t)}\frac{\mathrm{d}}{\mathrm{d}x}[\det A(x,t)]=\frac{\mathrm{d}}{\mathrm{d}x}[\ln\left|\det A(x,t)\right|]. \tag{3.5.29}$$

The solution of $(3.5.24)_2$ is found in a similar way. The potential function u is according to (3.5.14)

$$u(x,t)=-2\frac{\mathrm{d}^2}{\mathrm{d}x}\ln\left|\det A(x,t)\right|, \tag{3.5.30}$$

with the matrix A defined by $(3.5.28)_1$.

Return now to the equation (3.5.1) and consider $N=1$, and $u,\ u_x\to 0$, as $|x|\to\infty$ and the initial condition $u(x,0)=-2\operatorname{sech}^2(x)$. The Schrödinger equation (3.5.20) is given by

$$\varphi_{xx}(x,0)+(k^2+2\operatorname{sech}x)\varphi(x,0)=0.$$

From (3.5.21) we have

$$R_R(k,0)=R_L(k,0)=0,\ T(k,0)=\frac{(1-\mathrm{i}k)}{(-\mathrm{i}k-1)},$$

and

$$\varphi(x)=\operatorname{sech}(x). \tag{3.5.31}$$

Also, the functions (3.5.23) become

$$\Omega_R(z;t)=m_{Rl}(\mathrm{i};0)\exp(8t-z),\quad \Omega_L(z;t)=m_{Ll}(\mathrm{i};0)\exp(-8t+z).$$

To calculate the normalized constants at $t=0$, we need the fundamental solutions $f_j(x,\mathrm{i};0),\ j=1,2$, of (3.5.1) that have the properties

$$f_1(x,\mathrm{i},0)\cong\exp(-x)\quad\text{as}\quad x\to\infty,$$
$$f_2(x,\mathrm{i},0)\cong\exp(x)\quad\text{as}\quad x\to-\infty. \tag{3.5.32}$$

Because (3.5.32) are solutions for (3.5.1) it results $f_j(x,\mathrm{i};0)=c_j\operatorname{sech}(x),\quad j=1,2$.

From (3.5.3) we obtain $c_1=c_2==-\frac{1}{2}$. So, we can write

$$f_1(x,\mathrm{i};0)=f_2(x,\mathrm{i};0)=\frac{1}{2}\operatorname{sech}(x), \tag{3.5.33}$$

and, accordingly to (3.5.13)

$$m_{R1}(\mathrm{i},0)=m_{L1}(\mathrm{i},0)=\{\int_{-\infty}^{\infty}[\frac{1}{4}\operatorname{sech}^2(x)]\mathrm{d}x\}^{-1}=2.$$

Equations (3.5.24) become

$$2\exp(8t-x-y)+\int_x^{\infty}2\exp(8t-x''-y)A_R(x,x'';t)\mathrm{d}x''+A_R(x,y;t)=0,\quad x<y,$$

$$2\exp(-8t+x+y)+\int_{-\infty}^{x}2\exp(8t+x''+y)A_L(x,x'';t)\mathrm{d}x''+A_L(x,y;t)=0,\quad x>y.$$

To determine the solution u we solve one of the above equations and obtain

$$A_R(x,y,t)=-2\frac{\exp(8t-x-y)}{1+\exp(8t-2x)},$$

and them from (3.5.30)

$$u(x,t)=-2\mathrm{sech}(x-4t). \tag{3.5.34}$$

3.6 Multi-soliton solutions of the KdV equation

Two-soliton solutions for (3.5.1) are derived for $N=2$, and u, $u_x\to 0$ at $|x|\to\infty$, and the initial condition $u(x,0)=-6\mathrm{sech}^2(x)$, (Data and Tanaka). The Schrödinger equation is given by (3.5.20). From (3.5.21) we have

$$R_R(k,0)=R_L(k,0)=0,$$

$$T(k,0)=\frac{(1-\mathrm{i}k)(2-\mathrm{i}k)}{(-\mathrm{i}k-1)(-\mathrm{i}k-2)}.$$

The fundamental solutions $f_j(x,\mathrm{i};0)$, $f_j(x,2\mathrm{i};0)$, $j=1,2$, of (3.5.1) have the properties

$$f_j(x,\mathrm{i},0)\cong\exp(\mp x)\quad\text{for}\quad x\to\pm\infty,\ j=1,2, \tag{3.6.1a}$$

$$f_j(x,2\mathrm{i},0)\cong\exp(\mp 2x)\quad\text{for}\quad x\to\pm\infty,\ j=1,2, \tag{3.6.1b}$$

and then we have

$$f_j(x,\mathrm{i};0)=c_j^{\,i}\varphi(x,\mathrm{i};0),\quad j=1,2,$$

$$f_j(x,2\mathrm{i};0)=c_j^{\,2i}\varphi(x,2\mathrm{i};0),\quad j=1,2.$$

From (3.6.1) it results c_j

$$c_1^{\,i}=c_2^{\,i}=\frac{1}{6},\ c_1^{\,2i}=c_2^{\,2i}=\frac{1}{12}.$$

Therefore,

$$f_j(x,\mathrm{i};0) = \frac{1}{2}\tanh(x)\operatorname{sech}(x),$$
$$f_j(x,2\mathrm{i};0) = \frac{1}{4}\operatorname{sech}^2(x). \tag{3.6.2}$$

From (3.5.13) it results

$$m_{R1}(\mathrm{i},0) = m_{L1}(\mathrm{i},0) = \{\int_{-\infty}^{\infty}[\frac{1}{4}\tanh^2(x)\operatorname{sech}^2(x)]\mathrm{d}x\}^{-1} = 6\,,$$

$$m_{R2}(2\mathrm{i},0) = m_{L2}(2\mathrm{i},0) = \{\int_{-\infty}^{\infty}[\frac{1}{16}\operatorname{sech}^4(x)]\mathrm{d}x\}^{-1} = 12,$$

and (3.5.23) become

$$\Omega_R(z;t) = 6\exp(8t - z) + 12\exp(64t - 2z),$$

$$\Omega_L(z;t) = 6\exp(-8t + z) + 12\exp(-64t + 2z).$$

Next we solve the first Marchenko equation (3.5.24)

$$6\exp(8t - x - y) + 12\exp(64t - 2x - 2z) + A_R(x,y;t)\times$$
$$\times\int_{x}^{\infty}[2\exp(8t - x'' - y) + 12\exp(64t - 2x'' - y)]A_R(x,x'';t)\mathrm{d}x'' = 0, \tag{3.6.3}$$

for $x < y$. Assuming the solutions of the form (3.5.26)

$$A_R(x,y,t) = L_1^R(x,t)\exp(-y) + L_2^R(x,t)\exp(-2y)\,,$$

the equation (3.6.3) is reduced to (3.5.26), where

$$A(x,t) = \begin{bmatrix} 1+3\exp(8t-2x) & 2\exp(8t-3x) \\ 4\exp(64t-3x) & 1+3\exp(64t-4x) \end{bmatrix},$$

$$L^R(x,t) = \begin{bmatrix} L_1^R(x,t) \\ L_2^R(x,t) \end{bmatrix},\quad B(x,t) = \begin{bmatrix} 6\exp(8t-x) \\ 12\exp(64t-2x) \end{bmatrix}.$$

The solution of (3.5.26) is given by

$$L^R(x,t) = \frac{6}{\det A(x,t)}\begin{bmatrix} \exp(72t-5x) - \exp(8t-x) \\ -2[\exp(64t-2x) + \exp(72t-4x)] \end{bmatrix},$$

where

$$\det A(x,t) = 1 + 3\exp(8t-2x) + 3\exp(64t-4x) + \exp(72t-6x)\,.$$

So, we can write

$$A(x,t) = 6\frac{-\exp(8t-2x) - 2\exp(64t-4x) - \exp(72t-6x)}{1 + 3\exp(8t-2x) + 3\exp(64t-4x) + \exp(72t-6x)}.$$

Let us remark that this solution may be obtained also from (3.5.29).
Finally, the solution is given by

$$u(x,t) = -2\frac{\mathrm{d}}{\mathrm{d}x}A(x,x;t) = -12\frac{3+4\cosh(2x-8t)+\cosh(4x-64t)}{[3\cosh(x-28t)+4\cosh(3x-36t)]^2}. \tag{3.6.4}$$

This solution was derived for $t>0$, but it makes sense also for $t<0$. The solution consists of two waves. The taller wave catches the shorter one, coalesces to form a single wave at $t=0$, and then reappears to the right and moves away from the shorter wave as t increasing. The interaction of waves is not linear. The taller wave has moved forward, and the shorter one backward relative to the positions they would have reached if the interaction were linear (Drazin and Johnson). The nonlinear interaction of waves is characterized by the phase shifts. The solitons occur as $t\to\pm\infty$, and interact in this special way.

The asymptotic behavior of the solution for $t\to\infty$, is examined by introducing $\xi = x-16t$. The solution is written as

$$u(x,t) = -12\frac{3+4\cosh(2\xi+24t)+\cosh(4\xi)}{[3\cosh(\xi-12t)+\cosh(3\xi+12t)]^2}.$$

As $t\to\infty$, we obtain the solution

$$u(x,t) \approx -8\mathrm{sech}^2(2\xi - \frac{1}{2}\ln 3).$$

This wave is moving with a velocity of 16 units, the amplitude of 8 units, and a phase shift of $\frac{\ln 3}{2}$. As $t\to-\infty$, we have the wave

$$u(x,t) \approx -8\mathrm{sech}^2(2\xi + \frac{1}{2}\ln 3),$$

having a phase shift of $(-\frac{\ln 3}{2})$. For $\eta = x-4t$, the solution becomes

$$u(x,t) = -12\frac{3+4\cosh(2\eta)+\cosh(4\eta-48t)}{[3\cosh(\eta-24t)+\cosh(3\eta-24t)]^2},$$

which yields

$$u(x,t) \approx -2\mathrm{sech}^2(\eta + \frac{1}{2}\ln 3),\ t\to\infty,$$

and

$$u(x,t) \approx -2\mathrm{sech}^2(\eta - \frac{1}{2}\ln 3),\ t\to-\infty.$$

By approximating we can consider that

$$u(x,t) \approx -8\,\mathrm{sech}^2(2\xi - \frac{1}{2}\ln 3) - 2\mathrm{sech}^2(\eta + \frac{1}{2}\ln 3),\quad t\to\infty,$$

$$u(x,t) \approx -8\,\mathrm{sech}^2(2\xi + \frac{1}{2}\ln 3) - 2\,\mathrm{sech}^2(\eta - \frac{1}{2}\ln 3), \quad t \to -\infty .$$

The terms exponentially decreasing at $t \to \pm\infty$ were neglected.

Note that the solution at $+\infty$ consists of two solitons having the velocities 16 and 14 units, respectively, with the tallest and then fastest soliton moving forward by an amount $\varphi_0 = \frac{\ln 3}{2}$, and the shorter wave moving back by $\varphi_0 = -\frac{\ln 3}{2}$. The three-soliton solutions are derived from (3.6.1), with $N = 3$, and $u, u_x \to 0$, $|x| \to \infty$, $u(x,0) = -12\,\mathrm{sech}^2(x)$. As the previous example, from (3.5.21) we have

$$R_R(k,0) = R_L(k,0) = 0 ,$$

$$T(k,0) = \frac{(1-\mathrm{i}k)(2-\mathrm{i}k)(3-\mathrm{i}k)}{(-\mathrm{i}k-1)(-\mathrm{i}k-2)-\mathrm{i}k-3)},$$

and

$$\varphi(x,\mathrm{i};0) = P_3^1(\tanh x) = \frac{3}{2}[5\tanh^2(x) - 1]\,\mathrm{sech}(x),$$

$$\varphi(x,2\mathrm{i};0) = P_3^2(\tanh x) = 15\tanh(x)\,\mathrm{sech}^2(x),$$

$$\varphi(x,3\mathrm{i};0) = P_3^3(\tanh x) = -15\,\mathrm{sech}^3(x).$$

The fundamental solutions are

$$f_j(x,\mathrm{i};0) = \frac{1}{8}[5\tanh^2(x) - 1]\,\mathrm{sech}(x),$$

$$f_j(x,2\mathrm{i};0) = \frac{1}{2}\tanh(x)\,\mathrm{sech}^2(x), \quad f_j(x,3\mathrm{i};0) = \frac{1}{8}\mathrm{sech}^3(x).$$

According to (3.5.13) we have

$$\Omega_R(z;t) = 12\exp(8t - z) + 60\exp(64t - 2z) + 60\exp(216t - 3z),$$

$$\Omega_L(z;t) = 12\exp(-8t + z) + 60\exp(-64t + 2z) + 60\exp(-216t + 3z),$$

$$m_{R,L1}(\mathrm{i},0) = 12\ , \ m_{R,L2}(2\mathrm{i},0) = 60,\ m_{R,L3}(3\mathrm{i},0) = 60.$$

Further, it is sufficient to study the Marchenko equation $(3.5.24)_1$

$$12\exp(8t - x - y) + 60\exp(64t - 2x - 2z) + 60\exp(216t - 3x - 3z) + A_R(x,y;t) +$$

$$+\int_x^\infty [2\exp(8t - x'' - y) + 12\exp(64t - 2x'' - y) + 60\exp(216t - 3x'' - 3z)]A_R(x,x'';t)\mathrm{d}x'' = ($$

for $x < y$. Assuming the solutions of the form

$$A_R(x,y,t) = L_1^R(x,t)\exp(-y) + L_2^R(x,t)\exp(-2y) + L_3^R(x,t)\exp(-3y),$$

the above equation is reduced to (3.5.27), where

$$A(x,t)=\begin{bmatrix}1+6\exp(8t-2x) & 20\exp(8t-3x) & 15\exp(8t-4x)\\ 4\exp(64t-3x) & 1+15\exp(64t-4x) & 12\exp(64t-5x)\\ 3\exp(216t-4x) & 15\exp(216t-5x) & 1+10\exp(216t-6x)\end{bmatrix},$$

$$L^R(x,t)=\begin{bmatrix}L_1^R(x,t)\\ L_2^R(x,t)\\ L_3^R(x,t)\end{bmatrix},\ B(x,t)=\begin{bmatrix}12\exp(8t-x)\\ 60\exp(64t-2x)\\ 60\exp(216t-3x)\end{bmatrix},$$

$$\det A(x,t)=2\exp(144t-6x)[\cosh(144t-6x)+6\cosh(136t-4x)+ \\ +15\cosh(80t-2x)+10\cosh(72t).$$

On the basis of (3.4.30) we have

$$u(x,t)=-12\frac{M(x,t)}{N(x,t)},$$

where

$$M(x,t)=252+12\cosh(280t-10x)+100\cosh(8t-2x)+20\cosh(224t-8x)+ \\ +160\cosh(64t-4x)+60\cosh(216t-6x)+270\cosh(56t-2x)+ \\ +30\cosh(72t-6x)+80\cosh(208t-4x)+50\cosh(152t-2x),$$

$$N(x,t)=[\cosh(144t-6x)+6\cosh(136t-4x)+15\cosh(80t-2x)+10\cosh(72t)]^2.$$

Finally, we present the periodic solutions of the KdV equation

$$u_t+6uu_x+u_{xxx}=0,$$

written as a sum of solitons (Whitham 1984). The steadily progressing waves (trains of solitons) are given by

$$u=2k^2 f(\xi),\ \xi=kx-\omega t, \tag{3.6.5}$$

where $f(\xi)$ satisfies

$$f_{\xi\xi}=6B+(4-6A)f-6f^3,\ A=\frac{4}{6}(1-\frac{\omega}{k^3}), \tag{3.6.6}$$

with B a constant of integration. For $A=B=0$, we obtain a single soliton, $f=\operatorname{sech}^2(\xi-\xi_0)$, ξ_0 a constant. A row of these solitons spaced 2σ apart

$$f(\xi)=\sum_{-\infty}^{\infty}\operatorname{sech}^2(\xi-2m\sigma), \tag{3.6.7}$$

leads to an exact solution. Substitution of (3.6.7) into (3.6.6) yield

$$\left\{\sum_{-\infty}^{\infty}\operatorname{sech}^2(\xi-2m\sigma)\right\}^2-\sum_{-\infty}^{\infty}\operatorname{sech}^4(\xi-2m\sigma)=B-A\sum_{-\infty}^{\infty}\operatorname{sech}^2(\xi-2m\sigma).$$

This identity is true, and we have

$$A(\sigma) = 4\sum_{1}^{\infty}\frac{1}{\sinh^2(2j\sigma)}, \quad B(\sigma) = -\frac{1}{2}\frac{\mathrm{d}A(\sigma)}{\mathrm{d}\sigma}.$$

As referring to the inverse scattering method, we mention that the representation (3.6.7) may be viewed as another instance of the clean interaction of solitons. They are superimposed but keep their identity and do not destroy each other under the nonlinear coupling. Konno and Ito have studied the mechanism of interaction between solitons in 1987, by extending one of independent variables of the equations to complex and by observing how singularities of solutions behave in the complex plane. Following Konno and Ito, we choose t as the complex variable. In the complex t-plane one soliton solution is expressed as sum of poles, which are located equidistantly along the imaginary t-axis. In asymptotic regions two-soliton solution is expressed as a superposition of two separated solitons. Konno and Ito have studied the behavior of singularities during the collision. They introduced an auxiliary function and analyze its zeros which correspond to the poles of the solution at the same places in the complex plane.

To illustrate the Konno and Ito method, consider the KdV equation

$$u_t + 12uu_x + u_{xxx} = 0, \tag{3.6.8}$$

which admits the following one soliton solution

$$u(x,t) = \frac{k^2}{4}\operatorname{sech}^2\frac{1}{2}(kx - \beta t + \delta), \tag{3.6.9}$$

where $\beta = k^3$. The auxiliary function is

$$u(x,t) = \frac{\partial^2}{\partial x^2}\log\phi(x,t). \tag{3.6.10}$$

The double poles of the solution correspond with simple zeros of this function. For $\phi(x,t) = 0$, we have $\phi(x,t^*) = 0$, the zeros being symmetrically distributed with respect to the real t-axis. Consider only the upper half plane. For one soliton we have

$$\phi(x,t) = 1 + A\exp(kx - \beta t), \tag{3.6.11}$$

where $A = \exp\delta$ is a positive constant. We can obtain zeros of ϕ at

$$t_n(k) = t_{Rn}(k) + \mathrm{i}t_{In}(k), \quad n = 0, \pm 1, \pm 2, \ldots, \tag{3.6.12}$$

$$t_{Rn} = \frac{kx + \delta}{\beta}, \quad t_{Rn} = \frac{(2n+1)\pi}{\beta}. \tag{3.6.13}$$

By using the zeros, the soliton solution (3.6.9) is given by

$$u = -\frac{k^2}{\beta^2}\sum_{n=-\infty}^{\infty}\frac{1}{(t - t_n(k))^2}. \tag{3.6.14}$$

Therefore, we find that

1. Double poles are equidistantly located in parallel with the imaginary t-axis.
2. Real part $t_{Rn}(k)$ depends linearly on x and admits a trajectory of the soliton.
3. Poles located at smaller imaginary parts $t_{In}(k)$ affect more effectively the soliton amplitude.

For two-soliton solution the auxiliary function is

$$\begin{aligned}\phi(x,t) = 1 + A_1 \exp(k_1 x - \beta_1 t) + A_2 \exp(k_2 x - \beta_2 t) + \\ + A_3 \exp[(k_1 + k_2)x - (\beta_1 + \beta_2)t],\end{aligned} \tag{3.6.15}$$

with $\beta_j = k_j^3$, $j = 1,2$, and

$$A_3 = \frac{(k_1 - k_2)^2}{(k_1 + k_2)^2} A_1 A_2 . \tag{3.6.16}$$

As $x \to \pm\infty$ two solitons are separated in each other and locations of the zeros are given by the sum of contributions from two solitons. To increasing x the zeros are classified into two categories:

– A zero belonging to one soliton with k_1 changes asymptotically into one for another soliton with k_2 and vice versa, namely

$$t_{In}(k_1) \approx t_{In}(k_2) . \tag{3.6.17}$$

– The zeros preserve their identity even though they receive effect of nonlinear interaction in such a way as phase shift during interaction.

3.7 Boussinesq, modified KdV and Burgers equations

Fermi, Pasta and Ulam studied in 1955 the problem of a dynamical system of n identical particles of unit mass on a fixed end line with forces acting between particles (Dodd *et al.*). The motion equation of the particle n is

$$\ddot{Q}_n = f(Q_{n+1} - Q_n) - f(Q_n - Q_{n-1}) , \tag{3.7.1}$$

where Q_n is the displacement of the particle with respect to the equilibrium position, and $f(Q)$ is the interaction force, which may be of the forms

$$f(Q) = \gamma Q + \alpha Q^2 , \tag{3.7.2}$$

or

$$f(Q) = \gamma Q + \beta Q^3 , \tag{3.7.3}$$

with γ a constant, and α , β chosen in such a way that the maximum displacement of Q caused by the nonlinear term is small. Using initial sine-wave data, integrating the equation (3.7.1) with (3.7.2) or (3.7.3) yields to the conclusion that the equipartition of energy criterion failed. The energy is kept in the initial vibration mode and a few nearby modes, without spreading in all the normal modes of vibrations.

Using the McLaurin expansion

$$f(n+a)=\left[\exp(a\frac{\partial}{\partial n})\right]f(n),$$

where n is now a continuous variable, the equation (3.7.1) becomes (Dodd *et al.*)

$$Q=f[(\exp(\frac{\partial}{\partial n})-1)Q]-f[(1-\exp(-\frac{\partial}{\partial n}))Q],$$

with $Q(t,n)=Q_n(t)$. Expanding the function f as a McLaurin series we obtain

$$Q=f'(0)\left[\frac{\partial^2 Q}{\partial n^2}+\frac{1}{12}\frac{\partial^4 Q}{\partial n^4}+...\right]+\frac{1}{2!}f''(0)\left[2\frac{\partial Q}{\partial n}\frac{\partial^2 Q}{\partial n^2}+...\right]+$$
$$+\frac{1}{3!}f'''(0)\left[3\left(\frac{\partial Q}{\partial n}\right)^2\frac{\partial^2 Q}{\partial n^2}+...\right].$$

This equation can be written as

$$\ddot{P}=f'(0)\left[l^2\frac{\partial^2 P}{\partial x^2}+\frac{l^4}{12}\frac{\partial^4 P}{\partial x^4}+...\right]+\frac{1}{2!}f''(0)\left[2l^{r+3}\frac{\partial P}{\partial x}\frac{\partial^2 P}{\partial x^2}+...\right]+$$
$$+\frac{1}{3!}f'''(0)\left[3\left(\frac{\partial P}{\partial x}\right)^2\frac{\partial^2 P}{\partial x^2}+...\right],$$

if we note $x=nl$ and $P=Ql^{-r}$, where r is to be determined from the balance of the fourth derivative term with the nonlinear term. For the quadratic nonlinearity given by (3.7.2), it results $\gamma=f'(0)$, $2\alpha=f''(0)$ and $f'''(0)=0$. Therefore, we choose $r=1$, and taking $u(x,t)=\frac{\partial P}{\partial x}$ up to $O(l^4)$, we obtain the Boussinesq equation

$$\frac{\partial^2}{\partial x^2}\left(\frac{l^4}{12}\frac{\partial^2 u}{\partial x^2}+\alpha l^4 u^2+l^2\gamma u\right)=\frac{\partial^2 u}{\partial t^2}, \tag{3.7.4}$$

which describes the motion of waves in both directions. In the approximation $O(l^2)$, (3.7.4) reduces to the linear wave equation.

One solution of (3.7.4) is

$$u=\frac{1}{8\alpha}a^2\text{sech}^2\frac{1}{2}[ax-\omega t+\delta], \tag{3.7.5}$$

with

$$\omega^2=\gamma a^2 l^2+a^4 l^4/12. \tag{3.7.6}$$

The Boussinesq equation is known either in the form of the system of equations

$$u_t=v_x,\ v_t=\frac{1}{3}u_{xxx}+\frac{8}{3}uu_x,$$

and admits the Hamiltonian $\frac{1}{2}v$. The system has an infinite number of symmetries and conservation laws, also an infinite number of exact solutions.

The Boussinesq equation (3.7.6) can be reduced to a KdV equation if the waves are propagating in only one direction

$$\frac{l^3}{12}\frac{\partial^3 u^{(1)}}{\partial\xi^3} + 2\alpha l^3 u^{(1)}\frac{\partial u^{(1)}}{\partial\xi} + 2\sqrt{\gamma}\frac{\partial u^{(1)}}{\partial\tau} = 0\,. \tag{3.7.7}$$

Indeed, (3.7.7) is obtained by a scaling transformation on x and t

$$\xi = \varepsilon^p (x - ct),\quad \tau = \varepsilon^q t\,, \tag{3.7.8}$$

where ε is a small parameter, and by expanding u in powers of ε

$$u = \varepsilon u^{(1)} + \varepsilon^2 u^{(2)} + \ldots, \tag{3.7.9}$$

with $c = \sqrt{\gamma} l$, $p = 1/2$ and $q = 3/2$.

For a cubic nonlinearity given by (3.7.3), it is obtained the motion equation up to the approximation $O(l^4)$

$$\frac{\partial^2}{\partial x^2}\left(\frac{l^4}{12}\frac{\partial^2 u}{\partial x^2} + \beta l^4 u^3 + l^2 \gamma u\right) = \frac{\partial^2 u}{\partial t^2}\,. \tag{3.7.10}$$

This equation may be written under the form of a modified KdV

$$\frac{l^3}{12}\frac{\partial^3 u^{(1)}}{\partial\xi^3} + 3\beta l^3 \left(u^{(1)}\right)^2 \frac{\partial u^{(1)}}{\partial\xi} + 2\sqrt{\gamma}\frac{\partial u^{(1)}}{\partial\tau} = 0\,, \tag{3.7.11}$$

for $p=1$, $q=3$ and $c = \sqrt{\gamma} l$ with wave solutions traveling only in the direction $+\xi$

$$u = \frac{1}{\sqrt{6\beta}} a^2 \operatorname{sech}^2 \frac{1}{2}\left[a\xi - \omega\tau + \delta\right], \tag{3.7.12}$$

$$\omega^2 = \frac{a^3 l^3}{24\sqrt{\gamma}}\,. \tag{3.7.13}$$

With an appropriate changing of variable, the modified KdV equation (3.7.11) may be written under the normalized form

$$u_{xxx} + 3u^2 u_x + u_t = 0\,. \tag{3.7.14}$$

Miura (1976) has shown that the explicit nonlinear transformation $v = u^2 + \mathrm{i}u_x$, maps solutions of the modified KdV equation into solutions of the KdV equation.

Writing $u = \frac{f}{g}$, the equation (3.7.14) reduces to

$$g^2 = ff_{xx} - f_x^2\,,\ fg_{xxx} - 3f_x g_{xx} + 3f_{xx} g_x - f_{xxx} g + g_t f - g f_t = 0\,, \tag{3.7.15}$$

which admits two-soliton solutions

$$g = 2a_1 \exp\theta_1 + 2a_2 \exp\theta_2 + 2a_2 A \exp(2\theta_1 + \theta_2) + 2a_2 A \exp(\theta_1 + 2\theta_2),$$

$$f = 1 + \exp 2\theta_1 + \exp 2\theta_2 + 2(1 - A)\exp(\theta_1 + \theta_2) + A\exp(2\theta_1 + 2\theta_2), \qquad (3.7.16)$$

with

$$\theta_i = a_i x - a_i^3 t + \delta_i\,,\ \theta_i = a_i x - a_i^3 t + \delta_i\,.$$

Whitham obtained in 1984 the periodic solutions for (3.7.14) expressed as sums of solitons. Any steady progressive wave is given by

$$u = \sqrt{2}kf(\xi)\,,\ \xi = kx - \omega t\,, \qquad (3.7.17)$$

where $f(\xi)$ satisfies

$$f_{\xi\xi} = 2B + (2A+1)f - 2f^3\,,\ A = \frac{1}{2}(\frac{\omega}{k^3} - 1)\,, \qquad (3.7.18)$$

with B a constant of integration. For $A = B = 0$ we obtain a single soliton, $f = \operatorname{sech}(\xi - \xi_0)$. A row of these solitons spaced 2σ apart

$$f(\xi) = \sum_{-\infty}^{\infty} \operatorname{sech}(\xi - 2m\sigma)\,, \qquad (3.7.19)$$

leads to exact solution. Substitution of (3.7.19) into (3.7.18) yield

$$\left\{\sum_{-\infty}^{\infty} \operatorname{sech}(\xi - 2m\sigma)\right\}^3 - \sum_{-\infty}^{\infty} \operatorname{sech}^3(\xi - 2m\sigma) = B + A\sum_{-\infty}^{\infty} \operatorname{sech}(\xi - 2m\sigma)\,.$$

This identity is true, and $A = 6\sum_{1}^{\infty} \frac{\cosh(2j\sigma)}{\sinh^2(2j\sigma)}$, $B = 0$. Integrating (3.7.18) we obtain a Weierstrass equation with polynomial of four degrees

$$f_\xi^2 = -C + (2A+1)f^2 - f^4\,, \qquad (3.7.20)$$

writen as

$$f_\xi^2 = (f_0^2 - f^2)(f^2 - f_1^2)\,,\ f_0 \ge f \ge f\,.$$

The periodic solutions are expressed as

$$f = f_0 \operatorname{dn}(f_0\xi)\,. \qquad (3.7.21)$$

Therefore, (3.7.19) gives the Jacobi elliptic function dn in terms of a sum of sech functions. The modified KdV equation has also an infinite number of conserved quantities. Its Hamiltonian is $\frac{1}{2}u^2$.

Another remarkable equation is the nonlinear diffusion equation

$$u_t + uu_x = \delta u_{xx}\,, \qquad (3.7.22)$$

known as the Burgers equation. The diffusion coefficient δ is a positive and real constant. The Burgers equation (3.7.22) includes nonlinearity and dissipation, leading to a nonlinear version of the heat equation.

The simplest solution of (3.7.22) is the Taylor shock wave

$$u = a\delta(1 - \tanh\frac{1}{2}(ax - \delta a^2 t)) . \qquad (3.7.23)$$

As $x \to \infty$, we have $u \to 0$, and as $x \to -\infty$, $u \to 2a\delta$. The shock solution is continuous because the term δu_{xx} prevents the tendency of nonlinearity to form discontinuities. For $u(x,0) = f(x)$ and $\delta \to 0$, the solutions of (3.7.22) may be written as a functional

$$u = f(x - ut) , \qquad (3.7.24)$$

that can be determined by the characteristics method.

In 1931 Fay (see Whitham 1984) derived the following solution for Burgers equation ($\delta = 1$)

$$u(x,t) = -2\sum_{1}^{\infty} \frac{\sin nx}{\sinh nt} . \qquad (3.7.25)$$

Another two solutions are derived from the Cole–Hopf transformation

$$u = -\frac{2\phi_x}{\phi} , \qquad (3.7.26)$$

by taking ϕ to be initially a row of equally spaced functions. We have

$$\phi = 1 + 2\sum_{1}^{\infty}(-1)^n \exp(-n^2 t)\cos nx ,$$

$$\phi = \frac{1}{\sqrt{4\pi t}} \sum_{-\infty}^{\infty} \exp\{\frac{1}{4t}[-(x-(j-1)\pi]^2\} . \qquad (3.7.27)$$

The last form leads to the saw-tooth profile for small t. Parker found in 1980 (see Whitham 1984) another equivalent form, obtained as a certain superposition of equally spaced shocks of the form (3.7.23) moving with zero speed relative to a basic frame.

In the saw-tooth wave the shocks decay, and in order to remain periodic and bounded as $x \to \pm\infty$, we must add a term proportional to x. The latter adjustment is made from the elementary solution of the Burgers equation, $u = x/t$. Therefore, the Parker exact solution is given by

$$u = \frac{x}{t} - \frac{\pi}{t}\sum_{k} \tanh\frac{\pi(x - 2k\pi)}{2t} , \qquad (3.7.28)$$

where the summation is interpreted as $\lim_{N\to\infty} \sum_{k=-N}^{N}$.

3.8 The sine-Gordon and Schrödinger equations

We have seen that the sine-Gordon equation is related to the construction of pseudospherical surfaces. In 1967 Lamb obtained the sine-Gordon equation in the analysis of the propagation of ultrashort light pulses (see Coley *et al.*). He exploited the permutability theorem associated with the Bäcklund transformation to generate an analytic expression for pulse decomposition corresponding to the two-soliton solution.

Let us begin with a mechanical pendulum problem, consisted from N identically pendulum of mass M linked by torque springs (Dodd *et al.*). The total restoring torque exerted by the springs is composed of the torque due to the gravity and the torque due to the torque springs given by

$$\Gamma_i = -Mdg\sin\varphi_i + k(\varphi_{i+1} - 2\varphi_i - \varphi_{i-1}), \tag{3.8.1}$$

where d is the distance of the center of mass from the central axis, g the gravity acceleration constant, φ_i is the angle made by the i-th pendulum with the downward vertical, and k is the torque constant.

For an array of pendula having the same moment of inertia J, the Newtonian motion equations for the i pendulum is

$$J\omega_i = \Gamma_i, \tag{3.8.2}$$

where ω_i is the angular velocity of the i-th pendulum. Inserting (3.8.1) into (3.8.2) we obtain the equation

$$J\ddot{\varphi}_i = -Mdg\sin\varphi_i + k(\varphi_{i+1} - 2\varphi_i - \varphi_{i-1}). \tag{3.8.3}$$

To obtain a continuous model for the system of pendula, a limiting process is considered, by introducing new space and time variable

$$X = \sqrt{\frac{Mdg}{k}}\frac{x}{h},\quad T = \sqrt{\frac{Mdg}{J}}t,$$

where h is the distance between pendula. The equation (3.8.3) is reduced to the sine-Gordon equation

$$\varphi_{TT} - \varphi_{XX} + \sin\varphi = 0. \tag{3.8.4}$$

Another form of sine-Gordon equation is the nonlinear version of the known Klein–Gordon linear equation from the field theory, derived by Skyrme in 1958 (Perring and Skyrme)

$$\phi_{xx} - \phi_{tt} = m^2\phi, \tag{3.8.5}$$

where m is a constant, namely

$$\phi_{xx} - \phi_{tt} = m^2\sin\phi. \tag{3.8.6}$$

For new variables $\xi = m\gamma(x - vt) + \delta$ and $\gamma^2 = (1 - v^2)^{-1}$, the equation (3.8.6) becomes

$$\phi_{\xi\xi} = \sin\phi. \tag{3.8.7}$$

Multiplying the equation (3.8.7) by ϕ_ξ and integrating, we have

$$\phi'^2 = -\cos\phi + C .$$

Assuming the boundary conditions of the form $\phi \to 0(\mathrm{mod}\, 2\pi)$, the integration constant C is zero. Using $\cos\phi = 1 - 2\sin^2\frac{\phi}{2}$, it results from $d\phi = \phi_\xi d\xi$

$$\xi = \int \frac{d\phi}{\sqrt{2\sin^2\frac{\phi}{2} - 1}} = \ln\tan\frac{\phi}{4},$$

or

$$\tan\frac{\phi}{4} = \exp\xi . \tag{3.8.8}$$

It is easy to show from $\phi_{\xi\xi} = -2\,\mathrm{sech}\xi\,\tanh\xi$, $\sin\phi = -2\,\mathrm{sech}\xi\,\tanh\xi$, that the solution of (3.8.8) is given by

$$\phi = 4\arctan\exp[m\gamma(x - vt) + \delta], \tag{3.8.9}$$

with

$$\gamma^2 = (1 - v^2)^{-1} . \tag{3.8.10}$$

The solution (3.8.9) is named *kink* because it represents a twist in the variable $\phi(x,t)$, which takes the system from one solution $\phi = 0$ to an adjacent solution with $\phi = 2\pi$. Using the Bäcklund transform we can derive for the equation sine-Gordon multi-soliton solutions. In section 1.7 we have obtained the two-soliton solution ϕ_{12} (1.7.23)

$$\tan\frac{\phi_{12} - \phi}{4} = \frac{\mu_2 + \mu_1}{\mu_2 - \mu_1}\tan\frac{\phi_2 - \phi_1}{4},$$

where ϕ_1 and ϕ_2 are two solutions for the sine-Gordon equation (3.8.9)

$$\phi_1 = 4\arctan\exp[m\gamma(x - v_1 t) + \delta_1],$$

$$\phi_2 = 4\arctan\exp[m\gamma(x - v_2 t) + \delta_2],$$

generated from ϕ given by (3.8.9), by the Bäcklund transform (1.7.23) of parameters μ_1 and μ_2. If ϕ_1 is a solution of the type (3.8.9) and ϕ_{12}, ϕ_{22} two-soliton solution generated from ϕ_1, then a three-soliton solution ϕ_3 is given by

$$\tan\frac{\phi_3 - \phi_1}{4} = \frac{\mu_2 + \mu_1}{\mu_2 - \mu_1}\tan\frac{\phi_{12} - \phi_{22}}{4}.$$

Another equation, exhaustively investigated by both physicists and mathematicians, is the nonlinear Schrödinger equation (NLS)

$$\mathrm{i}\frac{\partial\phi}{\partial t}+\frac{\partial^2\phi}{\partial x^2}+\beta\phi\,|\,\phi\,|^2=0\,, \tag{3.8.11}$$

where ϕ is a complex function. Applying to (3.8.11) the Painlevé analysis with associated conditions $\phi\to 0$ as $|\,x\,|\to\infty$, we obtain a traveling soliton solution

$$\phi=a\frac{\sqrt{2}}{\beta}\exp\{\mathrm{i}\,[\frac{1}{2}bx-(\frac{1}{4}b^2-a^2)]\}\,\mathrm{sech}[a(x-bt)]\,, \tag{3.8.12}$$

where a and b are arbitrary constants.

The propagation of waves in dielectric wires is described by the NLS equation with an additional dissipative term

$$\mathrm{i}\frac{\partial\phi}{\partial t}+\frac{1}{2}\frac{\partial^2\phi}{\partial x^2}+\phi\,|\,\phi\,|^2+\mathrm{i}\varepsilon\phi=0\,, \tag{3.8.13}$$

where ε is a positive constant. When $\varepsilon=0$ the solutions have the form

$$\phi=r\exp\mathrm{i}(\theta+nt)\,, \tag{3.8.14}$$

with $r(x-ct)$ and $\theta(x-ct)$ real functions, and c,n real constants. Substitution of (3.8.14) into (3.8.9) yields

$$\theta'=\frac{1}{2}(c+\frac{A}{S}),\;\; S'^2=-2F(S)\,, \tag{3.8.15}$$

with

$$S=r^2\,,\;\; F(S)=S^3-2(n-\frac{1}{4}c^2)S^2+BS+C\,. \tag{3.8.16}$$

In (3.8.16), B,C are arbitrary integration constants. Analyzing the nature of the roots of the cubic equation in the right-hand side of $(3.8.16)_2$, we obtain the soliton solution

$$\phi=A\,\mathrm{sech}\,\alpha\exp\mathrm{i}\beta,\;\; \alpha=A(x-\gamma-ct),$$

$$\beta=cx-\frac{1}{2}(n^2-A^2)t+\delta,\;\; A^2=2(n-\frac{1}{4}c^2)>0\,, \tag{3.8.17}$$

and $c,\gamma\,,\delta$ constants.

If we attach to (3.8.13) an initial condition of the type (3.8.17) for $\varepsilon\neq 0$, we obtain

$$\frac{\mathrm{d}A}{\mathrm{d}t}=-2\varepsilon A,\;\; \frac{\mathrm{d}V}{\mathrm{d}t}=0,\;\; \frac{\mathrm{d}\gamma}{\mathrm{d}t}=0,\;\; \frac{\mathrm{d}\theta}{\mathrm{d}t}=2\varepsilon A^2t. \tag{3.8.18}$$

The amplitude of wave is decreasing with time as $A=A_0\exp(-2\varepsilon t)$, where A_0 is the initial amplitude. As the amplitude decreases in time, the width of the wave $1/A$ becomes smaller, in contrast with linear waves for which $A=A_0\exp(-\varepsilon t)$.

In the case (3.8.13) when the dissipative term $i\varepsilon\phi$ has a bigger weight then the nonlinear and dispersive terms, namely for $\varepsilon >> A^2$, the waves are damped. In this case the width of the waves remain unchangeable.

For another NLS equation

$$i\frac{\partial\phi}{\partial t} + \frac{\partial^2\phi}{\partial x^2} - \phi|\phi|^2 = 0, \quad (3.8.19)$$

by repeating the calculations, we obtain the soliton solutions

$$\phi = r\exp i(\theta + nt),$$

$$r^2(\xi) = m - 2k^2\operatorname{sech}^2 k\xi, \quad (3.8.20)$$

$$c\tan\theta(\xi) = -2k\tanh k\xi,$$

for any c, and

$$\xi = x - ct, \; n = -m, \; k = \frac{1}{2}(2m - c^2)^{1/2}, \; m > \frac{1}{2}c^2.$$

3.9 Tricomi system and the simple pendulum

Consider the following differential system of equations (Halanay, Arnold)

$$y_i' = f_i(x, y_1, y_2, ... y_n), \quad i = 1, 2, ..., n,$$

where the functions f_i, $i = 1, 2, ..., n$, are continuous and verify the Lipschitz conditions with respect to y_i, $i - 1, 2, ..., n$,

$$|f_i(y_1, y_2, .., y_k', .., y_n) - f_i(y_1, y_2, .., y_k'', .., y_n)| \leq A_{ik}(y_k' - y_k''),$$

for $k = 1, 2, ..., n$.

By imposing the initial conditions

$$y_i(x_0) = y_i^0,$$

the solutions y_i, $i = 1, 2, ..., n$, verify the Volterra equations

$$y_i(x) = y_i^0 + \int_{x_0}^{x} f_i[t, y_1(t), y_2(t), ... y_n(t)]\mathrm{d}t, \quad i = 1, 2, ..., n,$$

and can be recursively calculated by

$$y_i^{(m+1)}(x) = y_i^0 + \int_{x_0}^{x} f_i[t, y_1^{(m)}(t), y_2^{(m)}(t), ... y_n^{(m)}(t)]\mathrm{d}t, \; m = 0, 1, 2, ...$$

Consider now the Tricomi problem (Tricomi)

$$y_1' = y_2 y_3, \; y_2' = -y_1 y_3, \; y_3' = -m\, y_1 y_2, \quad (3.9.1)$$

and the initial conditions

$$y_1(0)=0\ ,\ y_2(0)=1\,,\ y_3(0)=1\,, \tag{3.9.2}$$

with $0\le m\le 1$, $y'=\dfrac{\mathrm{d}y}{\mathrm{d}x}$. The recursive relations become

$$y_1^{(m+1)}(x)=\int_0^x y_2^{(m)}(t)y_3^{(m)}(t)\mathrm{d}t\ ,\ y_2^{(m+1)}(x)=1-\int_0^x y_1^{(m)}(t)y_3^{(m)}(t)\mathrm{d}t\,,$$

$$y_3^{(m+1)}(x)=1-k^2\int_0^x y_1^{(m)}(t)y_2^{(m)}(t)\mathrm{d}t\,, \tag{3.9.3}$$

or

$$y_1^{(1)}=x\ ,\ y_2^{(1)}=1\,,\ y_3^{(1)}=1\,,$$

$$y_1^{(2)}=x\ ,\ y_2^{(2)}=1-\frac{x^2}{2!}\,,$$

$$y_3^{(2)}=1-m\frac{x^2}{2!}\,,$$

$$y_1^{(3)}=x-(1+m)\frac{x^3}{3!}+6m\frac{x^5}{5!}\,,$$

$$y_2^{(3)}=1-\frac{x^2}{2!}+3m\frac{x^4}{4!}\,,$$

$$y_3^{(3)}=1-m\frac{x^2}{2!}+3m\frac{x^4}{4!}\,,$$

$$y_1^{(4)}=x-(1+m)\frac{x^3}{3!}+12m\frac{x^5}{5!}+\ldots\,,$$

$$y_2^{(4)}=1-\frac{x^2}{2!}+(1+4m)\frac{x^4}{4!}+\ldots\,,$$

$$y_3^{(4)}=1-m\frac{x^2}{2!}+m(1+m)\frac{x^4}{4!}+\ldots\,,$$

$$y_1^{(5)}=x-(1+m)\frac{x^3}{3!}+(1+14m+m^2)\frac{x^5}{5!}+\ldots\,,$$

$$y_2^{(5)} = 1 - \frac{x^2}{2!} + (1+4m)\frac{x^4}{4!} - m(42+15m)\frac{x^6}{6!} + \ldots,$$

$$y_3^{(5)} = 1 - m\frac{x^2}{2!} + m(1+m)\frac{x^4}{4!} - m(15+42m)\frac{x^6}{6!} + \ldots.$$

In this way we can calculate the approximations $y_1^{(m)}, y_2^{(m)}, y_3^{(m)}$, $m = 0,1,2,\ldots$. For $m \to \infty$, the solutions of (3.9.1) turn in

$$y_1 = x - (1+m)\frac{x^3}{3!} + (1+14m+m^2)\frac{x^5}{5!} - \ldots \to \operatorname{sn} x,$$

$$y_2 = 1 - \frac{x^2}{2!} + (1+4m)\frac{x^4}{4!} - (1+44m+16m^2)\frac{x^6}{6!} + \ldots \to \operatorname{cn} x, \tag{3.9.4}$$

$$y_3 = 1 - m\frac{x^2}{2!} + m(1+m)\frac{x^4}{4!} - m(16+44m+m^2)\frac{x^6}{6!} + \ldots \to \operatorname{dn} x .$$

From (3.9.1), after a little manipulation, we have

$$2y_1y_1' + 2y_2y_2' = \frac{\mathrm{d}}{\mathrm{d}x}(y_1^2 + y_2^2) = 0,$$

$$2my_1y_1' + 2y_3y_3' = \frac{\mathrm{d}}{\mathrm{d}x}(my_1^2 + y_3^2) = 0, \tag{3.9.5}$$

and by integrating, we obtain

$$y_1^2 + y_2^2 - \text{const.},\ my_1^2 + y_3^2 - \text{const.} \tag{3.9.6}$$

On the basis of (3.9.2), relations (3.9.6) become

$$y_1^2 + y_2^2 = 1,\ m\,y_1^2 + y_3^2 = 1,$$

or

$$\operatorname{sn}^2 x + \operatorname{cn}^2 x = 1,\ m\operatorname{sn}^2 x + \operatorname{dn}^2 x = 1. \tag{3.9.7}$$

For arbitrary initial conditions

$$y_1(0) = y_1^0,\ y_2(0) = y_2^0,\ y_3(0) = y_3^0, \tag{3.9.8}$$

the solution of (3.9.1) may be written in a general form

$$y_i = \sum_{k=1}^{n} A_{ik}\operatorname{cn}^2(\alpha_i x + \beta_i). \tag{3.9.9}$$

Consider now the motion of the simple pendulum of mass m and the length l (Figure 3.9.1) described by (Teodorescu)

$$\ddot{\theta} + \omega^2 \sin\theta = 0,\ \omega^2 = \frac{g}{l}. \tag{3.9.10}$$

For small θ we have $\sin\theta \cong \theta$, and the equation (3.9.10) admits harmonic solutions of period $T = \frac{2\pi}{\omega}$

$$\theta = A\cos(\omega t + \varphi) . \tag{3.9.11}$$

The constants are determined from the initial conditions

$$\theta(0) = \alpha , \ \dot\theta(0) = \beta . \tag{3.9.12}$$

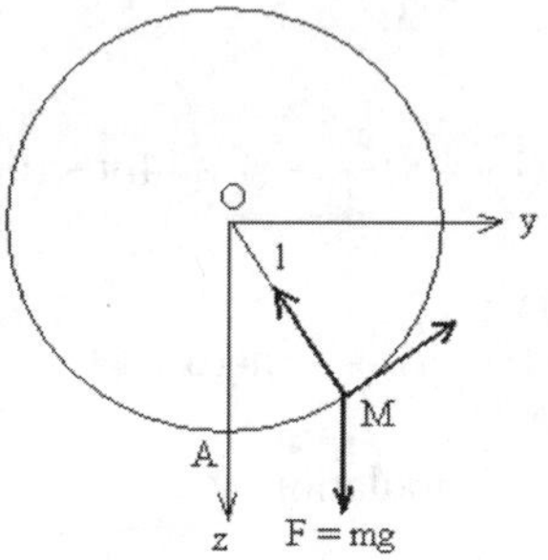

Figure 3.9.1 The simple pendulum.

It is preferable to consider two cases:

Case 1. $$\theta(0) = 0 , \ \dot\theta(0) = \frac{v_A}{l} = \beta , \tag{3.9.13}$$

with v_A , the speed of mass for $y = l$, and

Case 2. $$\theta(0) = \alpha , \ \dot\theta(0) = 0 , \tag{3.9.14}$$

with v_A the speed of a mass in free downfall from the distance a, without initial speed.

The energy theorem gives (Vâlcovici *et al.*)

$$v^2 - v_A^2 = 2g(y - l) , \ v = l\dot\theta ,$$

or

$$v^2 = 2g(y - a) , \ a = l - \frac{v_A^2}{2g} , \tag{3.9.15}$$

$$y \ge a , \ y = l\cos\theta , \ a < l . \tag{3.9.16}$$

When $a > -l$, we have the harmonic oscillations problem. It results $v_A < 2\sqrt{gl}$. Since $-l < a < l$, we write

$$a = l\cos\alpha , \ 0 < \alpha < \pi , \tag{3.9.17}$$

where the angle α is the amplitude of the motion. The equation (3.9.15) becomes

$$\dot{\theta}^2 = 2\omega^2(\cos\theta - \cos\alpha)\,,\ \cos\theta \geq \cos\alpha\,, \tag{3.9.18}$$

or

$$\dot{\theta}^2 = 4\omega^2(\sin^2\frac{\alpha}{2} - \sin^2\frac{\theta}{2})\,. \tag{3.9.19}$$

The solution of (3.9.19) and (3.9.13) is

$$\theta = 2\arcsin[\sqrt{m}\,\mathrm{sn}(\omega t)]\,, \tag{3.9.20}$$

with

$$m = \sin^2\frac{\alpha}{2} = (\frac{\beta}{2\omega})^2\,. \tag{3.9.21}$$

The solution (3.9.20) may be obtained directly from (3.9.10), noting $\sin\frac{\theta}{2} = \sqrt{m}\,\mathrm{sn}\omega t$, $\cos\frac{\theta}{2} = \sqrt{m}\,\mathrm{cn}\omega t$, that yields to $\dot{\theta} = 2\omega\sqrt{m}\,\mathrm{cn}\omega t$, and $\ddot{\theta} = 4\omega^2 m\,\mathrm{cn}^2\omega t$. The period of (3.9.20) is given by

$$T = 4\omega K\,, \tag{3.9.22}$$

where K is the complete elliptic integral. For small values of m we have

$$T = 2\pi\omega[1 + \sum_{n=1}^{\infty}(\frac{1\cdot 3\cdot 5\cdots(2n-1)}{4\cdot 2\cdot 6\cdots 2n})^2 m^n]\,. \tag{3.9.23}$$

Another way to derive the solution (3.9.10) and (3.9.14) is to define

$$\theta = 4\arctan U(t)\,. \tag{3.9.24}$$

The identity $\sin 4\varphi = \frac{4\tan\varphi(1-\tan^2\varphi)}{(1+\tan^2\varphi)^2}$, with $\varphi = \frac{\theta}{4}$, leads to another form of the equation (3.9.10)

$$\ddot{U} + \ddot{U}U^2 - 2U\dot{U}^2 + \omega^2 U(1-U^2) = 0\,,$$

which has the solution

$$U = A\exp\mathrm{i}\omega t\,. \tag{3.9.25}$$

Thus, the solution of (3.9.10) and (3.9.14) is represented by the real part of expression

$$\theta = 4\arctan(A\exp\mathrm{i}\omega t)\,,\ A = \tan\frac{\alpha}{4}\,. \tag{3.9.26}$$

This solution is known as the breather solution. The importance of this solution may lie in the fact that its rest energy varies from 16, the rest mass of two solitons, down to zero as ω tends to one.

For $a = -l$ it results in an asymptotical motion. We have $v_A = 2\sqrt{gl}$, and the equation (3.9.15) may be written under the form

$$\dot{\theta}^2 = 4\omega^2 \cos^2 \frac{\theta}{2}. \tag{3.9.27}$$

For (3.9.13) we have

$$\theta = 4\arctan(\exp \omega t) - \pi . \tag{3.9.28}$$

This solution can be derived from (3.9.10) and (3.9.13) by the Bäcklund transformation. So, consider u and v so that

$$\frac{1}{2}(u+v)_{,t} = \omega \sin\frac{u-v}{2}, \quad \frac{1}{2}(u-v)_{,t} = \omega \sin\frac{u+v}{2}. \tag{3.9.29}$$

For $v = 0$ it results $\frac{1}{2}u_{,t} = \omega \sin\frac{u}{2}$, and

$$t = \frac{1}{2\omega}\int_0^u \frac{\mathrm{d}u}{\sin u/2} = 2\log|\tan\frac{u}{4}| + C. \tag{3.9.30}$$

The solution (3.9.28) monotonically increases to π as $t \to \infty$, with $4\omega = \beta$.

When $a < -l$ we have a rotational motion, with $v_A > 2\sqrt{gl}$. The equation (3.9.15) becomes

$$\dot{\theta}^2 = 4\lambda^2(1 - m^* \sin^2 \frac{\theta}{2}), \tag{3.9.31}$$

with

$$\lambda^2 = \frac{g(l-a)}{2l^2} = \frac{1}{2}(\omega^2 - \omega_0^2), \; \omega_0^2 = \frac{ga}{l^2}, \quad 1 > m^* = \frac{2l}{l-a} > 0. \tag{3.9.32}$$

Solution of (3.9.31) and (3.9.13) is written as

$$\theta = 2\arcsin(\mathrm{sn}\lambda t), \; 2\lambda = \beta, \; m^{*2} = \frac{4\omega^2}{\beta^2}. \tag{3.9.33}$$

The solution (3.9.33) may be directly obtained from (3.9.10), by defining $\sin(\theta/2) = \mathrm{sn}\,\lambda t$, $\cos(\theta/2) = \mathrm{cn}\lambda t$, that yields $\dot{\theta} = 2\lambda \mathrm{dn}\,\lambda t$ and $\ddot{\theta} = -2\lambda^2 m^* \mathrm{cn}\,\lambda t\, \mathrm{sn}\,\lambda t$.

The rotation period is given by

$$T = \frac{\pi}{\lambda}[1 + (\frac{1}{2})^2 m^* + (\frac{1\cdot 3}{2\cdot 4})^2 m^{*2} + ...].$$

PART 2

APPLICATIONS TO MECHANICS

Chapter 4

STATICS AND DYNAMICS OF THE THIN ELASTIC ROD

4.1 Scope of the chapter

The theory of the thin elastic rod occupies an important position in the history of vibration theory. Wallis (1616–1703) and Joseph Sauveur (1653–1716) have observed that a stretched string can vibrate in parts with certain nodes at which no motion takes place, whereas a strong motion takes place at intermediate points, called 'loops'. The dynamical explanation of this vibration was provided by Daniel Bernoulli (1700–1782) in 1755. He stated the famous superposition principle of the coexistence in the vibrating string of a multitude of small oscillations at the same time.

The elastic line in which the resistance of a bent rod is assumed to arise from the extension and contraction of its longitudinal filaments was investigated by James Bernoulli (1667–1748). Daniel Bernoulli suggested to Leonhard Euler (1707–1783) that the differential equation of the elastica could be found by making the integral of the square of the curvature taken along the rod a minimum. Euler obtained on this suggestion the differential equation of the curve. J. L. Lagrange (1736–1813) and Lord Rayleigh (1842–1919) anticipated in their works the fundamental ideas of the modern topological and perturbation method foundered later by Poincaré and Lyapunov.

We must mention that the problem of the bending and twisting of thin rods was solved in an elegant analytical fashion by Love in 1926. He made a classification of the form of the rod, which is inflexional elastica and non-inflexional elastica. Since the thin elastic rod is an example of a solitonic medium, it is interesting to see that the inflectional elastica can be reobtained by using the soliton theory. Due to the large displacements in the fundamental equations, even if the strains are small, the nonlinear effects yield to the soliton solutions.

The study of solitons in the thin elastic rod is an exciting branch in mechanics. Besides the strings and rods there are many physical problems that can be treated as long thin elastic rods. In biology there are many 1D media such as DNA, RNA and α-helix of protein. In this chapter the basic laws of equilibrium and motion for a thin elastic rod are studied and solved, following the Tsuru and Munteanu works. The localized solutions are expressed by elliptic and hyperbolic functions. The solutions expressed by the hyperbolic functions are solitons. The rod deviates from a plane and has a 3-dimensional structure, changing its form as the torsion angle increases.

This chapter is referred to the publications of Love (1926), Tsuru (1986, 1987), Munteanu and Donescu (2002), Antman (1974) and Antman and Liu (1979).

4.2 Fundamental equations

Let us consider a thin elastic, homogeneous and isotropic rod of length l, straight and having a circular cross section of radius $a << l$ in its natural state. External forces and couples fix the ends of the bar. We suppose the rod deforms in space by bending and torsion. The rod occupies at time $t=0$ the region $\Omega_9 \subset R^3$. After motion takes place at time t, the rod occupies the region $\Omega(t)$.

We know the motion of the rod between $t=0$ and $t=t_1$ if and only if we know the mapping

$$S(0,t), \quad \forall t \in [0,t_1], \tag{4.2.1}$$

which takes a material point in Ω_0 at $t=0$ to a spatial position in $\Omega(t)$ at $t=t_1$.

The mapping (4.2.1) is single valued and possess continuous partial derivatives with respect to their arguments. The position of a material point in Ω_0 may be denoted by a rectangular fixed coordinate system $X \equiv (X,Y,Z)$ and the spatial position of the same point in $\Omega(t)$, by the moving coordinate system $x \equiv (x,y,z)$.

Following the current terminology, we shall call X the material or Lagrange coordinates and x the spatial or Euler coordinates. The origin of these coordinate systems is lying on the central axis of the rod. The motion of the rod carries various material points through various spatial positions. This is expressed by (Truesdell and Toupin, Şoós and Teodosiu, Şoós)

$$x = f_i(X,t), \quad i=1,2,3. \tag{4.2.2}$$

We take s to be the coordinate along the central line of the natural state. The orthonormal basis of the Lagrange coordinate system is denoted by (e_1,e_2,e_3), and the orthonormal basis of the Euler coordinate system by (d_1,d_2,d_3).

The basis $\{d_k\}$, $k=1,2,3$ is related to $\{e_k\}$, $k=1,2,3$ by the Euler angles θ, ψ and φ. These angles determine the orientation of the Euler axes relative to the Lagrange axes (Tsuru)

$$\begin{aligned} d_1 &= (-\sin\psi\sin\varphi + \cos\psi\cos\varphi\cos\theta)e_1 + \\ &+ (\cos\psi\sin\varphi + \sin\psi\cos\varphi\cos\theta)e_2 - \sin\theta\cos\varphi\, e_3, \end{aligned}$$

$$d_2 = (-\sin\psi\cos\varphi - \cos\psi\sin\varphi\cos\theta)e_1 + \\ +(\cos\psi\cos\varphi - \sin\psi\sin\varphi\cos\theta)e_2 + \sin\theta\sin\varphi e_3, \quad (4.2.3)$$

$$d_3 = \sin\theta\cos\psi e_1 + \sin\theta\sin\psi e_2 + \cos\theta\, e_3.$$

The Z-axis coincides with the central axis. The plane (xy) intersects the plane (XY) in the nodal line ON (Figure 4.2.1).

The motion of the rod is described by three vector functions

$$R \times R \ni (s,t) \to r(s,t), d_1(s,t), d_2(s,t) \in E^3 . \quad (4.2.4)$$

The material sections of the rod are identified by the coordinate s. The position vector $r(s,t)$ can be interpreted as the image of the central axis in the Euler configuration. The functions $d_1(s,t), d_2(s,t)$ can be interpreted as defining the orientation of the material section s in the Euler configuration. The function

$$d_3(s,t) = d_1(s,t) \times d_2(s,t), \quad (4.2.5)$$

represents the unit tangential vector along the rod and can be expressed as

$$d_3(\sin\theta\cos\psi, \sin\theta\sin\psi, \cos\theta) .$$

We introduce the strains y_1, y_2, y_3 by

$$r' = y_k d_k , \quad (4.2.6)$$

where $(')$ means the partial differentiation with respect to s.

Since $\{d_k\}$, $k = 1,2,3$, is orthonormal, there is a vector u such as

$$d_k' = u \times d_k . \quad (4.2.7)$$

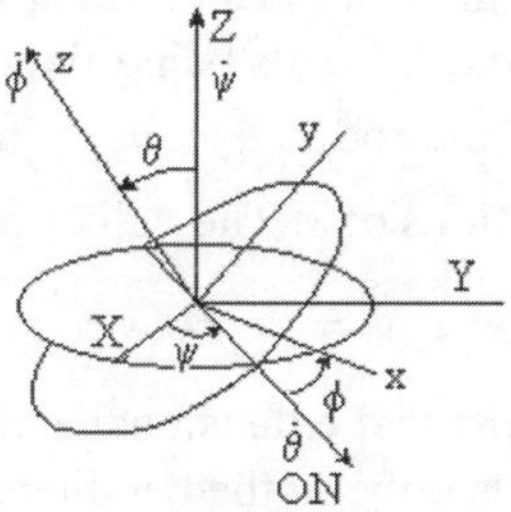

Figure 4.2.1 The Euler angles θ, ψ and φ .

The components of u with respect to the basis $\{d_k\}$ are

$$u_k = \frac{1}{2} e_{klm} d_l' \cdot d_m , \quad (4.2.8)$$

where e_{klm} the components of the alternating tensor. The relations (4.2.8) become

$$u_1 = d_{31}d'_{21} + d_{32}d'_{22} + d_{33}d'_{23} ,$$

$$u_2 = d_{11}d'_{31} + d_{12}d'_{32} + d_{13}d'_{33} ,$$

$$u_3 = d_{21}d'_{11} + d_{22}d'_{12} + d_{23}d'_{13} ,$$

where d_{ij}, $i = j = 1,2,3$, are the components of vectors d_i, i =1,2,3, given by (4.2.3).

Substitution of (4.2.3) into the above relations gives

$$u_1 = \theta' \sin\varphi - \psi' \sin\theta \cos\varphi ,$$

$$u_2 = \theta' \cos\varphi + \psi' \sin\theta \sin\varphi , \tag{4.2.9}$$

$$u_3 = \varphi' + \psi' \cos\theta .$$

These functions measure the bending and torsion of the bar. The functions u_k, $k = 1,2,3$, can be interpreted as the components of the angular velocity vector (the variation with respect to s) of the rotational motion of the moving system of coordinates relative to the fixed system of coordinates. If we substitute the differentiation with respect to s with the differentiation with respect to time we will obtain the components of the angular velocity vector defined by (4.2.9). The functions u_1 and u_2 represent the components of the curvature of the central line denoted by κ corresponding to the planes (yz) and (xz)

$$\kappa^2 = u_1^2 + u_2^2 = \theta'^2 + \psi'^2 \sin^2\theta , \tag{4.2.10}$$

and u_3 is the torsion of the bar denoted by τ

$$u_3 = \tau = \varphi' + \psi' \cos\theta . \tag{4.2.11}$$

In this way we consider the rod is rigid along the tangential direction and the total length of the rod l is invariant, the ends being fixed by external forces.

The full set of strains of the rod is $\{y_k, u_k\}$. In the natural state d_3 coincides with r', and d_k are constant functions of s. The values of the strains in the natural state are

$$y_1 = y_2 = 0, \quad y_3 = 1, \quad u_k = 0 . \tag{4.2.12}$$

In the following we assume that extensional and compression strains have the values $y_1 = y_2 = 0$, $y_3 = 1$ and focus only on the bending and torsion of the rod.

The link between the position vector $r = (x, y, z)$ and unit tangential vector d_3 is obtained from the first two relations of (4.2.12) and (4.2.6)

$$r' = d_3 . \tag{4.2.13}$$

From (4.2.13) we obtain

$$r = \int_0^s d_3 \mathrm{d}s , \tag{4.2.14}$$

or

$$x(s) = \int_0^s \cos\psi \sin\theta \mathrm{d}s\,,\quad y(s) = \int_0^s \sin\psi \sin\theta \mathrm{d}s\,,$$

$$z(s) = \int_0^s \cos\theta \mathrm{d}s\,. \tag{4.2.15}$$

To characterize the position of ends of the rod we introduce the vector D of the components $x(L), y(L), z(L)$

$$D = \int_0^l d_3 \mathrm{d}s\,. \tag{4.2.16}$$

The elastic energy of the deformed rod $\mathcal{U}$ is composed of the bending energy and the torsional energy (Landau and Lifshitz, Solomon)

$$\mathcal{U} = \frac{A}{2}\int_0^l \kappa^2 \mathrm{d}s + \frac{C}{2}\int_0^l \tau^2 \mathrm{d}s\,, \tag{4.2.17}$$

where κ and τ are given by (4.2.10) and (4.2.11).

The quantities A and C are the bending stiffness and respectively the torsional stiffness related to the Lamé constants λ, μ by

$$A = \frac{1}{4}\pi a^4 E, \quad C = \frac{1}{2}\pi a^4 \mu, \tag{4.2.18}$$

where $E = \dfrac{\mu(3\lambda + 2\mu)}{\lambda + \mu}$ is the Young's elastic modulus, and a is the radius of the cross section of the rod.

The elastic energy can be written in the form by using (4.2.10) and (4.2.11)

$$\mathcal{U} = \frac{A}{2}\int_0^l (\theta'^2 + \psi'^2 \sin^2\theta)\mathrm{d}s + \frac{C}{2}\int_0^l (\varphi' + \psi'\cos\theta)^2 \mathrm{d}s\,. \tag{4.2.19}$$

To write the equilibrium equations, the variation of the elastic energy $\mathcal{U}$ with respect to θ, φ and ψ is considered.

THEOREM 4.2.1 *The exact static equilibrium equations of the thin elastic rod with the ends fixed by the external force* $F = -\lambda$ *with* $\lambda = (\lambda_1, \lambda_2, \lambda_3)$ *are given by*

$$\begin{aligned} &A(\psi'^2 \sin\theta\cos\theta - \theta'') - C(\varphi' + \psi'\cos\theta)\psi'\sin\theta + \\ &+\lambda_1 \cos\theta\cos\psi + \lambda_2 \cos\theta\sin\psi - \lambda_3 \sin\theta = 0, \end{aligned} \tag{4.2.20}$$

$$\begin{aligned} &\frac{\partial}{\partial s}[A\psi'\sin^2\theta + C(\varphi' + \psi'\cos\theta)\cos\theta] - \\ &-\lambda_1 \sin\theta\sin\psi + \lambda_2 \sin\theta\cos\psi = 0, \end{aligned} \tag{4.2.21}$$

$$\frac{\partial}{\partial s}[-C(\varphi'+\psi'\cos\theta)]=0. \tag{4.2.22}$$

The end couples at $s=0$ *and* $s=1$ *are* $M_i(0)$, $i=1,2,3$, *and respectively* $M_i(l)$, $i=1,2,3$, *where*

$$M_1(s)=A\theta', \text{ the couples with respect to the nodal line ON,} \tag{4.2.23}$$

$$M_2(s)=A\psi'\sin^2\theta+C(\psi'\cos\theta+\varphi')\cos\theta, \text{ the couple with respect to Z-axis,} \tag{4.2.24}$$

$$M_3(s)=C(\varphi'\cos\theta+\psi'), \text{ the couple with respect to z-axis.} \tag{4.2.25}$$

Proof: The unknowns of the problem are Euler angles. The vector is assumed to be known. The position vector $r(x,y,z)$ is related to d_3 by (4.2.13). To take this constraint into account we introduce the functional $\mathfrak{I}$

$$\mathfrak{I}=\mathcal{U}+\lambda D, \tag{4.2.26}$$

where $\lambda=(\lambda_1,\lambda_2,\lambda_3)$ is the Lagrange multiplicator, and D is defined by (4.2.16).

Applying the variational Hamilton principle, the variation of $\mathfrak{I}$ is given by

$$\delta\mathfrak{I}=\delta\,\mathcal{U}+\lambda\delta D=0, \tag{4.2.27}$$

where δD is the variation of the end positions.

We write (4.2.27) under the form

$$\delta\mathfrak{I}=\delta\int_0^l L(\theta,\theta',\psi,\psi',\varphi,\varphi')\mathrm{d}s=0,$$

where the Lagrange function L is

$$L=\frac{A}{2}(\theta'^2+\psi'^2\sin^2\theta)+\frac{C}{2}(\psi'\cos\theta+\varphi')^2+ \tag{4.2.28}$$

$$+\lambda_1\sin\theta\cos\psi+\lambda_2\sin\theta\sin\psi+\lambda_3\cos\theta.$$

The variation of the functional with respect to θ, φ and ψ is

$$\delta\mathfrak{I}=\int_0^l\Big(\frac{\partial L}{\partial\theta}\delta\theta+\frac{\partial L}{\partial\theta'}\delta\theta'+\frac{\partial L}{\partial\psi}\delta\psi+\frac{\partial L}{\partial\psi'}\delta\psi'+\frac{\partial L}{\partial\varphi}\delta\psi+\frac{\partial L}{\partial\varphi}\delta\varphi'\Big)\mathrm{d}s=0.$$

Observing that $\delta\theta'=\frac{\partial}{\partial s}\delta\theta$, $\delta\psi'=\frac{\partial}{\partial s}\delta\psi$, $\delta\varphi'=\frac{\partial}{\partial s}\delta\varphi$, and integrating by parts the terms $\frac{\partial L}{\partial\theta'}\delta\theta', \frac{\partial L}{\partial\psi'}\delta\psi', \frac{\partial L}{\partial\varphi}\delta\varphi'$, we obtain

$$\delta\Im = \left|\frac{\partial L}{\partial\theta'}\delta\theta\right|_0^l + \left|\frac{\partial L}{\partial\psi'}\delta\psi\right|_0^l + \left|\frac{\partial L}{\partial\varphi'}\delta\varphi\right|_0^l +$$
$$+\int_0^l\left(\frac{\partial L}{\partial\theta} - \frac{\partial}{\partial s}\left(\frac{\partial L}{\partial\theta'}\right)\right)\delta\theta \mathrm{d}s + \int_0^l\left(\frac{\partial L}{\partial\psi} - \frac{\partial}{\partial s}\left(\frac{\partial L}{\partial\psi'}\right)\right)\delta\psi \mathrm{d}s + \tag{4.2.29}$$
$$+\int_0^l\left(\frac{\partial L}{\partial\varphi} - \frac{\partial}{\partial s}\left(\frac{\partial L}{\partial\varphi'}\right)\right)\delta\varphi \mathrm{d}s = 0.$$

Taking into account that

$$\delta\theta(0) = \delta\theta(l) = 0\,,\ \delta\psi(0) = \delta\psi(l) = 0\,,\ \delta\varphi(0) = \delta\varphi(l) = 0\,, \tag{4.2.30}$$

the first three terms in (4.2.29) are vanishing and we obtain the Lagrange equations

$$\frac{\partial L}{\partial\theta} - \frac{\partial}{\partial s}\left(\frac{\partial L}{\partial\theta'}\right) = 0,\quad \frac{\partial L}{\partial\psi} - \frac{\partial}{\partial s}\left(\frac{\partial L}{\partial\psi'}\right) = 0,\quad \frac{\partial L}{\partial\varphi} - \frac{\partial}{\partial s}\left(\frac{\partial L}{\partial\varphi'}\right) = 0. \tag{4.2.31}$$

The conditions (4.2.30) express that the functions $\theta+\delta\theta$, $\psi+\delta\psi$, $\varphi+\delta\varphi$ have the same values at $s = 0$ and $s = l$. The nonvanishing integrals must be zero for any variations $\delta\theta, \delta\psi, \delta\varphi$ and this is possible only if the integrands are identically zero.

Substituting (4.2.28) into (4.2.31) we obtain the differential equations (4.2.20)–(4.2.22).

If $F = (F_1, F_2, F_3)$ is the force applied to the rod's ends, with F_i, $i = 1,2,3$, the components of the force with respect to the fixed coordinate system (X,Y,Z), then this force is related to λ by

$$F = \frac{\partial U}{\partial D} = -\lambda\,. \tag{4.2.32}$$

Therefore, $-\lambda$ represents the external force that fixes the ends of the rod. This force is supposed to be known. The couples $M = (M_1, M_2 M_3)$ can be determined from

$$\delta\Im = \int_0^l \{(TS\ (4.2.20))\delta\theta + (TS\ (4.2.21))\delta\psi + (TS\,(4.2.22))\delta\varphi\}\mathrm{d}s +$$

$$+[A\theta'\delta\theta + [A\psi'\sin^2\theta + C(\varphi' + \psi'\cos\theta)\cos\theta]\delta\psi + C(\varphi' + \psi'\cos\theta)\delta\varphi]_0^l, \tag{4.2.33}$$

where TS are left-side parts of equations (4.2.20)–(4.2.22). Into equilibrium the integrand in (4.2.33) is zero

$$\delta\Im = [A\theta'\delta\theta + [A\psi'\sin^2\theta + C(\varphi' + \psi'\cos\theta)\cos\theta]\delta\psi + \\ + C(\varphi' + \psi'\cos\theta)\delta\varphi]_0^l. \tag{4.2.34}$$

From here we derive the couples at the ends of the rod with respect to ON and Z and z axes

$$M_1 = \frac{\partial\Im}{\partial\theta}\big|_{s=0\ or\ l} = \frac{\partial U}{\partial\theta}\big|_{s=0\ or\ l} = \big|A\theta'\big|_{s=0\ or\ l}\,,$$

$$M_2 = \frac{\partial \Im}{\partial \psi}|_{s=0 or l} = \frac{\partial U}{\partial \psi}|_{s=0 or l} = | A\psi' \sin^2\theta + C(\psi' \cos\theta + \varphi')\cos\theta |_{s=0 or l},$$

$$M_3 = \frac{\partial \Im}{\partial \psi}|_{s=0 or l} = \frac{\partial U}{\partial \psi}|_{s=0 or l} = | C(\varphi' \cos\theta + \psi')|_{s=0 or l}.$$

The equilibrium equations (4.2.20)–(4.2.22) are coupled nonlinear ordinary differential equations in unknown Euler angles.

Next, we see that the equation (4.2.22) can be solved

$$\varphi'(s) + \psi'(s)\cos\theta(s) = \alpha,$$

with α an integration constant.

From the torsion definition (4.2.11) we conclude that $\alpha = \tau$. So, the above relation becomes

$$\varphi'(s) + \psi'(s)\cos\theta(s) = \tau. \tag{4.2.35}$$

With (4.2.35), equations (4.2.20) and (4.2.21) can be written as

$$\begin{gathered} A(\psi'^2 \sin\theta\cos\theta - \theta'') - C\tau\psi' \sin\theta + \lambda_1 \cos\theta\cos\psi + \\ + \lambda_2 \cos\theta\sin\psi - \lambda_3 \sin\theta = 0, \end{gathered} \tag{4.2.36}$$

$$A(\psi'' \sin\theta + 2\psi'\theta'\cos\theta) - C\tau\theta' + \lambda_1 \sin\psi - \lambda_2 \cos\psi = 0. \tag{4.2.37}$$

We introduce

$$\lambda = \sqrt{\lambda_1^2 + \lambda_2^2}, \quad \psi_1 = \psi + \pi + \arctan\frac{\lambda_1}{\lambda_2}, \tag{4.2.38}$$

and write

$$\begin{gathered} \lambda_1 \cos\psi + \lambda_2 \sin\psi = -\lambda\sin\psi_1, \\ \lambda_1 \sin\psi - \lambda_2 \cos\psi = \lambda\cos\psi_1. \end{gathered} \tag{4.2.39}$$

Adding (4.2.36) multiplied by $2\theta'$ and (4.2.37) multiplied by $(-2\psi_1' \sin\theta)$ we obtain

$$\begin{gathered} -A\psi_1'^2 (\sin^2\theta)' - A(\psi_1'^2)' \sin^2\theta - A(\theta'^2)' + \lambda_3(\cos\theta)' - \\ -2\lambda\sin\psi_1(\sin\theta)' - 2\lambda\sin\theta(\sin\psi_1)' = 0. \end{gathered}$$

Dividing by 1/2 we have

$$[\frac{A}{2}(\theta'^2 + \psi_1'^2 \sin^2\theta) + \lambda\sin\theta\sin\psi_1 - \lambda_3\cos\theta]' = 0. \tag{4.2.40}$$

Integrating (4.2.40) with respect to s we find the bending energy density of the thin elastic rod

$$\frac{A}{2}\kappa^2 = -\lambda\sin\theta\sin\psi_1 + \lambda_3\cos\theta + C_0, \tag{4.2.41}$$

with C_0 an integration constant. In the case $\lambda = (0,0,\lambda_3)$ the equilibrium equations (4.2.35)–(4.2.37) become

$$\varphi'(s) + \psi'(s)\cos\theta(s) = \tau\,, \tag{4.2.42}$$

$$A(\psi'^2 \sin\theta\cos\theta - \theta'') - C\tau\psi' \sin\theta - \lambda_3 \sin\theta = 0\,, \tag{4.2.43}$$

$$A(\psi'' \sin\theta + 2\psi'\theta'\cos\theta) - C\tau\theta' = 0\,. \tag{4.2.44}$$

To write the motion equations, let us introduce the inertia of the rod characterized by the functions

$$R \ni s \to (\rho_0 A_0)(s), (\rho_0 I_1)(s), (\rho_0 I_2(s) \in (0,\infty)\,, \tag{4.2.45}$$

where ($\rho_0 A_0$) is the natural mass density per unit length, A_0 the area of the cross section, ($\rho_0 I_1$) the principal mass moment of inertia around the axis which is perpendicular to the central axis and ($\rho_0 I_2$) the principal mass moment of inertia around the central axis. We suppose to have

$$\rho = A_0\rho_0 = \pi a^2 \rho_0\,,\; k_1 = I_1\rho_0 = \frac{\pi a^4}{4}\rho_0,\; k_2 = I_2\rho_0 = \frac{\pi a^4}{2}\rho_0, \tag{4.2.46}$$

where ρ_0 is the mass density per unit volume, and I_1, I_2 geometrical moments of inertia around the axis, which is perpendicular to the central axis and respectively around the central axis. The kinetic energy K of the rod is a sum between the energy of the translational motion K_1, the energy of the rotational motion of the tangential vector K_2 and the energy of the rotational motion around the central axis K_3 (Tsuru)

$$K - K_1 + K_2 + K_3\,, \tag{4.2.47}$$

with

$$K_1 = \frac{\rho}{2}\int_0^l \dot{r}^2 \mathrm{d}s\,, \tag{4.2.48}$$

$$K_2 = \frac{k_1}{2}\int_0^l \dot{d}_3^{\,2} \mathrm{d}s = \frac{k_1}{2}\int_0^l (\Omega_1^2 + \Omega_2^2)\mathrm{d}s\,, \tag{4.2.49}$$

$$K_3 = \frac{k_2}{2}\int_0^l \Omega_3^2 \mathrm{d}s\,, \tag{4.2.50}$$

where the dot represents the differentiation with respect to time, and $\Omega(\Omega_1, \Omega_2, \Omega_3)$ is the vector of angular velocity of rotation

$$\begin{aligned} \Omega_1 &= -\dot{\psi}\sin\theta\cos\varphi + \dot{\theta}\sin\varphi, \\ \Omega_2 &= \dot{\psi}\sin\theta\sin\varphi + \dot{\theta}\cos\varphi, \\ \Omega_1 &= \dot{\psi}\cos\theta + \dot{\varphi}. \end{aligned} \tag{4.2.51}$$

These relations are analogous to (4.2.9). Using (4.2.7) we get for K_2 and K_3

$$K_2 = \frac{k_1}{2}\int_0^l (\dot{\psi}^2 \sin^2\theta + \dot{\theta}^2)\mathrm{d}s\,, \tag{4.2.52}$$

$$K_3 = \frac{k_2}{2}\int_0^l (\dot{\psi}\cos\theta + \dot{\varphi})^2 \mathrm{d}s\,. \tag{4.2.53}$$

We prove now the following theorem:

THEOREM 4.2.2 *The exact set of motion equations of the thin elastic rod with the ends fixed by the force* $F = -\lambda$ *with* $\lambda = (\lambda_1, \lambda_2, \lambda_3)$ *are*

$$-\rho\ddot{r} - \lambda' = 0\,, \tag{4.2.54}$$

$$\begin{aligned}
&k_1(\dot{\psi}^2 \sin\theta\cos\theta - \ddot{\theta}) - k_2(\dot{\varphi} + \dot{\psi}\cos\theta)\dot{\psi}\sin\theta - \\
&-A(\psi'^2 \sin\theta\cos\theta - \theta'') + C(\varphi' + \psi'\cos\theta)\psi'\sin\theta - \\
&-\lambda_1\cos\theta\cos\psi - \lambda_2\cos\theta\sin\psi + \lambda_3\sin\theta = 0,
\end{aligned} \tag{4.2.55}$$

$$\begin{aligned}
&-\frac{\partial}{\partial t}\{k_1\dot{\psi}\sin^2\theta + k_2(\dot{\varphi} + \dot{\psi}\cos\theta)\cos\theta\} + \\
&+\frac{\partial}{\partial s}\{A\psi'^2\sin^2\theta + C(\varphi' + \psi'\cos\theta)\cos\theta\} + \\
&\qquad +\lambda_1\sin\theta\sin\psi - \lambda_2\sin\theta\cos\psi = 0,
\end{aligned} \tag{4.2.56}$$

$$-k_2\frac{\partial}{\partial t}(\dot{\varphi} + \dot{\psi}\cos\theta) + C\frac{\partial}{\partial s}(\varphi' + \psi'\cos\theta) = 0\,. \tag{4.2.57}$$

Proof: The unknowns are the force vector $\boldsymbol{\lambda}$ and Euler angles. The position vector is related to Euler angles through the constraint (4.2.13).

The kinetic energy K is given by

$$K = \frac{\rho}{2}\int_0^L \dot{r}^2\mathrm{d}s + \frac{k_1}{2}\int_0^L \dot{d}_3^{\,2}\mathrm{d}s + \frac{k_2}{2}\int_0^L (\dot{\psi}\cos\theta + \dot{\varphi})^2\mathrm{d}s\,. \tag{4.2.58}$$

We introduce the Lagrangian

$$L = K - U\,, \tag{4.2.59}$$

and the action

$$I = \int_0^T \mathrm{d}t L\,. \tag{4.2.60}$$

To take account of the constraint (4.2.13) we introduce a new functional

$$I_c = I + L = I + \int_0^T \mathrm{d}t \int_0^l \mathrm{d}s\,(r' - d_3)\cdot\lambda\,. \tag{4.2.61}$$

We write this functional under the form

$$I_c = \int_0^T dt \int_0^l ds \, [\frac{\rho}{2}\dot{r}^2 + \frac{k_1}{2}\dot{d}_3^2 + \frac{k_2}{2}(\dot{\psi}\cos\theta + \dot{\varphi})^2 - \frac{A}{2}\kappa^2 - \frac{C}{2}\tau^2 + (r' - d_3)\lambda],$$

or

$$I_c = \int_0^T dt \int_0^l ds \, L(r, \dot{r}, r', \varsigma, \dot{\varsigma}, \varsigma'),$$

where ς is the vector of Euler angles $\varsigma(\theta, \psi, \varphi)$ and L the Lagrangian

$$L = \frac{\rho}{2}\dot{r}^2 + \frac{k_1}{2}\dot{d}_3^2 + \frac{k_2}{2}(\dot{\psi}\cos\theta + \dot{\varphi})^2 - \frac{A}{2}(\theta'^2 + \psi'^2 \sin^2\theta) - \frac{C}{2}(\varphi' + \psi'\cos\theta)^2 + (r' - d_3)\lambda. \tag{4.2.62}$$

The variation with respect to r and ς gives

$$\delta I_c = \int_0^T dt \int_0^l ds (\frac{\partial L}{\partial r}\delta r + \frac{\partial L}{\partial r'}\delta r' + \frac{\partial L}{\partial \dot{r}}\delta \dot{r} + \frac{\partial L}{\partial \varsigma}\delta \varsigma' + \frac{\partial L}{\partial \varsigma'}\delta \varsigma' + \frac{\partial L}{\partial \dot{\varsigma}}\delta \dot{\varsigma}) = 0.$$

We see that $\delta r' = \frac{\partial}{\partial s}\delta r$, $\delta \dot{r} = \frac{\partial}{\partial t}\delta r$, $\delta \varsigma' = \frac{\partial}{\partial s}\delta \varsigma, \delta \dot{\varsigma} = \frac{\partial}{\partial t}\delta \varsigma$, and by integrating through parts the terms $\frac{\partial L}{\partial r'}\delta r', \frac{\partial L}{\partial \dot{r}}\delta \dot{r}$ etc., we obtain

$$\begin{aligned} \delta I_c = & \left\| \frac{\partial L}{\partial r'}\delta r \right\|_0^l \Big|_0^T + \left\| \frac{\partial L}{\partial \dot{r}}\delta r \right\|_0^l \Big|_0^T + \left\| \frac{\partial L}{\partial \varsigma'}\delta \varsigma \right\|_0^l \Big|_0^T + \left\| \frac{\partial L}{\partial \dot{\varsigma}}\delta \varsigma \right\|_0^l \Big|_0^T + \\ & + \int_0^T dt \int_0^l ds \left(\frac{\partial L}{\partial r} - \frac{\partial}{\partial s}\left(\frac{\partial L}{\partial r'}\right) - \frac{\partial}{\partial t}\left(\frac{\partial L}{\partial \dot{r}}\right) \right) \delta r + \\ & + \int_0^T dt \int_0^l ds \left(\frac{\partial L}{\partial \varsigma} - \frac{\partial}{\partial s}\left(\frac{\partial L}{\partial \varsigma'}\right) - \frac{\partial}{\partial t}\left(\frac{\partial L}{\partial \dot{\varsigma}}\right) \right) \delta \varsigma = 0. \end{aligned} \tag{4.2.63}$$

According to

$$\delta r(0,0) = \delta r(0,T) = 0, \ \delta r(l,0) = \delta r(l,T) = 0,$$

$$\delta \varsigma(0,0) = \delta \varsigma(0,T) = 0, \ \delta \varsigma(l,0) = \delta \varsigma(l,T) = 0, \tag{4.2.64}$$

the first four terms in (4.2.63) are vanishing. The condition for integrands to be identically zero lead to the Lagrange equations

$$\frac{\partial L}{\partial r} - \frac{\partial}{\partial s}\left(\frac{\partial L}{\partial r'}\right) - \frac{\partial}{\partial t}\left(\frac{\partial L}{\partial \dot{r}}\right) = 0, \quad \frac{\partial L}{\partial \varsigma} - \frac{\partial}{\partial s}\left(\frac{\partial L}{\partial \varsigma'}\right) - \frac{\partial}{\partial t}\left(\frac{\partial L}{\partial \dot{\varsigma}}\right) = 0. \tag{4.2.65}$$

By substituting (4.2.62) into (4.2.65) we obtain the motion equations (4.2.54)–(4.2.57).

Differentiating both sides of (4.2.54) with respect to s, and using (4.2.13) the equation (4.2.54) becomes

$$-\rho\ddot{d}_3 - \lambda'' = 0. \tag{4.2.66}$$

The motion equations (4.2.54)–(4.2.57) are coupled nonlinear partial differential equations in unknown Euler angles and the vector function λ which characterizes the external force applied to the ends of the rod to maintain it fixed.

We have to supplement the motion equations with initial conditions

$$\begin{gathered}\lambda(s,0) = \lambda_0(s) = -\rho v^2 d_3(s,0) + (0,0,\lambda_3),\\ \theta(s,0) = \theta_0(s),\ \ \psi(s,0) = \psi_0(s),\\ \varphi(s,0) = \varphi_0(s).\end{gathered} \tag{4.2.67}$$

4.3 The equivalence theorem

This section focuses on the relation between the equilibrium equations and the motion equations of the thin elastic rod (Tsuru). The purpose is to determine the conditions when the motion equations can be reduced to the equilibrium equations. In connection to this, the following theorem holds:

THEOREM 4.3.1 *Given* λ, θ, ψ *and* φ *as functions only of the variable* $\xi = s - vt$ *and suppose that*

$$\lambda = \zeta d_3 + (0,0,\lambda_3), \tag{4.3.1}$$

with $\zeta = -\rho v^2$ *and* $d_3 = (\sin\theta\cos\psi, \sin\theta\sin\psi, \cos\theta)$. *In these conditions the motion equations (4.2.54)–(4.2.57) are equivalent to the equilibrium equations (4.2.20)–(4.2.22) for*

$$A - k_1 v^2 \to A,\ \ C - k_2 v^2 \to C,\ \ \xi \to s. \tag{4.3.2}$$

Proof: We note by prime the differentiation with respect to the new variable $\xi = s - vt$. The system of equations (4.2.54)–(4.2.57) become

$$-\rho v^2 r'' = \lambda', \tag{4.3.3}$$

$$\begin{gathered}k_1 v^2(\psi'^2 \sin\theta\cos\theta - \theta'') - k_2 v^2(\varphi' + \psi'\cos\theta)\psi'\sin\theta - \\ -A(\psi'^2\sin\theta\cos\theta - \theta'') + C(\varphi' + \psi'\cos\theta)\psi'\sin\theta - \\ -\lambda_1\cos\theta\cos\psi - \lambda_2\cos\theta\sin\psi + \lambda_3\sin\theta = 0,\end{gathered} \tag{4.3.4}$$

$$-\frac{\partial}{\partial\xi}\{-vk_1\psi'\sin^2\theta - vk_2(\varphi'+\psi'\cos\theta)\cos\theta + \\ +A\psi'^2\sin^2\theta + C(\varphi'+\psi'\cos\theta)\cos\theta\} + \lambda_1\sin\theta\sin\psi - \\ -\lambda_2\sin\theta\cos\psi = 0, \tag{4.3.5}$$

$$-k_2\frac{\partial}{\partial\xi}[-v\varphi' - v\psi'\cos\theta + C(\varphi'+\psi'\cos\theta)] = 0\,. \tag{4.3.6}$$

Rearranging the terms in (4.3.3)–(4.3.6) we have

$$-\rho v^2 r' = \lambda + \hat{c}\,, \tag{4.3.7}$$

$$(A-k_1v^2)(\psi'^2\sin\theta\cos\theta - \theta'') - \\ -(C-k_2v^2)(\varphi'+\psi'\cos\theta)\psi'\sin\theta + \\ +\lambda_1\cos\theta\cos\psi + \lambda_2\cos\theta\sin\psi - \lambda_3\sin\theta = 0, \tag{4.3.8}$$

$$(A-k_1v^2)(\psi''\sin\theta + 2\psi'\theta'\cos\theta) - \\ (C-k_2v^2)\theta'(\varphi'+\psi'\cos\theta) + \\ +\lambda_1\sin\psi \; -\lambda_2\cos\psi = 0, \tag{4.3.9}$$

$$\varphi'(s) + \psi'(s)\cos\theta(s) = \tau\,, \tag{4.3.10}$$

with $\hat{c}$ an integration constant.

Taking into account (4.2.13) the equation (4.3.8) becomes

$$-\rho v^2 d_3 = \lambda + \hat{c}\,,$$

or

$$\boldsymbol{\lambda} = (-\rho v^2\sin\theta\cos\psi - \hat{c}_1,\;\; -\rho v^2\sin\theta\sin\psi - \hat{c}_2,\;\; -\rho v^2\cos\theta - \hat{c}_3)\,. \tag{4.3.11}$$

From (4.3.1) we have

$$\boldsymbol{\lambda} = (\zeta\sin\theta\cos\psi,\;\; \zeta\sin\theta\sin\psi,\;\; \zeta\cos\theta + \lambda_3)\,. \tag{4.3.12}$$

It can be seen that equations (4.3.12) and (4.3.13) are equivalent if

$$\zeta = -\rho v^2\,,\;\; \hat{c} = (0,0,-\lambda_3)\,. \tag{4.3.13}$$

With these conditions, the motion equations (4.2.54)–(4.2.57) become

$$(A-k_1v^2)(\psi'^2\sin\theta\cos\theta - \theta'') - \\ -(C-k_2v^2)(\varphi'+\psi'\cos\theta)\psi'\sin\theta - \lambda_3\sin\theta = 0, \tag{4.3.14}$$

$$(A-k_1v^2)(\psi''\sin\theta + 2\psi'\theta'\cos\theta) - \\ -(C-k_2v^2)\theta'(\varphi'+\psi'\cos\theta) = 0, \tag{4.3.15}$$

$$\varphi'(s) + \psi'(s)\cos\theta(s) = \tau\,. \tag{4.3.16}$$

Hence, the equations (4.2.54)–(4.2.57) coincide with equilibrium equations (4.2.35)–(4.2.37). We note that the quantity λ_3 is known. The constant *v*, which defines the variable ξ can be determined from $(4.2.67)_1$.

4.4 Exact solutions of the equilibrium equations

Consider the case when the force $\lambda = (0, 0, \lambda_3)$ with $\lambda_1 = \lambda_2 = 0$ is parallel to the *Z*-axis of the Lagrangian system of equations. Let us consider the equilibrium equations (4.2.42)–(4.2.44) and introduce $\lambda_1 = \lambda_2 = 0$

$$A(\psi'^2 \sin\theta\cos\theta - \theta'') - C\tau\psi' \sin\theta - \lambda_3 \sin\theta = 0\,, \tag{4.4.1}$$

$$A(\psi'' \sin\theta + 2\psi'\theta' \cos\theta) - C\tau\theta' = 0\,, \tag{4.4.2}$$

$$\varphi'(s) + \psi'(s)\cos\theta(s) = \tau\,. \tag{4.4.3}$$

Multiplying both sides of (4.4.2) with $\sin\theta$ we get

$$A\sin^2\theta\psi'' + A(\sin^2\theta)'\psi' + C\tau(\cos\theta)' = 0\,,$$

or

$$A(\sin^2\theta\psi')' + C\tau(\cos\theta)' = 0\,,$$

integrating we have

$$A\sin^2\theta\psi' + C\tau\cos\theta + \beta = 0\,, \tag{4.4.4}$$

with β an integration constant. In the virtue of (4.4.4) we obtain

$$\psi' = -\frac{C\tau\cos\theta + \beta}{A\sin^2\theta}. \tag{4.4.5}$$

Substituting (4.4.5) into (4.4.1) and multiplying the resulting equation with $2\theta'$ we obtain by integration

$$A\theta'^2 - \frac{(C\tau\cos\theta + \beta)^2}{A\sin^2\theta} + 2C\tau\cos\theta\frac{C\tau\cos\theta + \beta}{A\sin^2\theta} - 2\lambda_3\cos\theta + \gamma = 0\,,$$

with γ an integration constant. The above equation can be written as

$$A\theta'^2 + \frac{C^2\tau^2\cos^2\theta - \beta^2}{A\sin^2\theta} - 2\lambda_3\cos\theta + \gamma = 0\,. \tag{4.4.6}$$

Substituting $u = \cos\theta$ into (4.4.6) we obtain the differential equation

$$u'^2 = \frac{1-u^2}{A}\left\{-\frac{C^2\tau^2u^2 - \beta^2}{A(1-u^2)} + 2\lambda_3 u - \gamma\right\}, \tag{4.4.7}$$

or

$$\frac{1}{2}u'^2 = f(u), \tag{4.4.8}$$

$$f(u) = -\frac{1}{A}[\lambda_3 u^3 - \frac{1}{2}(\gamma - \frac{C^2\tau^2}{A})u^2 - \lambda_3 u + \frac{1}{2}(\gamma - \frac{\beta^2}{A})] .$$

We have obtained a Weierstrass equation (1.4.14) with a polynomial of third order. The torsion τ and the integration constants γ and β are determined from the boundary conditions

$$\psi(0) = \psi(l) = \psi_0 ,$$

$$\theta(0) = \theta(l) = \theta_0 ,$$

$$\tau(0) = \tau(l) = \tau_0 . \tag{4.4.9}$$

The qualitative nature of the solutions of (4.4.8) for arbitrary values of the constants can be studied by elementary analysis. We write the equation (4.4.8) in the form

$$\frac{1}{2}u'^2 = au^3 + bu^2 - au + c \equiv f(u) , \tag{4.4.10}$$

with

$$a = -\frac{\lambda_3}{A} \neq 0, \quad b = \frac{1}{2A}(\gamma - \frac{C^2\tau^2}{A}), \quad c = -\frac{1}{2A}(\gamma - \frac{\beta^2}{A}) . \tag{4.4.11}$$

We are looking for general bounded waves of permanent form. We have $u'^2 \geq 0$. So, u varies monotonically until u' vanishes. The graph of f for different values of the constants a, b and c has six possible forms (Figure 4.4.1). Since $u'^2 - f(u) \geq 0$, solutions will occur only in the intervals shaded in the figure.

If u_1 is a simple zero for f, from Taylor's formula we have

$$f(u) = f(u_1) + (u - u_1)f'(u_1) + O\{(u - u_1)^2\} ,$$

or, because $f(u_1) = 0$

$$u'^2 = 2f'(u_1)(u - u_1) + O\{(u - u_1)^2\} , \text{ for } u \to u_1 . \tag{4.4.12}$$

Therefore,

$$u(s) = u(s_1) + (s - s_1)u'(s_1) + \frac{1}{2}(s - s_1)^2 u''(s_1) + O\{(s - s_1)^3\} , \tag{4.4.13}$$

where $u(s_1) = u_1$. From $u'^2 = 2f(u)$ we have $u'^2(s_1) = 2f(u(s_1)) = 0$, and $u'(s_1) = 0$.

Differentiating both sides of $u'^2 = 2f(u)$, it results $u''(s) = f'(u(s))$ and $u''(s_1) = f'(u_1)$. From (4.4.13) we get

$$u = u_1 + \frac{1}{2}f'(u_1)(s - s_1)^2 + O\{(s - s_1)^3\} , \text{ for } s \to s_1 . \tag{4.4.14}$$

It follows that u has a local minimum or maximum u_1 at s_1, as $f'(u_1)$ is positive or negative, respectively. Similarly, if u_1 is a double zero of f we have

$$f(u) = f(u_1) + (u-u_1)f'(u_1) + \frac{1}{2}(u-u_1)^2 f''(u_1) + O\{(u-u_1)^3\},$$

or

$$u'^2 = f''(u_1)(u-u_1)^2 + O\{(u-u_1)^3\}, \quad \text{for} \quad u \to u_1. \tag{4.4.15}$$

The quantity u' is real only if $f''(u_1) > 0$ (Figure 4.4.1b) and in this case we have

$$u' = \sqrt{f''(u_1)}(u-u_1). \tag{4.4.16}$$

Relation (4.4.16) yields

$$\frac{u'}{u-u_1} = \sqrt{f''(u_1)},$$

and integrating with respect to s we have

$$\ln(u-u_1) = \pm\sqrt{f''(u_1)}s + \ln C,$$

where

$$u - u_1 = \mathrm{C}\ \exp(\pm\sqrt{f''(u_1)}s. \tag{4.4.17}$$

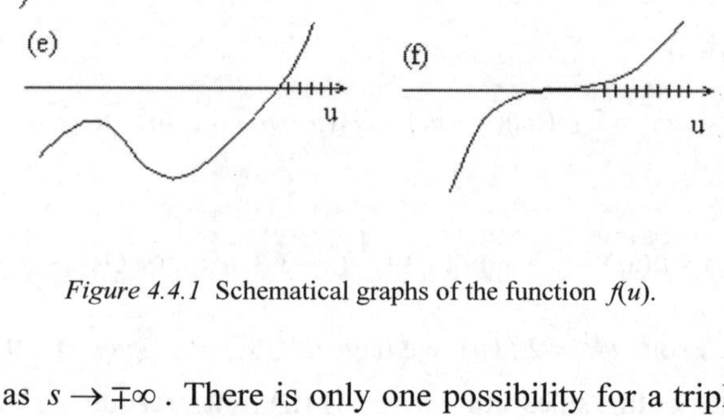

Figure 4.4.1 Schematical graphs of the function *f*(*u*).

Thus, $u \to u_1$ as $s \to \mp\infty$. There is only one possibility for a triple zero $u_1 = -\dfrac{b}{3a}$ (Figure 4.4.1f). The exact solution is

$$u(s) = -\frac{b}{3a} + \frac{2}{(s-s_0)^2}, \tag{4.4.18}$$

where s_0 is an arbitrary constant. This solution is unbounded at $s = s_0$.

Consider now the cases a, d, e, and the right-hand part of the case c shown in Figure 4.4.1. If $u' > 0$, then $f > 0$ for all $s > 0$ and $u \to +\infty$, as $s \to +\infty$. If $u'(0) < 0$, then u will decrease until it reaches $u = u_1$. In this case u_1 is a simple zero and so u' changes sign and once again $u \to +\infty$ as $s \to +\infty$. So, for these four cases we have unbounded solutions.

Let us see the case from Figure 4.4.1b. The solution u has a simple zero at u_3 and a double zero at u_1. The solution has a minimum at $u = u_3$ and attains $u = u_1$ as $s \to \pm\infty$. Here we have a solitary wave solution with a amplitude $u_3 - u_1 < 0$ (Figure 4.4.2a)

$$\begin{aligned} u &= u_3 + (u_2 - u_3)\sin^2\phi = u_3 + (u_2 - u_3)(1 - \mathrm{cn}^2\phi) = \\ &= u_3 + (u_2 - u_3)(1 - \mathrm{cn}^2\phi) = u_3 + (u_2 - u_3)\mathrm{sn}^2\phi = \\ &= u_2 - (u_2 - u_3)\mathrm{cn}^2\left(\sqrt{\frac{|a|}{2}}(u_1 - u_3)(s - s_3), m\right). \end{aligned}$$

Finally we consider the cases in Figure 4.4.1c. The solution u has simple zeros, that is a local maximum at u_2 and a local minimum at u_3. Thus u' changes sign at these points and since the behavior near them is algebraic, consecutive points $u = u_2$, $u = u_3$, will be a finite distance apart. The solution will oscillate between u_2 and u_3 with a finite period (Figure 4.4.2b).

The period can be determined from $u' = \sqrt{2f(u)}$ as

$$2\int_{u_3}^{u_2} \frac{du}{u'} = 2\int_{u_3}^{u_2} \frac{du}{\sqrt{2f(u)}}. \tag{4.4.19}$$

We have

$$\int_{u_3(s_3)}^{u_2(s_2)} \frac{du}{u'} = \int_{u_3(s_3)}^{u_2(s_2)} \frac{u'ds}{u'} = \int_{s_3}^{s_2} ds = \int_{u_3}^{u_2} \frac{du}{\sqrt{2f(u)}}.$$

From the above formula we have implicitly the solution

$$s = s_3 + \int_{u_3}^{u} \frac{du}{\pm\sqrt{2f(u)}}, \tag{4.4.20}$$

where $u(s_3) = u_3$. The sign + corresponds to the case $u' > 0$ and the sign – to the case $u' < 0$.

Solutions (4.4.20) are known as *the cnoidal solutions* because they are described by the cosine and sine Jacobean elliptic functions (Freeman).

These solutions were found by Korteweg & de Vries in 1895. Finally, the single bounded solutions of the equation (4.4.8) are the cnoidal solutions (the solitary solution is a particular case of the cnoidal solution).

We can write the solution (4.4.20) as (Abramowitz and Stegun)

$$s = s_3 + \int_{u_3}^{u} \frac{du}{\pm\sqrt{2a(u-u_1)(u-u_2)(u-u_3)}}, \tag{4.4.21}$$

with $u_3 < u_2 < u_1$ three distinct and real roots of the equation $f(u) = 0$.

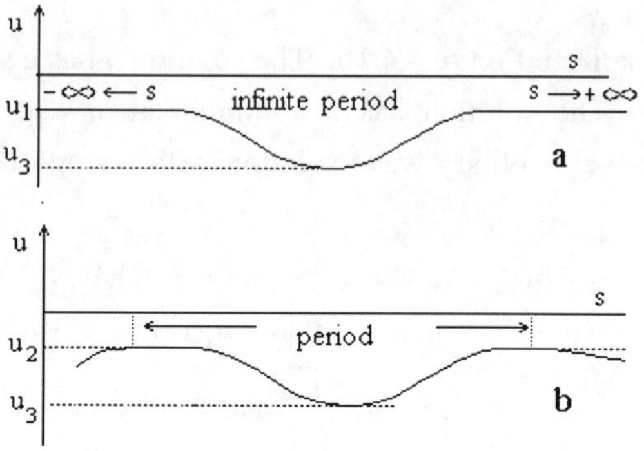

Figure 4.4.2 a) Soliton solution, b) Cnoidal solution.

The integral (4.4.21) may be reduced to an elliptic integral of the first kind

$$v = \sqrt{\frac{|a|}{2}}(u_1 - u_3)(s - s_3), \tag{4.4.22}$$

with $m = \dfrac{u_2 - u_3}{u_1 - u_3}$, $0 \le m \le 1$, the modulus of the elliptic functions. The solution is

$$\begin{aligned} u &= u_3 + (u_2 - u_3)\sin^2\phi = u_3 + (u_2 - u_3)(1 - \mathrm{cn}^2\phi) = \\ &= u_3 + (u_2 - u_3)(1 - \mathrm{cn}^2\phi) = u_3 + (u_2 - u_3)\mathrm{sn}^2\phi = \\ &= u_2 - (u_2 - u_3)\mathrm{cn}^2(\sqrt{\frac{|a|}{2}}(u_1 - u_3)(s - s_3), m). \end{aligned} \tag{4.4.23}$$

We have

$$v = \int_0^{\phi} \frac{d\theta}{\sqrt{1 - m\sin^2\theta}}. \tag{4.4.24}$$

The period of $\cos\phi$ is 2π, so the period of the function cn is

$$4\int_0^{\pi/2} \frac{d\theta}{\sqrt{1 - m\sin^2\theta}} = 4K(m). \tag{4.4.25}$$

In fact $K(1) = \infty$, then the period of the function sech v is infinite. Parameters u_2 and u_3 determine the amplitude of the cnoidal functions. The parameter s_3 determines the phase. The period of the solution u is

$$2K(m)\sqrt{\frac{2}{u_1 - u_3}}. \tag{4.4.26}$$

The case of three distinct and real roots.

We will prove the following theorem in the case when the equation $f(u) = 0$ has three distinct and real roots.

THEOREM 4.4.1 *Given $u_3 < u_2 < u_1$ the distinct and real roots of the cubic equation $f(u) = 0$, the equilibrium equations (4.4.1)–(4.4.3) have a unique solution for the Euler angles*

$$\begin{aligned} u &= u_2 - (u_2 - u_3)\mathrm{cn}^2\left(\sqrt{\frac{|\lambda_3|}{2A}}(u_1 - u_3)(s - s_3), m\right) = \\ &= u_2 - (u_2 - u_3)\mathrm{cn}^2[w(s - s_3), m], \end{aligned} \tag{4.4.27}$$

with $m = \dfrac{u_2 - u_3}{u_1 - u_3}$ *and* $w = \sqrt{\dfrac{|\lambda_3|}{2A}(u_1 - u_3)}$,

$$\begin{aligned} \psi = \frac{1}{4A^2w^2}\{&-\frac{\beta + C\tau}{1 - u_3}\Pi[w(s - s_3), \frac{u_2 - u_3}{1 - u_3}, m] - \\ &-\frac{\beta - C\tau}{1 + u_3}\Pi[w(s - s_3), \frac{u_2 - u_3}{1 + u_3}, m]\}, \end{aligned} \tag{4.4.28}$$

$$\begin{aligned} \varphi = -\frac{\tau(C - A)}{A}s + \frac{1}{4A^2w^2}\{&\frac{\beta + C\tau}{1 - u_3}\Pi[w(s - s_3), \frac{u_2 - u_3}{1 - u_3}, m] - \\ &-\frac{\beta - C\tau}{1 + u_3}\Pi(w(s - s_3), \frac{u_2 - u_3}{1 + u_3}, m)\}, \end{aligned} \tag{4.4.29}$$

where $\Pi(x, z, m)$ is the normal elliptic integral of the third kind

$$\Pi(x, z, m) = \int_0^x \frac{\mathrm{d}y}{1 - z\,\mathrm{sn}^2(y, m)}. \tag{4.4.30}$$

Proof: The solution of (4.4.7) is given by (4.4.23) that implies (4.4.27). In the virtue of (4.4.3) we have

$$\psi' = -\frac{C\tau u + \beta}{A(1 - u^2)}. \tag{4.4.31}$$

Decomposing in simple fractions, we rewrite (4.4.31) as

$$\psi' = -\frac{\beta + C\tau}{2A(1 - u)} - \frac{\beta - C\tau}{2A(1 + u)}. \tag{4.4.32}$$

From (4.4.27) we have

$$1-u=1-u_3-(u_2-u_3)\mathrm{sn}^2[w(s-s_3),m],$$

$$1+u=1+u_3+(u_2-u_3)\mathrm{sn}^2[w(s-s_3),m], \tag{4.4.33}$$

or

$$\frac{1-u}{1-u_3}=1-\frac{u_2-u_3}{1-u_3}\mathrm{sn}^2[w(s-s_3),m],$$

$$\frac{1+u}{1+u_3}=1+\frac{u_2-u_3}{1+u_3}\mathrm{sn}^2[w(s-s_3),m]. \tag{4.4.34}$$

We introduce (4.4.34) into (4.4.32) and integrate with respect to s. The solution (4.4.28) is straightly obtained. Substituting (4.4.31) into (4.4.3) we get

$$\varphi'=\tau-\psi'\cos\theta=\frac{\tau(C-A)u^2+\beta u+A\tau}{A(1-u^2)}. \tag{4.4.35}$$

Decomposing in simple fractions, we have for (4.4.35)

$$\varphi'=-\frac{\tau(C-A)}{A}+\frac{\beta+C\tau}{2A(1-u)}-\frac{\beta-C\tau}{2A(1+u)}. \tag{4.4.36}$$

Similarly, the solution (4.4.29) is obtained. We now prove another theorem.

THEOREM 4.4.2 *In conditions given by the theorem 4.4.1, the components of the position vector are*

$$x=R\sin(\psi+\Delta)=R\sin\Phi\ , \tag{4.4.37}$$

$$y=R\cos(\psi+\Delta)=R\cos\Phi\,, \tag{4.4.38}$$

$$z=u_1 s-\frac{(u_1-u_3)}{w}\mathrm{E}[\mathrm{am}(w(s-s_3)),m]\,, \tag{4.4.39}$$

with

$$R=\frac{1}{\lambda_3}\sqrt{2|\lambda_3|Au+C^2\tau^2A\gamma\beta^2}\ , \tag{4.4.40}$$

$$\Delta=-\arctan\left(\frac{A}{\beta u+C\tau}\sqrt{2f(u)}\right), \tag{4.4.41}$$

$$\Phi=\frac{\beta}{2A}s+C_0\Pi(ws,C_1,m)\,, \tag{4.4.42}$$

where $f(u)$ is given by (4.4.8) and C_0 , C_1 are constants, and

$$E(am(v),m) = E(v,m) = \int_0^{\phi} \sqrt{1 - m\sin^2\theta}\,d\theta = \int_0^{v} dn^2(y,m)dy\,, \qquad (4.4.43)$$

is the elliptic integral of the first kind.

Proof: The components x and y of the position vector are obtained from $(4.2.15)_{1,2}$. Multiplying (4.4.1) by $\sin\psi$ and (4.4.2) by $\cos\theta\cos\psi$ and subtracting we have

$$\frac{d}{ds}[-A\theta'\sin\psi - A\psi'\sin\theta\cos\theta\cos\psi + C\tau\sin\theta\cos\psi] = \\ = \lambda_3\sin\theta\sin\psi = \lambda_3 d_{32}, \qquad (4.4.44)$$

where d_{32} is the y component of the unit tangential vector.

Multiplying (4.4.1) by $\cos\psi$ and (4.4.2) by $\cos\theta\sin\psi$ and summing we have

$$\frac{d}{ds}[A\theta'\cos\psi - A\psi'\sin\theta\cos\theta\sin\psi + C\tau\sin\theta\sin\psi] = \\ = -\lambda_3\sin\theta\cos\psi = -\lambda_3 d_{31}, \qquad (4.4.45)$$

where d_{31} is the x component of the unit tangential vector.

Integrating with respect to s (4.4.44) and (4.4.45) we have

$$x = \int_0^s d_{31}ds = -\frac{1}{\lambda_3}[A\theta'\cos\psi - A\psi'\sin\theta\cos\theta\sin\psi + C\tau\sin\theta\sin\psi] + x_0\,, \qquad (4.4.46)$$

$$y = \int_0^s d_{32}ds = \frac{1}{\lambda_3}[-A\theta'\sin\psi - A\psi'\sin\theta\cos\theta\cos\psi + C\tau\sin\theta\cos\psi] + y_0\,, \qquad (4.4.47)$$

with x_0, y_0 integration constants.

Choosing $x_0 = y_0 = 0$ and substituting (4.4.5) and (4.2.6) into (4.4.46) and (4.4.47) we have

$$x = \frac{1}{\nu}\sqrt{2|\lambda_3|Au + C^2\tau^2 + A\gamma + \beta^2}\Big[\frac{(\beta u + C\tau)\sin\psi}{\sqrt{(\beta u + C\tau)^2 + 2f(u)A^2}} - \\ - \frac{A\sqrt{2f(u)}\cos\psi}{\sqrt{(\beta u + C\tau)^2 + 2f(u)A^2}}\Big],$$

$$y = \frac{1}{\nu}\sqrt{2|\lambda_3|Au + C^2\tau^2 + A\gamma + \beta^2}\Big[\frac{(\beta u + C\tau)\cos\psi}{\sqrt{(\beta u + C\tau)^2 + 2f(u)A^2}} + \\ + \frac{A\sqrt{2f(u)}\sin\psi}{\sqrt{(\beta u + C\tau)^2 + 2f(u)A^2}}\Big].$$

Denoting

$$R=\frac{1}{\lambda_3}\sqrt{2|\lambda_3|Au+C^2\tau^2+A\gamma+\beta^2},$$

the above relations become

$$x=R\left[\frac{\sin\psi}{\sqrt{1+\frac{2f(u)A^2}{(\beta u+C\tau)^2}}}-\frac{A\frac{\sqrt{2f(u)}}{\beta u+C\tau}\cos\psi}{\sqrt{1+\frac{2f(u)A^2}{(\beta u+C\tau)^2}}}\right], \tag{4.4.48}$$

$$y=R\left[\frac{\cos\psi}{\sqrt{1+\frac{2f(u)A^2}{(\beta u+C\tau)^2}}}+\frac{A\frac{\sqrt{2f(u)}}{\beta u+C\tau}\sin\psi}{\sqrt{1+\frac{2f(u)A^2}{(\beta u+C\tau)^2}}}\right]. \tag{4.4.49}$$

We mention that R is always real. The quantity under the radical is always positive due to the condition $u'^2>0$. Using the formulae $\cos(-\arctan x)=\frac{1}{\sqrt{1+x^2}}$ and $\sin(-\arctan x)=-\frac{x}{\sqrt{1+x^2}}$, we denote

$$\Delta=-\arctan\left(\frac{A}{\beta u+C\tau}\sqrt{2f(u)}\right),$$

and obtain

$$x=R(\sin\psi\cos\Delta+\cos\psi\sin\Delta),\quad y=R(\cos\psi\cos\Delta-\sin\psi\sin\Delta),$$

that yield (4.4.37) and (4.4.38).

From (4.4.5), (4.4.7), (4.4.9) and (4.4.41) we have

$$\Phi'=\psi'+\Delta'=-\frac{\beta}{2A}-\frac{\beta(\beta^2+A\gamma-C^2\tau^2)+2\lambda_3AC\tau}{2A(2\lambda_3Au-\beta^2-A\gamma+C^2\tau^2)},$$

that yields (4.4.42) by integration with respect to s, where

$$C_0=\frac{-\beta(\beta^2+A\gamma-C^2\tau^2)-2\lambda_3AC\tau+C^3\tau^3}{2A(2\lambda_3Au_2-\beta^2-A\gamma+C^2\tau^2)},$$

$$C_1=\frac{2\lambda_3A(u_2-u_3)}{2\lambda_3Au_2-\beta^2-A\gamma+C^2\tau^2}. \tag{4.4.50}$$

The component z of the position vector r is computed from (4.2.15)$_3$

$$z=\int_0^s u\mathrm{d}s=\int_0^s[u_2-(u_2-u_3)\mathrm{cn}^2(w(s-s_3),m)]\mathrm{d}s=$$

$$= u_1 s - \frac{(u_1 - u_3)}{w}\mathrm{E}(\mathrm{am}(w(s - s_3), m), \tag{4.4.51}$$

where $E(am(v), m)$ is the elliptic integral of the second kind (4.4.43).

We used here the formula

$$E(am(v), m) = m_1 v + m\int_0^v \mathrm{cn}^2(w, m)\mathrm{d}w\,, \tag{4.4.52}$$

where $m + m_1 = 1$ and $m = \dfrac{u_2 - u_3}{u_1 - u_3}$.

We see in the virtue of (4.4.37)–(4.4.39) that the rod is confined on a plane when $C\tau = 0$ and $\beta = 0$. In this case the shape of the rod is called the 'elastica' by Love. In the general case when $C\tau \neq 0$ and $\beta \neq 0$ the shapes of the rod are very complicated and the rod is not confined in a plane, the deformation being spatial. Tsuru has obtained various shapes of the rod analytically and numerically.

The case of a simple root and a double root. Let us consider the case

$$\beta = C\tau, \quad \gamma = \frac{C^2\tau^2}{A} - 2\lambda_3\,. \tag{4.4.53}$$

This implies that $a = b$, $c = -a$ and $f(u)$ from (4.4.12) becomes $f(u)$ = $a(u+1)^2(u-1)$. Therefore, $f(u)$ has a simple root and a double root. We prove the theorem:

THEOREM 4.4.3 *Given* $u_1 = u_2 = -1$, $u_3 = 1$ *the roots of the cubic equation* $f(u) = 0$, *the Euler angles are uniquely determined from the equilibrium equations (4.4.1)–(4.4.3)*

$$u(s) = -1 + 2\frac{|\lambda_3|}{A}\mathrm{sech}^2\sqrt{\frac{|\lambda_3|}{A}}s\,, \tag{4.4.54}$$

$$\psi = \frac{-C\tau s}{2A} + \arctan\left(\frac{4A}{C\tau}\tanh\left(-\sqrt{\frac{|\lambda_3|}{A}}s\right)\right), \tag{4.4.55}$$

$$\varphi = \frac{\tau(2A - C)s}{2A} + \arctan\left(\frac{4A}{C\tau}\tanh\sqrt{\frac{|\lambda_3|}{A}}s\right). \tag{4.4.56}$$

Proof: The solution of (4.4.8) is expressed in terms of hyperbolic functions. Indeed, consider the solution in the form

$$u(s) = A_1\mathrm{sech}^2 B_1 s + C_1\,.$$

Substituting this solution into (4.4.8) and taking account that $u' = -2A_1B_1\,\mathrm{sech}^2 B_1 s\,\sinh s$, we obtain by balancing the powers of the function $\mathrm{sech}\,z$

$$C_1 = -1,\ A_1 = 2\frac{|\lambda_3|}{A},\ B_1 = \sqrt{\frac{|\lambda_3|}{A}} > 0 .$$

So, the solution of (4.2.8) is

$$u(s) = -1 + 2\frac{|\lambda_3|}{A}\operatorname{sech}^2\sqrt{\frac{|\lambda_3|}{A}}s .$$

The minimum of u is -1 at $s = \pm\infty$. The maximum of u is $(-1+2\frac{|\lambda_3|}{A})$ as $s = 0$. The curve is asymptotically straight along Z-axis. The deformation is localized around $s = 0$. Using (4.4.31) we have

$$\psi' = -\frac{C\tau}{A(1-u)} . \tag{4.4.57}$$

Substituting (4.4.54) into (4.4.57) we get

$$\psi' = -\frac{C\tau}{2A\tanh^2\sqrt{\frac{|\lambda_3|}{A}}\ s} . \tag{4.4.58}$$

Integrating with respect to s we have (4.4.55). Substituting (4.4.53) into (4.4.32) we obtain

$$\varphi' = \frac{u\tau(C-A)+\tau A}{A(1-u)} , \tag{4.4.59}$$

or

$$\varphi' = \frac{C\tau}{A(1-u)} + \frac{\tau(A-C)}{A} . \tag{4.4.60}$$

Finally, integrating (4.4.60) with respect to s gives (4.4.56).
We prove the following theorem:

THEOREM 4.4.4 *Given the conditions of the theorem 4.4.3 the components of the position vector are*

$$x = \frac{\sqrt{2|\lambda_3|A(u+1)}}{\lambda_3}\sin\left(\frac{C\tau s}{2A}\right) , \tag{4.4.61}$$

$$y = \frac{\sqrt{2|\lambda_3|A(u+1)}}{\lambda_3}\cos\left(\frac{C\tau s}{2A}\right), \tag{4.4.62}$$

$$z = -s + \frac{4A}{\lambda_3}\tanh(2s) . \tag{4.4.63}$$

Proof: Substituting (4.4.53) into $(4.4.50)_1$ we obtain

$$C_0 = 0 . \tag{4.4.64}$$

From (4.4.42) we have

$$\Phi = \frac{C\tau}{2A} s . \tag{4.4.65}$$

The relations (4.4.61) and (4.4.63) are derived from (4.4.37) and (4.4.38). From (4.4.39) it results (4.4.63).

For graphical illustration we consider the values $E = 194200 \times 10^6$ Pa, $\nu_1 = 0.28$, $\rho = 7876 \text{ kg} / \text{m}^3$, $l = 10$ m, and $a = 2 \times 10^{-3}$ m.

The shape of the rod of infinite length into which the central line is deformed is called *elastica*.

Figure 4.4.3 displays four shapes of elastica for $\tau = 0$ and different set of values ($\beta = 0.3$, $\lambda_3 = 0.4, \gamma = 0.2$), ($\beta = 0.7$, $\lambda_3 = 0.2, \gamma = 0.1$), ($\beta = 0.3$, $\lambda_3 = 0.1, \gamma = 0.1$) and ($\beta = 0.9$, $\lambda_3 = 0.4$, $\gamma = 0.3$) from left to right. These shapes are similar to the shapes of elastica found by Love in 1926.

The case $\tau \neq 0$ is illustrated in Figure 4.4.4. We see that for $\tau \neq 0$ the rod deviates from a plane and has a three-dimensional structure. This structure is simpler for small values of τ and more complicated when τ increases.

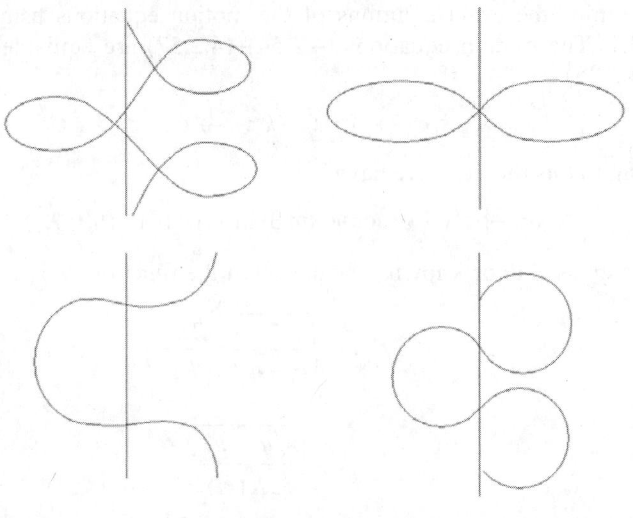

Figure 4.4.3 Shapes of *elastica* of Love for $\tau = 0$ $(k = 0)$.

In Figure 4.4.4, the following sets of parameters were considered from left to right ($\tau = 0.2$, $\beta = 0.3$, $\lambda_3 = 0.4, \gamma = 0.2$), ($\tau = 0.3$, $\beta = 0.7$, $\lambda_3 = 0.2, \gamma = 0.1$), ($\tau = 0.4$, $\beta = 0.3$, $\lambda_3 = 0.1, \gamma = 0.1$) and ($\tau = 0.5$, $\beta = 0.9$, $\lambda_3 = 0.4$, $\gamma = 0.3$).

In this case, the shape of the rod consists of a single loop or a series of loops lying altogether in space.

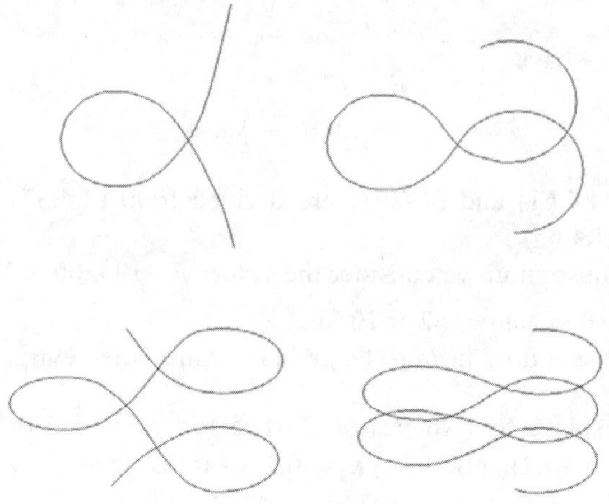

Figure 4.4.4 Four shapes of *elastica* for $\tau \neq 0$.

4.5 Exact solutions of the motion equations

We determine the exact solutions of the motion equations using the equivalence theorem 4.3.1. The motion equations (4.2.54)–(4.2.57) are equivalents to equilibrium equations (4.2.35)–(4.2.37) if

$$A - k_1 v^2 \to A, \quad C - k_2 v^2 \to C, \quad \xi \to s.$$

According to this theorem, we have

$$\lambda = -\rho v^2 (\sin\theta\cos\psi, \sin\theta\sin\psi, \cos\theta) + (0, 0, \lambda_3),$$

where λ_3 is supposed to be known. The period of the function u is given by (4.4.26)

$$2K(m)\sqrt{\frac{2}{u_1 - u_3}} = \frac{2\pi}{k}, \tag{4.5.1}$$

$$k = \frac{\rho\sqrt{u_1 - u_3}}{\sqrt{2}K(m)}. \tag{4.5.2}$$

The frequency of a cnoidal wave is given by

$$\omega = k\frac{b}{a} = -2k(u_1 + u_2 + u_3). \tag{4.5.3}$$

The theorems demonstrated in the static case are valid also in the dynamic case. We suppose λ, θ, ψ and φ are functions of variables $\xi = s - vt$. We have $\lambda = \zeta d_3 + (0, 0, \lambda_3)$, $\zeta = -\rho v^2$, $d_3 = (\sin\theta\cos\psi, \sin\theta\sin\psi, \cos\theta)$ and λ_3 a constant.

The case of three distinct and real roots.

THEOREM 4.5.1 *Given* $u_3 < u_2 < u_1$ *the distinct and real roots of the cubic equation* $f(u) = 0$, *the motion equations (4.2.54)–(4.2.57) have a unique solution for the Euler angles*

$$\begin{aligned} u &= u_2 - (u_2 - u_3)\mathrm{cn}^2(\sqrt{\frac{|\lambda_3|}{2A}(u_1 - u_3)}(\xi - \xi_3), m) = \\ &= u_2 - (u_2 - u_3)\mathrm{cn}^2[w(\xi - \xi_3), m], \end{aligned} \tag{4.5.4}$$

with $m = \dfrac{u_2 - u_3}{u_1 - u_3}$ *and* $w = \sqrt{\dfrac{|\lambda_3|}{2A}(u_1 - u_3)}$,

$$\begin{aligned} \psi = \frac{1}{4(A - k_1 v^2)^2 w^2}\{&\frac{\beta + (C - k_2 v^2)\tau}{1 - u_3}\Pi[w(\xi - \xi_3), \frac{u_2 - u_3}{1 - u_3}, m] - \\ &-\frac{\beta - (C - k_2 v^2)\tau}{1 + u_3}\Pi[w(\xi - \xi_3), \frac{u_2 - u_3}{1 + u_3}, m]\}, \end{aligned} \tag{4.5.5}$$

$$\begin{aligned} \varphi = -\frac{\tau[C - A - (k_2 + k_1)v^2]}{A - k_1 v^2}\xi + \frac{1}{4(A - k_1 v^2)^2 w^2}\{\frac{\beta + (C - k_2 v^2)\tau}{1 - u_3} \times \\ \times \Pi[w(\xi - \xi_3), \frac{u_2 - u_3}{1 - u_3}, m] - \frac{\beta - (C - k_2 v^2)\tau}{1 + u_3}\Pi(w(\xi - \xi_3), \frac{u_2 - u_3}{1 + u_3}, m)\}, \end{aligned} \tag{4.5.6}$$

where $\Pi(x, z, m)$ is the normal elliptic integral of the third kind.

THEOREM 4.5.2 *In conditions given by the theorem 4.5.1, the components of the position vector are*

$$x = R\sin(\psi + \Delta) = R\sin\Phi, \tag{4.5.7}$$

$$y = R\cos(\psi + \Delta) = R\cos\Phi, \tag{4.5.8}$$

$$z = u_1\xi - \frac{(u_1 - u_3)}{w}\mathrm{E}[\mathrm{am}(w(\xi - \xi_3)), m], \tag{4.5.9}$$

with

$$R = \frac{1}{\lambda_3}\sqrt{2|\lambda_3|(A - k_1 v^2)u + (C - k_2 v^2)^2\tau^2 + (A - k_1 v^2)\gamma + \beta^2}, \tag{4.5.10}$$

$$\Delta = -\arctan\left(\frac{A}{\beta u + C\tau}\sqrt{2f(u)}\right), \tag{4.5.11}$$

$$\Phi = \frac{\beta}{2A}\xi + C_0\Pi(w\xi, C_1, m), \tag{4.5.12}$$

where f(u) is given by (4.4.8), and C_0 , C_1 *are constants, and* $E(am(v),m)$ *is the elliptic integral of the first kind.*

Unknowns ξ_3, v, m, γ, λ_3, τ and β are determined from initial conditions (4.2.67) and boundary conditions (4.4.9).

The case of a simple root and a double root.

THEOREM 4.5.3 *Given* $u_1 = u_2 = -1$, $u_3 = 1$ *the roots of the cubic equation* $f(u) = 0$, *the Euler angles are uniquely determined from the motion equations (4.2.54)–(4.2.57)*

$$u(\xi) = -1 + 2\frac{|\lambda_3|}{A - k_1 v^2}\operatorname{sech}^2\sqrt{\frac{|\lambda_3|}{A - k_1 v^2}}\xi, \tag{4.5.13}$$

$$\psi = \frac{-(C - k_2 v^2)\tau\xi}{2(A - k_1 v^2)} + \arctan[\frac{4(A - k_1 v^2)}{(C - k_2 v^2)\tau}\tanh(-\sqrt{\frac{|\lambda_3|}{A - k_1 v^2}}\xi)], \tag{4.5.14}$$

$$\varphi = \frac{\tau[2A - C + v^2(k_2 - 2k_1)]\xi}{2(A - k_1 v^2)} + \arctan[\frac{4(A - k_1 v^2)}{(C - k_2 v^2)\tau}\tanh\sqrt{\frac{|\lambda_3|}{A - k_1 v^2}}\xi]. \tag{4.5.15}$$

THEOREM 4.5.4 *Given the conditions of the theorem 4.5.3 the components of the position vector are*

$$x = \frac{\sqrt{2|\lambda_3|(A - k_1 v^2)(u + 1)}}{\lambda_3}\sin\left(\frac{(C - k_2 v^2)\tau\xi}{2(A - k_1 v^2)}\right), \tag{4.5.16}$$

$$y = \frac{\sqrt{2|\lambda_3|(A - k_1 v^2)(u + 1)}}{\lambda_3}\cos\left(\frac{(C - k_2 v^2)\tau\xi}{2(A - k_1 v^2)}\right), \tag{4.5.17}$$

$$z = -\xi + \frac{4(A - k_1 v^2)}{\lambda_3}\tanh(2\xi). \tag{4.5.18}$$

Finally, let us remember that the equation (4.4.8) has the form

$$\frac{1}{2}u'^2 = f(u), \tag{4.5.19}$$

with

$$f(u) = -\frac{1}{A - k_1 v^2}[\lambda_3 u^3 - \frac{1}{2}(\gamma - \frac{(C - k_2 v^2)^2\tau^2}{A - k_1 v^2})u^2 - \lambda_3 u + \frac{1}{2}(\gamma - \frac{\beta^2}{A - k_1 v^2})]. \tag{4.5.20}$$

Chapter 5

VIBRATIONS OF THIN ELASTIC RODS

5.1 Scope of the chapter

The first treatment of the partial differential equations of wave motion was made by D'Alembert in 1750 in connection with the vibrations of thin rods. Daniel Bernoulli, Euler, Riccati, Poisson, Cauchy, Lord Rayleigh (1877) and also Strehlke (1833), Lissajous (1833) and Seebeck (1852) are foremost among those who have advanced knowledge in this problem. The vibrating rod has been a point of research in physics and mathematics, at last centuries up to the present day. Its current importance is associated with the theory of vibrating solitons.

In this chapter the vibrations of thin elastic rods are studied. In section 5.2 the small transverse and torsional vibrations of the initial deformed rod are considered. The large relative displacements can occur even when the strains are small. The nonlinear effect caused by the geometrical constraint yields to soliton solutions. The transverse vibrations of a helical rod are presented in section 5.3. The rod vibrates in space by bending and twisting. The vibrations are studied around the strained position of the rod which satisfies the static equilibrium equations given by the theorem 4.2.1.

In section 5.4 it is shown that for a special class of media that do not remember the interaction process (DRIP media), the waves admit the solitonic features, but in contrast to solitons, the waves distort as they propagate by an amount that is not altered by the interaction. The only effect of the interaction is to alter the arrival time of their fronts at any point. The vibrations of a heterogeneous string are considered in the last section.

Before proceeding, we mention the material we referred in this chapter: Love (1926), Landau and Lifshitz (1968), Lewis (1980), Synge (1981), Seymour and Varley (1982), Tsuru (1986, 1987), Munteanu and Donescu (2002).

5.2 Linear and nonlinear vibrations

Consider the transverse vibrations of a rod confined on a plane with no torsional displacements. In this case the Eulerian angles φ and ψ are zero in the motion equations (4.2.54)–(4.2.57). So, we have

$$-\rho\ddot{r} - \lambda' = 0\,, \tag{5.2.1}$$

$$-k_1\ddot{\theta} + A\theta'' - \lambda_1\cos\theta + \lambda_3\sin\theta = 0\,, \tag{5.2.2}$$

$$\lambda_2 \sin\theta = 0 \,. \tag{5.2.3}$$

The unit tangential vector along the rod is given by

$$d_3 = (\sin\theta, \quad 0, \quad \cos\theta) \,. \tag{5.2.4}$$

Substitution of (5.2.4) into (4.2.66) gives

$$\rho\dot{\theta}^2 \sin\theta - \rho\ddot{\theta}\cos\theta - \lambda_1'' = 0 \,, \tag{5.2.5}$$

$$\rho\dot{\theta}^2 \cos\theta + \rho\ddot{\theta}\sin\theta - \lambda_3'' = 0 \,. \tag{5.2.6}$$

Equation (5.2.2) becomes

$$-k_1\ddot{\theta} + A\theta'' - \lambda_1 \cos\theta + \lambda_3 \sin\theta = 0 \,. \tag{5.2.7}$$

For $\theta = \varepsilon << 1$, and neglecting terms of the order ε^2, equations (5.2.5)–(5.2.7) can be written as (Tsuru)

$$-\rho\ddot{\varepsilon} - \lambda_1'' = 0 \,, \tag{5.2.8}$$

$$\lambda_3'' = 0 \,, \tag{5.2.9}$$

$$\lambda_1 = -k_1\ddot{\varepsilon} + A\varepsilon'' \,. \tag{5.2.10}$$

Substitution (5.2.10) into (5.2.8) leads to the transverse vibrational motion

$$\rho\ddot{\varepsilon} - k_1\ddot{\varepsilon}'' + A\varepsilon'''' = 0 \,. \tag{5.2.11}$$

A wave solution of this equation is given by

$$\varepsilon = \varepsilon_0 \exp[\mathrm{i}(ks - \omega t)] \,, \tag{5.2.12}$$

with ε_0 a constant. Substituting (5.2.12) into (5.2.11) we have

$$\omega^2 = \frac{Ak^4}{\rho + k_1 k^2} \,. \tag{5.2.13}$$

The group velocity v_g of this wave is

$$v_g = \frac{\mathrm{d}\omega}{\mathrm{d}k} = \sqrt{\frac{A}{\rho}} \frac{k(2+\beta k^2)}{(1+\beta k^2)^{3/2}} \,, \quad \beta = \frac{k_1}{\rho} \,. \tag{5.2.14}$$

Let us consider now the torsional motion characterized by $\theta = \psi = 0$ (Zachmann). The motion equations (4.2.54)–(4.2.57) yields the known wave equation of motion for small linear vibrations

$$-k_2\ddot{\varphi} + C\varphi'' = 0 \,, \tag{5.2.15}$$

that admits solutions of the form

$$\varphi = \varphi_0 \exp[\mathrm{i}(ks - \omega t)] \,, \tag{5.2.16}$$

and the dispersion relation

$$\omega = \sqrt{\frac{C}{k_2}} k = \sqrt{\frac{\mu}{\rho_0}} k . \qquad (5.2.17)$$

For large displacements, even when the strains are small, the nonlinear terms in the motion equations (4.2.54)–(4.2.57) cannot be neglected. In this case we must consider the geometric constraints expressed by $r' = d_3, |d_3| = 1$.

Consider firstly the plane motion of the rod, in the absence of torsion ($\psi = \varphi = 0$).

The motion equations are given by (5.2.5)–(5.2.7). We may assume that the functions $\theta(s,t)$ and $\lambda(s,t)$ are functions of the variable $\xi = s - vt$.

Denoting with prime the differentiation with respect to ξ, the motion equations (5.2.5) –(5.2.7) become

$$\rho v^2 \theta'^2 \sin\theta - \rho v^2 \theta'' \cos\theta - \lambda_1'' = 0 , \qquad (5.2.18)$$

$$\rho v^2 \theta'^2 \cos\theta + \rho v^2 \theta'' \sin\theta - \lambda_3'' = 0 , \qquad (5.2.19)$$

$$(-k_1 v^2 + A)\theta'' - \lambda_1 \cos\theta + \lambda_3 \sin\theta = 0 . \qquad (5.2.20)$$

Note that the equations (5.2.18) and (5.2.19) can be written under the form

$$\rho v^2 (\theta' \cos\theta)' + \lambda_1'' = 0 ,$$

$$\rho v^2 (\theta' \sin\theta)' - \lambda_3'' = 0 .$$

Integrating twice with respect to ξ, the above equations are

$$\lambda_1 = -\rho v^2 \sin\theta + \gamma_1 \xi + \delta_1 , \qquad (5.2.21)$$

$$\lambda_3 = \rho v^2 \cos\theta + \gamma_3 \xi + \delta_3 , \qquad (5.2.22)$$

where $\gamma_1, \gamma_3, \delta_1$ and δ_3 are constants.

By imposing the boundary conditions

$$\theta' = 0, \quad \dot{r} = 0 \text{ as } s \to \pm\infty , \; t \to \infty , \qquad (5.2.23)$$

it results $\gamma_1 = \gamma_3 = 0$.

Inserting (5.2.21) and (5.2.22) into (5.2.20) we have

$$(A - k_1 v^2)\theta'' - \delta_1 \cos\theta + \delta_3 \sin\theta = 0 . \qquad (5.2.24)$$

Defining

$$\delta_1 = \sqrt{\delta_1^2 + \delta_3^2} \sin\alpha , \; \delta_3 = \sqrt{\delta_1^2 + \delta_3^2} \cos\alpha ,$$

the equation (5.2.24) becomes the known simple pendulum equation

$$\frac{\mathrm{d}^2\theta}{\mathrm{d}\xi^2} + \omega_0^2 \sin\theta = 0 , \qquad (5.2.25)$$

where we denoted $\theta \to \theta + \alpha$, and

$$\omega_0^2 = \frac{\sqrt{\delta_1^2 + \delta_3^2}}{A - k_1 v^2}. \tag{5.2.26}$$

The exact solution of (5.2.25) is expressed in terms of elliptic or hyperbolic functions, depending on the initial conditions. For the conditions (5.2.23), the solution of (5.2.25) is

$$\sin\frac{\theta}{2} = \tanh(\omega_0 \xi). \tag{5.2.27}$$

From (5.2.27) it results

$$\cos\frac{\theta}{2} = \sqrt{1 - \tanh^2(\omega_0 \xi)} = \operatorname{sech}(\omega_0 \xi). \tag{5.2.28}$$

Thus, with the aid of the formulae

$$\sin\theta = 2\tanh(\omega_0\xi)\operatorname{sech}(\omega_0\xi),\ \cos\theta = \operatorname{sech}^2(\omega_0\xi) - \tanh^2(\omega_0\xi), \tag{5.2.29}$$

the equations (5.2.21) and (5.2.22) become

$$\lambda_1 = -2\rho v^2 \tanh(\omega_0\xi)\operatorname{sech}(\omega_0\xi) + \delta_1,$$

$$\lambda_3 = -2\rho v^2 [\operatorname{sech}^2(\omega_0\xi) - \tanh^2(\omega_0\xi)] + \delta_3. \tag{5.2.30}$$

When the ends of the bar are fixed, the equation (5.2.1) yields

$$\begin{aligned} \rho\ddot{x} + \lambda_1' &= 0, \\ \rho\ddot{z} + \lambda_3' &= 0. \end{aligned} \tag{5.2.31}$$

From (5.2.31) and (5.2.30) we obtain the following solutions

$$x = \frac{2}{\omega_0}\operatorname{sech}[\omega_0(s - vt)], \tag{5.2.32}$$

$$z = -s + \frac{2}{\omega_0}\tanh[\omega_0(s - vt)]. \tag{5.2.33}$$

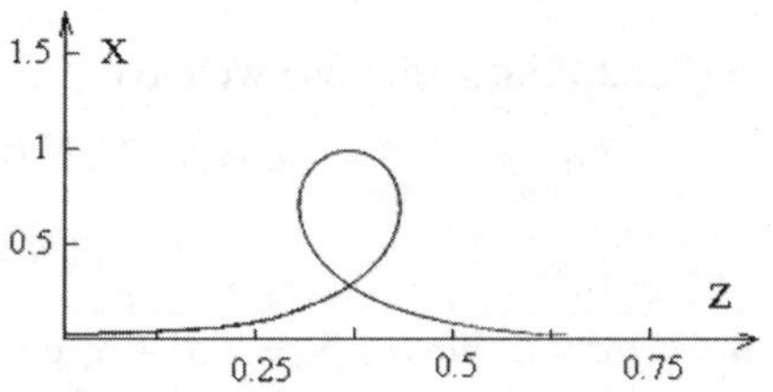

Figure 5.2.1 Shape of the elastica.

The solution (5.2.32) represents a soliton. The solution (5.2.33) is a kink and represents a twist in the variable z. The shape of elastica, into which the central line is deformed, is a single loop and it is represented in Figure 5.2.1 ($\omega_0 = 8$). The shape of the kink (5.2.33) is represented in Figure 5.2.2.

Since the waves can move in the rod in both directions, a head-on collision is possible. Konno and Ito have studied nonlinear interactions between solitons for KdV and Boussinesq equations. In 1962, Perring and Skyrme investigate the head-on collision of two kinks with equal but opposite velocity. The kink traveling to the left is called an anti-kink. Using the same definition, an anti-soliton is a soliton traveling to the left. It is interesting to investigate the behavior of waves during their mutual interaction.

To understand the effect of the collision between waves into the motion of the rod, let us consider next two initial conditions for the equations (5.2.31)

$$x_1 = \frac{2}{\omega_0}\operatorname{sech}[\omega_0(s - v_1 t)],$$

$$z_1 = -s + \frac{2}{\omega_0}\tanh[\omega_0(s - v_1 t)], \tag{5.2.34}$$

$$x_2 = \frac{2}{\omega_0}\operatorname{sech}[\omega_0(-s - v_2 t + \delta)],$$

$$z_2 = s + \Delta + \frac{2}{\omega_0}\tanh[\omega_0(-s - v_2 t + \delta)],$$

that represent two waves traveling with the velocities $v_1 > v_2$, in opposite directions.

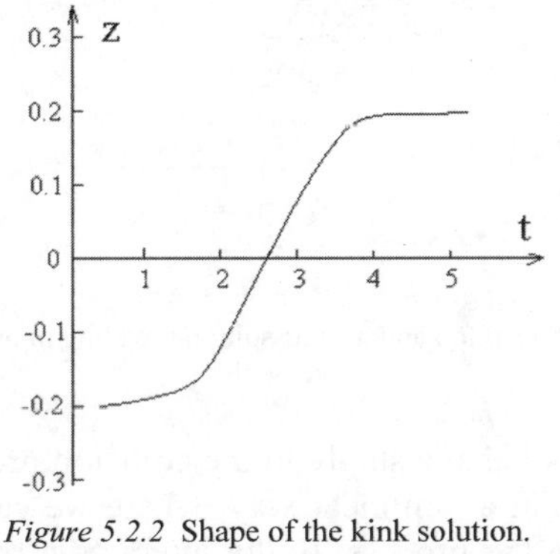

Figure 5.2.2 Shape of the kink solution.

The result of integrating the equation $(5.2.34)_1$ is shown in Figure 5.2.3, and the result of integrating the equation $(5.2.34)_2$ in Figure 5.2.4, for $v_1 = 0.5$, $v_2 = 0.35$, $\Delta =$ 0.75, $\delta = 1.75$ and $\omega_0 = 2.44$.

The two solitons are separated before the collision, then coalesce in the interaction zone and separate again afterwards, without change of velocity and shape, but with a small phase shift. In the collision zone there is no linear superposition. In contrast to the elastic soliton collision, the collision of kinks is inelastic because after collision, a small amount of energy is left in the form of oscillations.

The two kinks keep in some way their structure, but the oscillations indicate a dissipation of energy. After collision the velocities of kinks are $v_1' = 0.46$ and $v_2' = 0.33$. The coalescing of two kinks in the interaction zone (s,t) is represented in Figure 5.2.5 for two cases: $v_1 = 0.5$, $v_2 = 0.35$, and $v_1 = v_2 = 1$. In the second case, after collision the kinks velocities are $v_1' = v_2' = 0.87$.

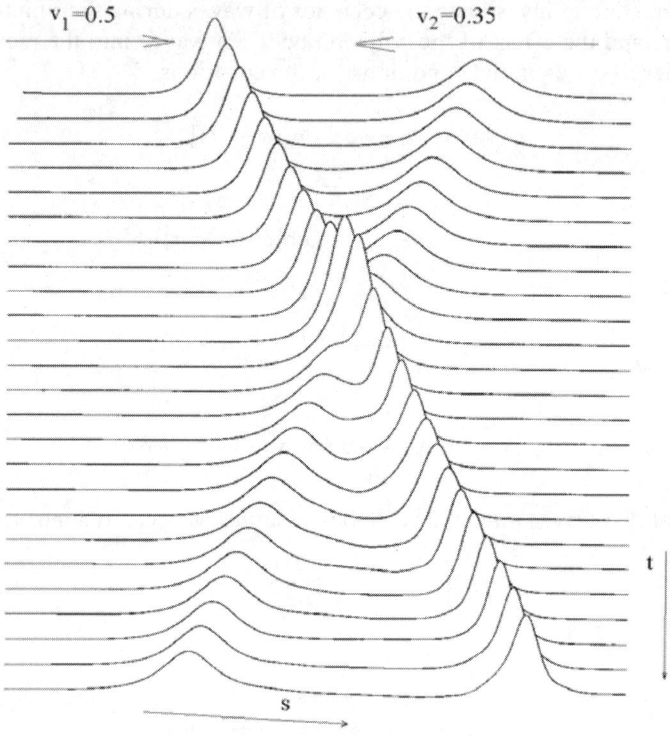

Figure 5.2.3 Collision of a soliton and an anti-soliton traveling in opposite directions with different velocities.

The solitons and kinks behave stably in the collision process even if the interaction between them takes place in a nonlinear way, where we cannot apply the principle of the linear superposition to the process. In the interaction region the kinks behave in a complicated way. A projection of the collision zone on the plane (s,t) can be seen in Figure 5.2.5.

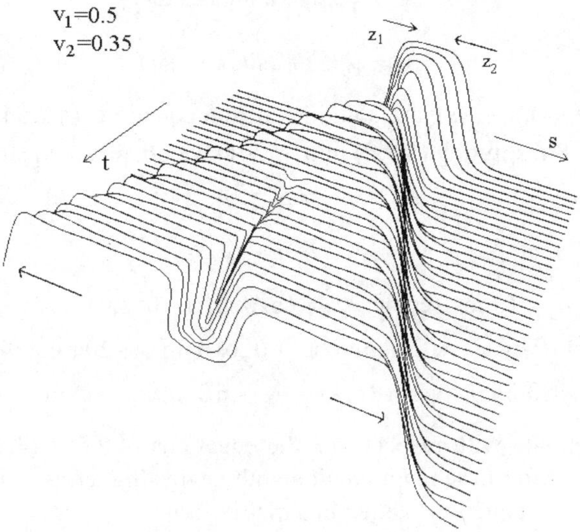

Figure 5.2.4 Collision of a kink and anti-kink traveling in opposite directions with different velocities.

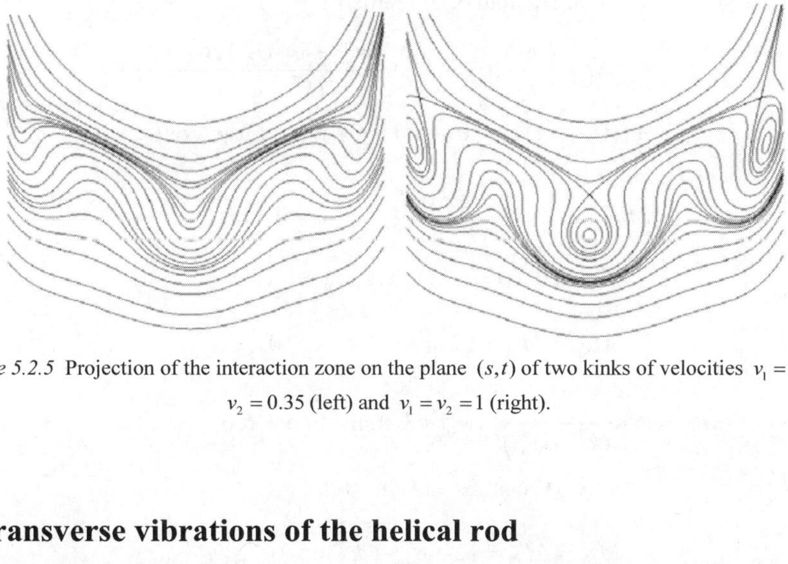

Figure 5.2.5 Projection of the interaction zone on the plane (s,t) of two kinks of velocities $v_1 = 0.5$, $v_2 = 0.35$ (left) and $v_1 = v_2 = 1$ (right).

5.3 Transverse vibrations of the helical rod

Let us consider a rod that vibrates around the strained position, which satisfies the static equilibrium equations given by the theorem 4.2.1 for $\lambda = (0,0,\lambda)$.

We begin with differentiating the equation (4.2.54) with respect to *s* (Tsuru)

$$-\rho\ddot{d}_3 - \lambda'' = 0\,. \tag{5.3.1}$$

Suppose that the strained rod has a helical form. The Euler angles are written as

$$\theta = \theta_0(s) + \varepsilon\cos(kx - \omega t)\,,$$

$$\psi = \psi_0(s) + \varepsilon \sin(ks - \omega t)\,, \tag{5.3.2}$$

$$\varphi = \varphi_0(s) + \varepsilon \sin(kx - \omega t)\,,$$

where $\theta_0(s)$, $\psi_0(s)$, $\varphi_0(s)$ are solutions of the equations (4.2.54)–(4.2.57), and ε a small parameter. Substituting (5.3.2) into (5.3.1) we obtain an equation in λ

$$\lambda'' = -\rho \ddot{d}_3\,, \tag{5.3.3}$$

with

$$\begin{gathered}\ddot{d}_3 = (\ddot{\theta}\cos\theta\cos\psi - 2\dot{\theta}\dot{\psi}\cos\theta\sin\psi - \dot{\theta}^2\sin\theta\cos\psi - \\ -\ddot{\psi}\sin\theta\sin\psi - \dot{\psi}^2\sin\theta\cos\psi,\quad \ddot{\theta}\cos\theta\sin\psi + 2\dot{\theta}\dot{\psi}\cos\theta\cos\psi - \\ -\dot{\theta}^2\sin\theta\sin\psi + \ddot{\psi}\sin\theta\cos\psi - \dot{\psi}^2\sin\theta\sin\psi,\quad -\ddot{\theta}\sin\theta - \dot{\theta}^2\cos\theta).\end{gathered} \tag{5.3.4}$$

Then, differentiate with respect to s the equations (4.2.55)–(4.2.57) and insert λ'' given by (5.3.3). After little manipulation with neglecting terms of third order in ε, we obtain the vibrations equation written in a matrix form

$$M\varepsilon = 0\,, \tag{5.3.5}$$

where M is a 3×3 symmetric matrix of elements

$$\begin{gathered}M_{11} = \frac{\omega^2}{k^2}\{\rho + k_1 k^2 - \frac{(\psi_0'^4 - 3k^2\psi_0'^2)\cos^2\theta_0}{(k^2 - \psi_0'^2)^2}\} + \\ +\{(C-2A)\cos^2\theta_0 + (A-C)\}\psi_0'^2 + C\tau\psi_0\cos\theta_0 - Ak^2,\end{gathered} \tag{5.3.6}$$

$$\begin{gathered}M_{12} = M_{21} = -\frac{2k\rho\omega^2}{(k^2 - \psi_0'^2)^2}\psi_0'\cos\theta_0\sin\theta_0 + \\ +k\sin\theta_0\{(C-2A)\cos\theta_0\psi_0' + C\tau,\end{gathered} \tag{5.3.7}$$

$$M_{13} = M_{31} = (2k_1\omega^2 - Ck^2)\cos\theta_0\,, \tag{5.3.8}$$

$$\begin{gathered}M_{22} = \omega^2\{\frac{\rho}{(k^2 - \psi_0'^2)^2}(\psi_0'^2 + k^2)\sin^2\theta_0 + k_1(\cos^2\theta_0 + 1)\} - \\ -k^2(C\cos^2\theta_0 + A\sin^2\theta_0),\end{gathered} \tag{5.3.9}$$

$$M_{23} = M_{32} = (2k_1\omega^2 - Ck^2)\cos\theta_0\,, \tag{5.3.10}$$

$$M_{33} = 2k_1\omega^2 - Ck^2\,, \tag{5.3.11}$$

and ε a column vector

$$\varepsilon = (\varepsilon, \varepsilon, \varepsilon)'\,. \tag{5.3.12}$$

In (5.3.6)–(5.3.11) we have used the relations

$$k_2 = 2k_1\,,\quad \varphi_0' = \tau - \psi_0'\cos\theta_0\,. \tag{5.3.13}$$

The dispersion relations are calculated from

$$\det M = 0\,, \tag{5.3.14}$$

where the determinant of M is a cubic polynomial of ω^2.

Consider the case of transverse vibrations. The characteristic equation is given by (Tsuru)

$$\rho^2[(1-u_0^2)\psi_0'^2 + k^2(1+k_1)]\omega^4 - \rho k^2[Au_0^2\psi_0'^2(\psi_0'^2 + 13k^2 - k_1(u_0^2+1)) -$$

$$-C\tau u_0\psi_0'(\psi_0'^2 + 5k^2) + 2A(k^4 - k^2\psi_0'^2 + k_1k^2 - 2k_1u_0]\omega^2 -$$

$$-k^4(\psi_0'^2 - k^2)[A^2u_0^2\psi_0'^2 - 3AC\tau u_0\psi_0' + A(\psi_0'^2 - k^2 + 2k_1u_0) + C^2\tau^2] = 0, \tag{5.3.15}$$

where $u_0 = \cos\theta_0 \neq \pm 1$.

The equation (5.3.15) is a cubic polynomial in ω^2. We note with $(\omega^+)^2$ and $(\omega^-)^2$ the roots of (5.3.15) which are functions of *k*. Numerical investigations show that for $\psi_0'^2 < k^2$, the roots $(\omega^+)^2$ and $(\omega^-)^2$ are positive, and for $\psi_0'^2 \geq k^2$ the root $(\omega^+)^2$ is positive and $(\omega^-)^2$ is negative.

The waves are stable for real values of the angular frequency ω. Next, we consider only the case $\psi_0'^2 < k^2$.

The initial strain is determined by the parameters u_0, ψ_0' and τ. We represent graphically the variation of $(\omega^+)^2$ and $(\omega^-)^2$ with respect to *k*, for different values of u_0, ψ_0' and τ. These are the dispersion curves, calculated for A =2.4404 Nm^2, C =1.9066 Nm^2 and ρ =7876 kg/ m^3.

Let us choose u_0 =0.3 and $\psi_0' \in [0,1,2,3]$. Figure 5.3.1 corresponds to $\tau = 0$, and Figure 5.3.2, to $\tau = 0.5$. Then we take $\tau \in [0,1,2,3]$ and $\psi_0' = 1$. Figure 5.3.3 corresponds to u_0 =0.3, and Figure 5.3.4 to $u_0 = 0.7$. From these figures we observe that the dispersion relations depend strongly on all parameters.

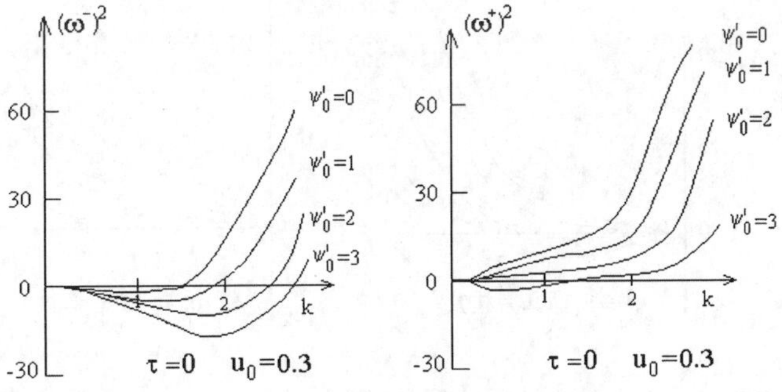

Figure 5.3.1 Dispersion relations of the transverse vibrations of a helical rod for different values of ψ_0'.

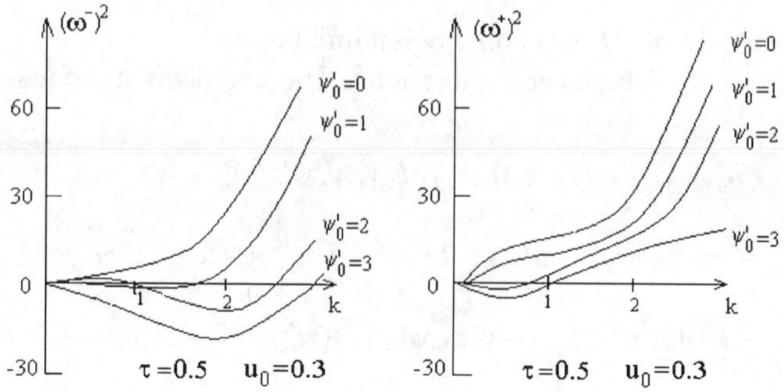

Figure 5.3.2 Dispersion relations of the transverse vibrations of a helical rod for different values of ψ'_0.

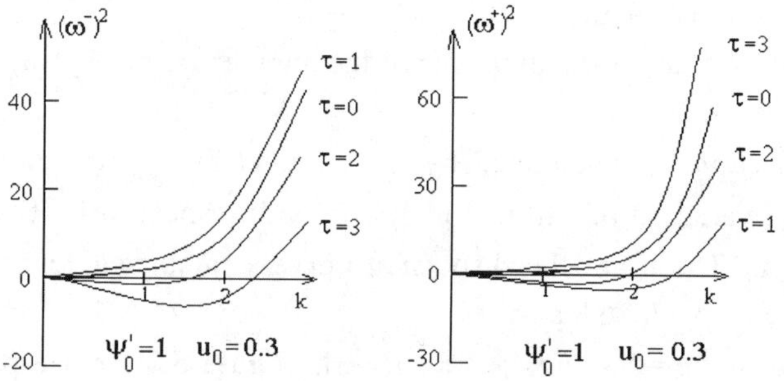

Figure 5.3.3 Dispersion relations of the transverse vibrations of a helical rod for different values of τ

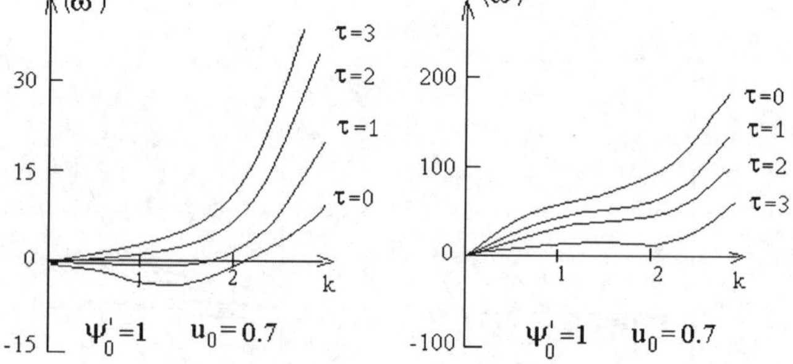

Figure 5.3.4 Dispersion relations of the transverse vibrations of a helical rod for different values of τ

Finally, we represent in Figs 5.3.5 and 5.3.6 the shape of the vibrating rod at $t = 35$ and $t = 124$, for $\psi_0' = 2$, $\tau = 0.5$, $u_0 = 0.2$, $\lambda_3 = 0.4$. The initial value of φ_0' is calculated from (5.3.13).

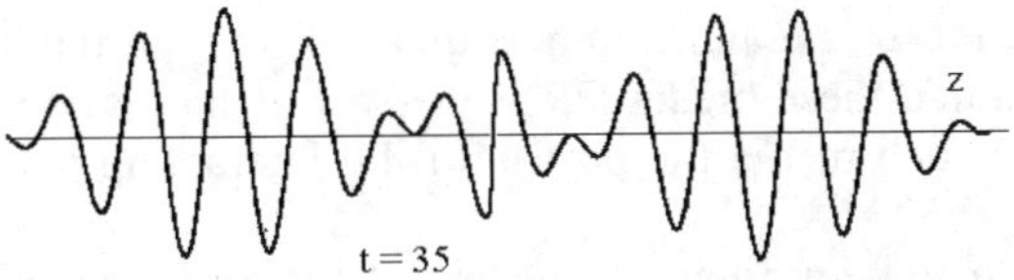

Figure 5.3.5 Shape of the vibrating rod at $t = 35$.

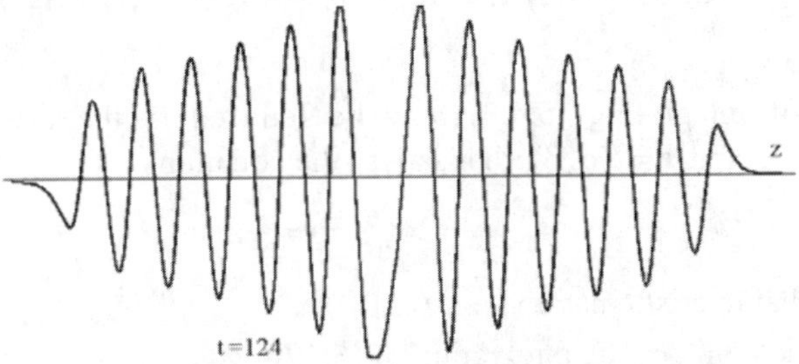

Figure 5.3.6 Shape of the vibrating rod at $t = 124$.

5.4 A special class of DRIP media

Fermi, Pasta and Ulam studied in 1955 the oscillations of a heterogeneous string which is governed by nonlinear wave equations of the form

$$y_{tt} = A^2(y_x, y_t) y_{xx}, \tag{5.4.1}$$

$$\frac{\mathrm{d}A}{\mathrm{d}y_x} = A^{3/2}(\mu + \nu A), \tag{5.4.2}$$

where $y(x,t)$ is the physical displacement, $A(y_x, y_t)$ a positive function representing the local speed of propagation, and μ, ν the material constants, and x ranges from $-\infty$ to $+\infty$. In particular, when $A = A(x)$, the equation (5.4.1) is applied to all physical systems essentially involving only one space-dimension and onetime-dimension. For the transverse vibrations of a string, we have $A^2 = T/m$ where T is the constant tension and m the mass per unit length function of x, for the compressional vibrations of an isotropic laminated elastic solid in which the density and elastic constants are functions of x only, $A^2 = (\lambda + 2\mu)/\rho$, and for the transverse vibrations of a laminated solid, $A^2 = \mu/\rho$.

In 1982 Seymour and Varley studied the equation (5.4.1) and have shown that, for certain functions $A(y_x)$ that satisfies (5.4.2), the solutions of (5.4.1) are simple waves whose profiles can be specified arbitrarily, that have the properties: they interact between themselves like solitons, being unaffected by the interaction, but in contrast to solitons, distort as they propagate by an amount that is not affected by the interaction.

Since the interaction phenomenon for waves of arbitrary shape and amplitude is a property of the transmitting medium rather than of the particular wave profiles, Seymour and Varley named these media DRIP media (media that do not remember the interaction process). In such media the profile of the interacting waves is not affected by such interactions.

The equation (5.4.1) can be written as a system of first order equations

$$u_t = A^2(e,u)e_x\,,\ u_x = e_t\,, \tag{5.4.3}$$

or under the form

$$u_t + Au_x = A(e_t + Ae_x)\,,\ u_t - Au_x = -A(e_t - Ae_x)\,, \tag{5.4.4}$$

where $u = y_t$ and $e = y_x$.

The problem of integrating (5.4.3) may be reduced to the determining of the functions $x = x(\alpha,\beta)$ and $t = t(\alpha,\beta)$ that satisfy the equations

$$x_\beta = At_\beta\,,\ x_\alpha = -At_\alpha\,, \tag{5.4.5}$$

where (α,β) are characteristic parameters (α,β).

When A depends only on e, integration of (5.4.6) yields

$$u = G(\beta) - F(\alpha)\,,\ c(e) = G(\beta) + F(\alpha)\,,\ c(e) = \int_0^e A(s)\mathrm{d}s\,. \tag{5.4.6}$$

Consequently, (5.4.5) must be regarded as

$$x_\beta = A(c)t_\beta\,,\ x_\alpha = -A(c)t_\alpha\,, \tag{5.4.7}$$

where A is a function of $c = G(\beta) + F(\alpha)$. The equation (5.4.4) becomes

$$u_\beta = Ae_\beta\,,\ u_\alpha = -Ae_\alpha\,. \tag{5.4.8}$$

The equations are very difficult to be integrated, but there are some exceptions when either $G = 0$ or $F = 0$. For $G = 0$, relations (5.4.6) and (5.4.7) yields

$$c = -u = F(\alpha)\,,\ x - A(F)t = \alpha\,, \tag{5.4.9}$$

that describe a right-traveling simple wave. For $F = 0$, relations (5.4.6) and (5.4.7) imply that

$$c = u = G(\beta)\,,\ x + A(G)t = \beta\,, \tag{5.4.10}$$

that describe a left-traveling simple wave. Both solutions represent waves moving with the velocity $A(c)$ into a uniform region where u and c are constant.

Let us consider now the class of DRIP media, that include a large variety of elastic–plastic materials, composite materials, gases, foams, and so on.

DEFINITION 5.4.1 (Seymour and Varley) *A DRIP medium is a nondispersive medium that transmits waves that do not remember the interaction process. The waves are of arbitrary shape and amplitude and distort in time but by an amount that is not affected by the interactions. A DRIP medium is defined by the condition that* $A(c)$ *satisfies an equation of state of the form*

$$\frac{dA}{dc} = \mu A^{1/2} + \nu A^{3/2}, \tag{5.4.11}$$

where μ, ν *are material constants.*

The equation of state (5.4.11) contains two arbitrary parameters that is an advantage to be used for a model of a wide class of different media.

The interaction of this type occurs when two solitons collide. The difference is that solitons are represented by waves of permanent form whose profiles are specific.

We will show that for a DRIP medium the equations (5.4.7) can be integrated to obtain a simple representation for $x(\alpha,\beta)$ and $t(\alpha,\beta)$.

THEOREM 5.4.1 (Seymour and Varley) *The general solutions of (5.4.7) are given by*

$$x(\alpha,\beta) = [l(\beta) - r(\alpha)]A^{1/2} + \mu(R(\alpha) - L(\beta)),$$

$$t(\alpha,\beta) = [l(\beta) + r(\alpha)]A^{-1/2} + \nu(R(\alpha) + L(\beta)), \tag{5.4.12}$$

where

$$L'(\beta) = lG'(\beta), \quad R'(\beta) = rF'(\alpha), \tag{5.4.13}$$

and $A(\alpha,\beta)$ *is determined from* $c - G(\beta) + F(\alpha)$ *and (5.4.11), and the functions* F, G, R, L, r, l *are determined from initial and boundary conditions.*

Proof. The proof of this theorem belongs to Seymour and Varley. Return to (5.4.7) and eliminate x. Thus, $t(\alpha,\beta)$ will satisfy the equation

$$\frac{\partial}{\partial\alpha}(At_\beta) + \frac{\partial}{\partial\beta}(At_\alpha) = 0, \tag{5.4.14}$$

or, if we take into consideration that $c = G(\beta) + F(\alpha)$

$$\frac{\partial}{\partial\alpha}(2A^{1/2}t_\beta) + (A'(c)A^{-1/2}t_\alpha)G'(\beta) = 0. \tag{5.4.15}$$

This equation can be easily integrated with respect to α. Using (5.4.11), integration of (5.4.15) yields

$$2A^{1/2}t_\beta + (\mu t - \nu x)G'(\beta) = g'(\beta), \tag{5.4.16}$$

where g is an unspecified function. From (5.4.16) and $(5.4.7)_1$ we obtain

$$\frac{\partial}{\partial\beta}\left[\frac{A^{1/2}}{\mu + \nu A}(\mu t - \nu x)\right] - \frac{1}{2}\frac{\mu - \nu A}{\mu + \nu A}g'(\beta) = 0. \tag{5.4.17}$$

In a similar way

$$\frac{\partial}{\partial\alpha}\left[\frac{A^{1/2}}{\mu+\nu A}(\mu t+\nu x)\right]-\frac{1}{2}\frac{\mu-\nu A}{\mu+\nu A}f'(\alpha)=0\,, \tag{5.4.18}$$

where f is an unspecified function. Relations (5.4.17) and (5.4.18) leads to

$$\mu t-\nu x=\frac{1}{f'(\alpha)}\frac{\mu+\nu A}{A^{1/2}}\frac{\partial\phi}{\partial\alpha}\,,\ \mu t+\nu x=\frac{1}{g'(\beta)}\frac{\mu+\nu A}{A^{1/2}}\frac{\partial\phi}{\partial\beta}\,, \tag{5.4.19}$$

where

$$\frac{\partial^2\phi}{\partial\alpha\partial\beta}=\frac{1}{2}\omega(c)f'(\alpha)g'(\beta)\,,\ \omega(c)=\frac{\mu-\nu A(c)}{\mu+\nu A(c)}\,. \tag{5.4.20}$$

On the basis of (5.4.11) it results that the function $\omega(c)$ defined above verifies the equation

$$\omega''(c)+\mu\nu\omega(c)=0\,. \tag{5.4.21}$$

Denoting $q=\dfrac{1}{f'(\alpha)}\dfrac{\partial\phi}{\partial\alpha}$ and $p=\dfrac{1}{g'(\beta)}\dfrac{\partial\phi}{\partial\beta}$, we obtain from (5.4.20)

$$\frac{\partial q}{\partial G}=\frac{1}{2}\omega(c)g'(G)\,,\ \frac{\partial p}{\partial F}=\frac{1}{2}\omega(c)f'(F)\,, \tag{5.4.22}$$

where it is convenient to regard (q,p) as functions of (F,G).

Integrating (5.4.22) we obtain the solutions

$$q=L'(G)\omega(c)-L(G)\omega'(c)+r(F)\,,$$

$$p=R'(F)\omega(c)-R(F)\omega'(c)+l(G)\,, \tag{5.4.23}$$

where

$$L''+\mu\nu L=\frac{1}{2}g'\,,\ R''+\mu\nu R=\frac{1}{2}f'\,. \tag{5.4.24}$$

In order that (x,t) given by (5.4.19) and (5.4.23) to satisfy the equations (5.4.7) it results that the functions l and r must be given by

$$l(G)=L'(G)\,,r(F)=R'(F)\,. \tag{5.4.25}$$

From (5.4.19), (5.4.20) and (5.4.23) it follows the solutions (5.4.12). The relation (5.4.13) is obtained from (5.4.25).

When A is independent of c, $A=A_L$, we see that (5.4.7) can be integrated

$$x-A_Lt=\alpha\,,\ x+A_Lt=\beta\,, \tag{5.4.26}$$

and, therefore, u and c are given by

$$u = G(x + A_L t) - F(x - A_L t),$$

$$c = G(x + A_L t) + F(x - A_L t). \qquad (5.4.27)$$

These representations of the linear theory are interpreted as the superposition of two nondistorting and noninteracting waves that move in opposite directions with speed A_L.

These formulae can be obtained from (5.4.12) and (5.4.13) by taking $A(c) = -\frac{\mu}{\nu} = A_L$.

Now, we attach to (5.4.7) the initial conditions

$$u(x,0) = u_0(x),\ c(x,0) = c_0(x),\ -\infty < x < \infty. \qquad (5.4.28)$$

Normalizing (α, β) so that at $t = 0$ to have $\alpha = x$ and $\beta = x$, it follows from (5.4.4)

$$F(x) = \frac{1}{2}[c_0(x) - u_0(x)],\ G(x) = \frac{1}{2}[c_0(x) + u_0(x)]. \qquad (5.4.29)$$

From (5.4.12) we obtain the solutions

$$x = [l(x) - r(x)]A_0^{1/2}(x) + \mu(R(x) - L(x)),$$

$$0 = [l(x) + r(x)]A_0^{-1/2}(x) + \nu(R(x) + L(x)), \qquad (5.4.30)$$

where $A_0(x)$ is determined from $c_0(x)$ by (5.4.11). The equations (5.4.30) and (5.4.13) and (5.4.29) determine the functions R, L, r, l.

5.5 Interaction of waves

We present the Seymour–Varley method for analysis of the interaction of any two waves that are traveling in opposite directions in DRIP media. The waves meet and interact and then emerge from the interaction region unchanged by the interaction. Following this method, we consider an interaction region in the plane (x,t) composed of five regions in which $u \neq 0$ and $c \neq 0$, as shown in Figure 5.5.1. At $t = 0$ the right-traveling wave (RW) occupies the region $-x_l \le x \le -x_f$, $x_l, x_f > 0$, and the left-traveling wave (LW) the region $0 < x_f \le x \le x_r$. The waves move into a region where $u = c = 0$. Suppose $u_0(x)$ and $c_0(x)$ are continuous. Suppose that at $t = 0$ we have $F(x) = f(x)$ for $-x_l \le x \le -x_f$, and zero otherwise, and $G(x) = g(x)$ for $x_f \le x \le x_r$, and zero otherwise. In region $\mathrm{I_R}$ we have $G = 0$, $F \neq 0$, $L = 0$; in $\mathrm{I_L}$, $F = 0$, $G \neq 0$, $R = 0$; in region II, $F \neq 0$, $G \neq 0$; in region $\mathrm{III_R}$, $G = 0$, $F \neq 0$, $L = L_R$ and in region $\mathrm{III_L}$, $F = 0$, $G \neq 0$, $R = R_L$.

We have also $f(-x_l) = f(-x_f) = g(x_f) = g(x_r) = 0$. The function $A(c)$ is supposed to be defined as

$$A(c) = \tilde{A}(x) = \begin{cases} A(f(x)), & \text{for } -x_l \leq x \leq -x_f, \\ A(g(x)), & \text{for } \quad x_f \leq x \leq x_r, \\ A(0) = A_0, & \qquad \text{otherwise.} \end{cases} \tag{5.5.1}$$

We also see that for $|x| \leq x_f$ we have

$$R = L = 0\,,\; l = r = \frac{x}{2A_0^{1/2}}\,, \tag{5.5.2}$$

and for $-x_l \leq x \leq -x_f$

$$L = 0\,,\; R = \tilde{A}^{1/2}(x)\frac{I(x) - I(-x_f)}{\mu + \nu\tilde{A}(x)}\,,$$

$$l = \frac{1}{2}[I(-x_f) - I(x) + \frac{x}{\tilde{A}^{1/2}(x)}]\,,$$

$$r = \frac{(\mu - \nu\tilde{A}(x))R(x) - x}{2\tilde{A}^{1/2}(x)}\,, \tag{5.5.3}$$

where

$$I(x) = \frac{x}{\tilde{A}^{1/2}(x)} - \int_0^x \frac{\mathrm{d}s}{\tilde{A}^{1/2}(s)}\,. \tag{5.5.4}$$

Also, we have for $x_f \leq x \leq x_r$

$$R = 0\,,\; L = \tilde{A}^{1/2}(x)\frac{I(x_f) - I(x)}{\mu + \nu\tilde{A}(x)}\,,$$

$$r = \frac{1}{2}[I(x) - I(x_f) - \frac{x}{\tilde{A}^{1/2}(x)}]\,,$$

$$l = \frac{(\mu - \nu\tilde{A}(x))L(x) + x}{2\tilde{A}^{1/2}(x)}\,, \tag{5.5.5}$$

and for $x \geq x_r$

$$R = 0\,,\; L = L_r = A_0^{1/2}\frac{I(x_f) - I(x_r)}{\mu + \nu A_0}\,, \tag{5.5.6}$$

$$l - \frac{(\mu - \nu A_0)L_r}{2A_0^{1/2}} = -r - \frac{(\mu + \nu A_0)L_r}{2A_0^{1/2}} = \frac{x}{2A_0^{1/2}}\,. \tag{5.5.7}$$

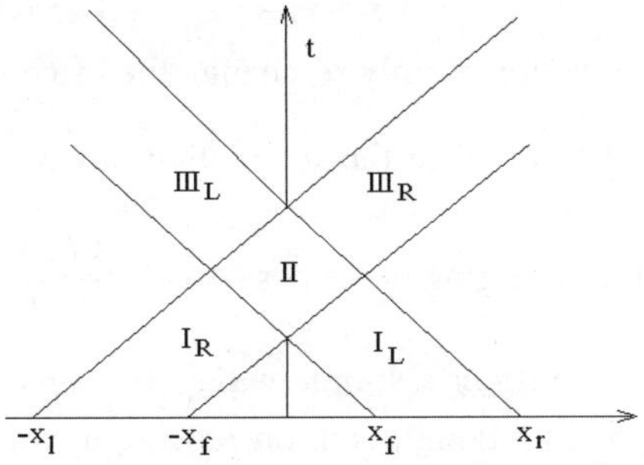

Figure 5.5.1 The plane (x,t) of the wave interaction

In the region I_R, the RW is a simple wave, in the region II it interacts with the LW, and in the region III_R it is again a simple wave. In a similar way, in the region I_L, the LW is a simple wave, in the region II it interacts with the RW, and in the region III_L it is again a simple wave. The important role of this analysis consists in that a simple wave in III_R is the same as the wave that is in I_R m and that the simple wave in III_L is the same as that in I_L. To show this, let us analyze the RW.

Since $G = 0$ in I_R and III_R, the relation (5.5.1) gives

$$A = \tilde{A}(\alpha),\ -x_l \le \alpha \le -x_f\,. \tag{5.5.8}$$

In the region I_R, the solutions (5.4.12) become

$$x = [l(\beta) - r(\alpha)]A^{1/2} + \mu R(\alpha)\,,$$

$$t = [l(\beta) + r(\alpha)]A^{-1/2} + \nu R(\alpha)\,. \tag{5.5.9}$$

Eliminating $l(\beta)$ in the relations (5.5.9) we have

$$x - \tilde{A}(\alpha)t = \alpha\,. \tag{5.5.10}$$

Mention that in the relation (5.5.10), A is given by (5.5.8), $l(\beta)$ by (5.5.2) and $R(\alpha), r(\alpha)$ by (5.5.3). In the region III_R, $R(\alpha), r(\alpha)$ are given by (5.5.3), and $l(\beta)$ and $L(\beta) = L_r$ are given by (5.5.7). Thus, here the solutions (5.4.12) are identical with (5.5.9) if x and t are replaced by

$$\tilde{x} = x + \mu L_r\,,\ \tilde{t} = x - \mu L_r\,. \tag{5.5.11}$$

Therefore, in the region III_R we have a RW given by

$$\tilde{x} - \tilde{A}(\alpha)\tilde{t} = \alpha\,. \tag{5.5.12}$$

So, by comparing (5.5.1) and (5.5.12) we see that the emerging wave is not affected by interaction, it being identical with waves produced by the initial conditions $t = \nu L_r$,

$F(x) = f(x + \mu L_r)$ for $-x_l - \mu L_r \le x \le -x_f - \mu L_r$, and zero otherwise, and $G(x) = 0$ for all x. The effect of interaction is only to change the effective origin of x and t in the original wave. If the RW is before the interaction, a centered wave with $A = \frac{x + x_f}{t}$, then after interaction, the emerging wave has $A = \frac{x + \mu L_r + x_f}{t - \nu L_r}$.

To illustrate this, we consider a simple wave profile $\tilde{A}(x) = \operatorname{sech} x$. This is the case of interaction of two pulses having a soliton profile, traveling in a DRIP medium. The calculations are performed numerically and presented in Figure 5.5.2. In Figure 5.5.2 we see that, in contrast to known theory of solitons interaction, these pulses travel in opposite directions, interact and emerge unaffected by the interaction. In the interaction region no coupling between waves is visible. This suggests that the waves may be regarded individually. Speaking from a physical viewpoint, this interaction requires that the energy of each field is carried individually without any transfer of energy between fields. This property may be of the transmitting medium rather than of the particular wave profiles.

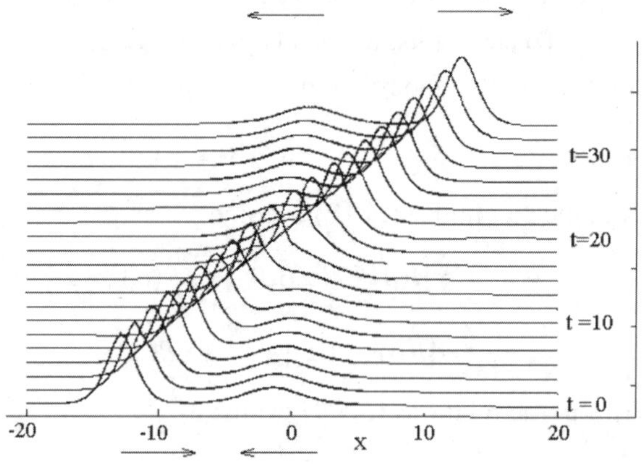

Figure 5.5.2 Profiles of waves against x for several t

5.6 Vibrations of a heterogeneous string

Synge studied in 1981 the vibrations of a heterogeneous string and has shown that the solutions can be obtained by applying a linear integral operator to a Cauchy function of position and time furnished by initial conditions. In the following we present the Synge method applied to the partial differential equation

$$y_{tt} = A^2(x) y_{xx}, \tag{5.6.1}$$

where $A(x)$ is a positive function.

Synge method straighten the characteristics of (5.6.1) by using the transformation

$$x \to u(x) = \int_0^x \frac{dz}{A(z)}, \tag{5.6.2}$$

that yields

$$y_{tt} - y_{uu} + \frac{c_u}{c} y_u = 0, \tag{5.6.3}$$

where $c(u(x)) = A(x)$ is transformed local speed. In the Cartesian coordinates (u,t), the characteristics are straight lines inclined to the axis at 45°. By a change of variable

$$y(u,t) = v(u,t)\phi(u), \tag{5.6.4}$$

the equation (5.6.3) becomes, on division by ϕ

$$v_{tt} - v_{uu} = 2kv_u + 2hv, \tag{5.6.5}$$

with

$$2k = 2\frac{\phi_u}{\phi} - \frac{c_u}{c}, \quad 2h = 2\frac{\phi_{uu}}{\phi} - \frac{\phi_u}{\phi}\frac{c_u}{c}. \tag{5.6.6}$$

For $\phi = \sqrt{c}$, we have $k = 0$, and equation (5.6.5) reduces to

$$v_{tt} - v_{uu} = 2hv. \tag{5.6.7}$$

Integrating (5.6.7) over the triangle PAB (Figure 5.6.1) where $P(u,t)$ is a generic point, we have

$$\iint (v_{uu} - v_{tt} + 2hv)\mathrm{d}u\mathrm{d}t = \int (v_u \mathrm{d}t + v_t \mathrm{d}u) + 2\iint hv\,\mathrm{d}u\,\mathrm{d}t = 0,$$

or

$$v(P) = \frac{1}{2}[v(A) + v(B) + \int_{AB} v_t\,\mathrm{d}u\,\mathrm{d}t] + \iint hv\,\mathrm{d}u\,\mathrm{d}t. \tag{5.6.8}$$

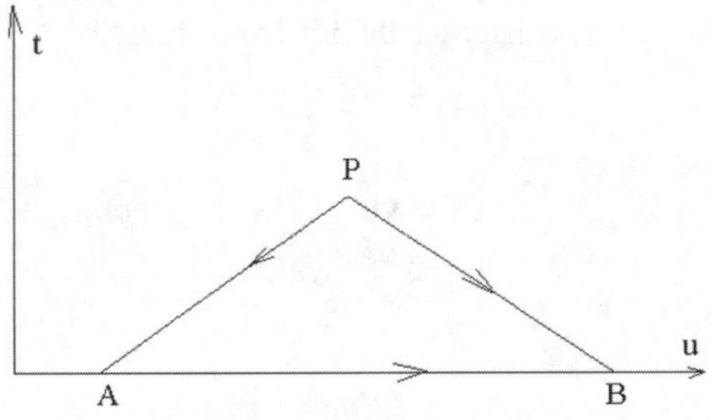

Figure 5.6.1 The triangle PAB in the plane (u,t).

Introduce now the Cauchy function determined by the initial conditions for v and v_t for $t=0$

$$C(u,t)=\frac{1}{2}[v(u-t,0)+v(u+t,0)+\int_{u-1}^{u+1} v_t(z)\,\mathrm{d}z], \qquad (5.6.9)$$

which satisfies the equation

$$C_{tt}-C_{uu}=0. \qquad (5.6.10)$$

Then introduce a linear integral operator H to be applied to function $v(u,t)$

$$Hv(u,t)=\iint h(u_1)v(u_1,t_1)\mathrm{d}u_1\,\mathrm{d}t_1. \qquad (5.6.11)$$

With (5.6.9) and (5.6.11), we obtain from (5.6.8) the integral equation for the solution $v(u,t)$

$$v(u,t)=C(u,t)+Hv(u,t). \qquad (5.6.12)$$

To solve (5.6.12) we write it as $C(u,t)=(1-H)v(u,t)$, and then multiply it by $(1-H)^{-1}$, and obtain an infinite sequence of operations applied to a specified Cauchy function, that is

$$v(u,t)=(1-H)^{-1}C=(1+H+H^2+\ldots)C. \qquad (5.6.13)$$

The series converges if C and h are bounded in absolute values. This solves completely the problem of vibrations of a heterogeneous string.

Next, we present the Lewis method to solve the equation (5.6.1) for the initial data

$$y(x,0)=f(x),\ y_t(x.0)=g(x). \qquad (5.6.14)$$

The energy E of the string is given by

$$E=\frac{1}{2}\int_{-\infty}^{\infty}\frac{y_u^2+y_t^2}{c(u)}\mathrm{d}u. \qquad (5.6.15)$$

Since the energy is conserved, the flow on phase-space, induced by the propagation of a wave according to (5.6.3) conserves the inner product

$$<v_1,v_2>=(\tilde{c}^{-1}v_1,\tilde{c}^{-1}v_2),$$

$$\tilde{c}=\begin{pmatrix}\sqrt{c(x)} & 0\\ 0 & \sqrt{c(x)}\end{pmatrix}, \qquad (5.6.16)$$

where

$$(v_1,v_2)=\frac{1}{2}\int_{-\infty}^{\infty}[q_1(u)q_2(u)+p_1(u)p_2(u)]\mathrm{d}u.$$

Here, v is a point in the phase-space (y_u,y_t)

$$v = \begin{pmatrix} q \\ p \end{pmatrix}, \; q(u) = y_u(u,0), \; p(u) = y_t(u,0). \tag{5.6.17}$$

Therefore, from (6.4.3) we have the equation governing the flow

$$\frac{\mathrm{d}}{\mathrm{d}t} v(t) = Lv(t), \tag{5.6.18}$$

where

$$v = \begin{pmatrix} y_u \\ y_t \end{pmatrix}, \; L = \begin{pmatrix} 0 & \partial_u \\ \partial_u - 2\gamma & 0 \end{pmatrix}, \; 2\gamma(u) = \frac{c_u}{c}. \tag{5.6.19}$$

We calculate $v(t)$ by a sequence of linear transformations that reduce $v(t)$ to a perturbation of the pulse. The operator L is screw-symmetric with respect to the inner product (5.6.16) and so there exists a one-parameter group $V(t)$ of orthogonal transformation determined by

$$\frac{\mathrm{d}}{\mathrm{d}t} V(t) = LV(t), \; V(0) = 1, \tag{5.6.20}$$

so that $v(t) = V(t)v$ is a solution of (5.6.18).

We based the following on the fact that $\frac{\mathrm{d}}{\mathrm{d}t}\exp(tL) = L\exp(tL)$, where

$$\exp L = E + L + \frac{L^2}{2!} + \frac{L^3}{3!} + \ldots = \lim_{m\to\infty}(E + \frac{L}{m})^m,$$

with E the unit matrix, and $\det(\exp L) = \exp(\mathrm{tr}L)$. Since $\mathrm{tr}L = 0$, we have $\det(\exp L) = 1$. Also we have $\exp(-t\partial x) = T(t)$, with $T(t)$ the right translation by t

$$[T(t)f](x) = f(x-t),$$

and $\exp(t\partial x) = T'(t)$, with

$$T'(t) = T(-t) = f(x+t),$$

the left translation by t.

The method is based on the decomposition of the phase-space (y_u, y_t) into a pair of complementary subspaces. This induces a decomposition of each initial datum into a forward-propagating part and a backward-propagating part.

In the homogeneous case $(\gamma = \frac{c_u}{2c} = 0)$, the equation (5.6.3) is reduced to $y_{tt} - y_{uu} = 0$ and the solution is expressed as a sum of two waves $f(x-t)$ and $f(x+t)$ that propagate independently. In the heterogeneous case both pulses are coupled by $\gamma \neq 0$ considered as a perturbation.

So, we take

$$V(t) = \tilde{c}R^{-1}\tilde{V}(t)R\tilde{c}^{-1}, \; \tilde{V}(t) = \exp(tL), \tag{5.6.21}$$

$$L = \tilde{c}R^{-1}\tilde{L}R\tilde{c}^{-1},$$

$$\tilde{L} = \begin{pmatrix} -\partial u & -\gamma \\ \gamma & \partial u \end{pmatrix},$$

$$R = \begin{pmatrix} \frac{1}{\sqrt{2}} & -\sqrt{2} \\ \frac{1}{\sqrt{2}} & \sqrt{2} \end{pmatrix}. \tag{5.6.22}$$

It results

$$\tilde{V}(t) = R\tilde{c}^{-1}V(t)\tilde{c}R^{-1}, \ \tilde{L} = R\tilde{c}^{-1}L\tilde{c}R^{-1}. \tag{5.6.23}$$

We see that $\tilde{L}$ can be written as a sum of two operators to separate the contribution of the coupling term $\gamma \neq 0$

$$\tilde{L} = \tilde{L}_0 + \Gamma, \ \tilde{L}_0 = \begin{pmatrix} -\partial u & 0 \\ 0 & \partial u \end{pmatrix}, \ \Gamma = \begin{pmatrix} 0 & -\gamma \\ \gamma & 0 \end{pmatrix}. \tag{5.6.24}$$

In the homogeneous case we have $\Gamma = 0$ and

$$\tilde{V}(t) = \tilde{U}(t), \quad \tilde{U}(t) = \begin{pmatrix} T(t) & 0 \\ 0 & T'(t) \end{pmatrix}, \tag{5.6.25}$$

where $T(t)$ is right translation by t

$$[T(t)f](s) = f(s-t), \tag{5.6.26}$$

and $T'(t) = T(-t)$ is left translation by t.

The initial conditions (5.6.14) can be written under the form

$$y(u,0) = \varphi(u), \ y_t(u,0) = \psi(u). \tag{5.6.27}$$

For $\Gamma = 0$ the solution of $y_{tt} - y_{uu} = 0$ is written as D'Alembert formula

$$y(u,t) = \frac{1}{2}[\varphi(u+t) + \varphi(u-t)] + \frac{1}{2}\int_{u-t}^{u+t} \psi(z)\mathrm{d}z. \tag{5.6.28}$$

Our aim is to obtain a similar formula for the inhomogeneous case $\Gamma \neq 0$. For this we use the well-known perturbation formula (Kato)

$$\tilde{V}(t) = \tilde{U}(t) + \int_0^t \tilde{U}(t-s)\Gamma\tilde{V}(s)\mathrm{d}s. \tag{5.6.29}$$

From this we can obtain an infinite series for $\tilde{V}(t)$ by an iteration scheme

$$\tilde{V}^{(n+1)}(t) = \tilde{U}(t) + \int_0^t \tilde{U}(t-s)\Gamma\tilde{V}^{(n)}(s)\mathrm{d}s,$$

$$\tilde{V}^{(0)}(t) = \tilde{U}(t) . \tag{5.6.30}$$

We take into account that $\tilde{V}(t) = \exp(tL)$ maps forward-going data into forward-going data and backward-going data.

So, we write

$$\tilde{V}(t) = \begin{pmatrix} \tilde{V}_{FF}(t) & \tilde{V}_{FB}(t) \\ \tilde{V}_{BF}(t) & \tilde{V}_{BB}(t) \end{pmatrix} . \tag{5.6.31}$$

Here, $\tilde{V}_{FF}$ maps forward-going data into forward-going data, $\tilde{V}_{FB}$ maps forward-going data into backward-going data, $\tilde{V}_{BF}$ maps backward-going data into forward-going data and $\tilde{V}_{BB}$ maps backward-going data into backward-going data.

From (5.6.25) and (5.6.30) we obtain for $\tilde{V}_{FF}$

$$\tilde{V}_{FF}(t) = T(t) - \int_0^t \int_0^{t_1} T(t-t_1)\gamma T'(t_1 - t_2)\gamma T(t_2)\mathrm{d}t_1 \mathrm{d}t_2 + \ldots . \tag{5.6.32}$$

The first term in (5.6.32) is simply translating a forward-going datum into a forward direction. The integrant $T(t-t_1)\gamma T'(t_1 - t_2)\gamma T(t_2)$ translates a forward-going datum in the forward direction from time zero to time t_2 when it is reflected. On reflection it is multiplied by the local reflection coefficient γ , then translated backwards from time t_2 to time t_1, when it is reflected again, multiplied by γ and translated forwards from time t_1 to time t . So, the second term represents the contribution to the forward-going disturbance from all possible double reflections. The following terms consider third reflections and so on.

Knowing this, it is easy to write

$$\tilde{V}_{FB}(t) = \int_0^t T'(t-t_1)\gamma T(t_1)\mathrm{d}t_1 -$$
$$- \int_0^t \int_0^{t_1} \int_0^{t_2} T'(t-t_1)\gamma T(t_1 - t_2)\gamma T'(t_2 - t_3)\gamma T(t_3)\mathrm{d}t_1 \mathrm{d}t_2 \mathrm{d}t_3 + \ldots , \tag{5.6.33}$$

$$\tilde{V}_{BF}(t) = -\int_0^t T'(t-t_1)\gamma T(t_1)\mathrm{d}t_1 -$$
$$- \int_0^t \int_0^{t_1} \int_0^{t_2} T(t-t_1)\gamma T'(t_1 - t_2)\gamma T(t_2 - t_3)\gamma T'(t_3)\mathrm{d}t_1 \mathrm{d}t_2 \mathrm{d}t_3 + \ldots , \tag{5.6.34}$$

$$\tilde{V}_{BB}(t) = T'(t) - \int_0^t \int_0^{t_1} T'(t-t_1)\gamma T(t_1 - t_2)\gamma T'(t_2)\mathrm{d}t_1 \mathrm{d}t_2 + \ldots . \tag{5.6.35}$$

Now, from (5.6.31) and (5.6.32)–(5.6.35) we have

$$V(t)=\begin{pmatrix} V_{11}(t) & V_{12}(t) \\ V_{21}(t) & V_{22}(t) \end{pmatrix}, \tag{5.6.36}$$

with

$$\begin{aligned}
V_{11}(t) &= \frac{1}{2}\sqrt{c}[V_{FF}(t)+V_{BB}(t)+V_{FB}(t)+V_{BF}(t)]\frac{1}{\sqrt{c}}, \\
V_{12}(t) &= \frac{1}{2}\sqrt{c}[V_{BB}(t)-V_{FF}(t)+V_{FB}(t)-V_{BF}(t)]\frac{1}{\sqrt{c}}, \\
V_{21}(t) &= \frac{1}{2}\sqrt{c}[V_{BB}(t)-V_{FF}(t)+V_{BF}(t)-V_{FB}(t)]\frac{1}{\sqrt{c}}, \\
V_{22}(t) &= \frac{1}{2}\sqrt{c}[V_{FF}(t)+V_{BB}(t)-V_{FB}(t)-V_{BF}(t)]\frac{1}{\sqrt{c}}.
\end{aligned} \tag{5.6.37}$$

Taking account of the initial data (5.6.27) we have

$$\begin{aligned}
y_t(u,t) &= \frac{1}{2}\sqrt{c(u)}[V_{BB}(t)-V_{FF}(t)+V_{BF}(t)-V_{FB}(t)]\frac{1}{\sqrt{c(u)}}\varphi'(u)+ \\
&+\frac{1}{2}\sqrt{c(u)}[V_{BB}(t)+V_{FF}(t)-V_{BF}(t)-V_{FB}(t)]\frac{1}{\sqrt{c(u)}}\psi(u),
\end{aligned} \tag{5.6.38}$$

and

$$\begin{aligned}
y(u,t) &= \varphi(u)+\frac{1}{2}\sqrt{c(u)}\int_0^t [V_{BB}(t)-V_{FF}(t)+V_{BF}(t)-V_{FB}(t)]\frac{1}{\sqrt{c(u)}}\varphi'(u)\mathrm{d}t_1 + \\
&+\frac{1}{2}\sqrt{c(u)}\int_0^t [V_{BB}(t)+V_{FF}(t)-V_{BF}(t)-V_{FB}(t)]\frac{1}{\sqrt{c(u)}}\psi(u)\mathrm{d}t_1.
\end{aligned} \tag{5.6.39}$$

When $c'(u)=0$, (5.6.39) becomes

$$y(u,t)=\varphi(u)+\frac{1}{2}\int_0^t [T'(t_1)-T(t_1)]\varphi'(u)\mathrm{d}t_1+\frac{1}{2}\int_0^t [T'(t_1)+T(t_1)]\psi(u)\mathrm{d}t_1. \tag{5.6.40}$$

After integration by parts we obtain D'Alembert formula.

Chapter 6

THE COUPLED PENDULUM

6.1 Scope of the chapter

In the first four sections, the inverse scattering transform is applied to solve the nonlinear equations that govern the motion of two pendulums coupled by an elastic spring. The theta-function representation of the solutions is describable as a linear superposition of Jacobi elliptic functions (cnoidal vibrations) and additional terms, which include nonlinear interactions among the vibrations. Comparisons between the cnoidal and LEM solutions are performed. Finally, an interesting phenomenon is put into evidence with consequences for dynamics of the coupled pendula.

The reason why a pendulum is chosen is that its dynamics is rich and complex and its equations are strongly related to the Weierstrass equation with a polynomial of higher order, that admits an analytical solution represented by a sum of a linear and a nonlinear superposition of cnoidal vibrations.

The modal interactions and the modal trading of energy were studied in the early 1950s by Fermi, Pasta and Ulam. In the section 6.5 we present the results of Davies and Moon concerning the modal interaction, characterized by highly localized waves, in an experimental structure. These waves are similar to Toda solitons. The extension of the Toda interacting equation is given by Toda–Yoneyama equation, which is presented in the last section.

This chapter is refers to the works by Donescu (2000, 2003), Munteanu and Donescu (2002), Davies and Moon (2001) and Yoneyama (1986).

6.2 Motion equations. Problem E1

Figure 6.2.1 shows a coupled pendulum consisting of two straight rods O_1Q_1, O_2Q_2 of masses M_1, M_2, lengths $O_1Q_1 = O_2Q_2 = a$, and mass centers C_1, C_2 with $O_1C_1 = l_1$, $O_2C_2 = l_2$ and $O_1O_2 = l$. The rods are linked together by an elastic spring Q_1Q_2, $Q_1 \in O_1C_1$, $Q_2 \in O_2C_2$ characterized by an elastic constant k. The elastic force in the spring is given by $k\,|\,O_1O_2 - Q_1Q_2\,|$. The kinetic energy T of the system is

$$T = \frac{1}{2}(I_1\dot{\theta}_1^2 + I_2\dot{\theta}_2^2)\,, \tag{6.2.1}$$

where θ_1 and θ_2 are the displacement angles in rapport to the verticals, I_1 is the mass moment of inertia of O_1O_2 with respect to C_1 and I_2 is the mass moment of inertia of O_3O_4 with respect to C_2. The elastic potential is written as (Teodorescu)

$$U = g(M_1 l_1 \cos\theta_1 + M_2 l_2 \cos\theta_2) - \frac{k}{2}(O_1O_2 - Q_1Q_2)^2, \tag{6.2.2}$$

where

$$\begin{aligned} Q_1Q_2^2 &= [O_1O_2 + a(\sin\theta_2 - \sin\theta_1)]^2 + a^2(\cos\theta_2 - \cos\theta_1)^2 = \\ &= O_1O_2^2 + 2aO_1O_2(\sin\theta_2 - \sin\theta_1) + 2a^2[1 - \cos(\theta_2 - \theta_1)]. \end{aligned} \tag{6.2.3}$$

The generalized force is

$$Q_1 = F \cdot \frac{\partial O_1 C_1}{\partial \theta_1} = \frac{A}{l_1}\cos\omega t(-\sin\theta_1 i + \cos\theta_1 j)(-l_1 \sin\theta_1 i + l_1 \cos\theta_1 j),$$

where i, j are unit normal vectors and

$$Q_1 = A\cos\omega t. \tag{6.2.4}$$

From Lagrange equations, where $L = T + U$

$$\begin{aligned} \frac{\mathrm{d}}{\mathrm{d}t}\left(\frac{\partial L}{\partial \dot\theta_1}\right) - \frac{\partial L}{\partial \theta_1} &= Q_1, \\ \frac{\mathrm{d}}{\mathrm{d}t}\left(\frac{\partial L}{\partial \dot\theta_2}\right) - \frac{\partial L}{\partial \theta_2} &= 0, \end{aligned} \tag{6.2.5}$$

we derive the motion equations of the pendulum

$$\begin{cases} I_1\ddot\theta_1 + M_1 g l_1 \sin\theta_1 + \dfrac{k}{2}\dfrac{\partial}{\partial\theta_1}(O_1O_2 - Q_1Q_2)^2 = A\cos\omega t, \\ I_2\ddot\theta_2 + M_2 g l_2 \sin\theta_2 + \dfrac{k}{2}\dfrac{\partial}{\partial\theta_2}(O_1O_2 - Q_1Q_2)^2 = 0, \end{cases} \tag{6.2.6}$$

with g the gravitational acceleration. Equations (6.2.6) are coupled and nonlinear.

Substitution (6.2.3) into (6.2.6) gives

$$\begin{cases} I_1\ddot\theta_1 + M_1 g l_1 \sin\theta_1 - kH[-al\cos\theta_1 - a^2\sin(\theta_2 - \theta_1)] = A\cos\omega t, \\ I_2\ddot\theta_2 + M_2 g l_2 \sin\theta_2 - kH[al\cos\theta_2 + a^2\sin(\theta_2 - \theta_1)] = 0, \end{cases} \tag{6.2.7}$$

where

$$\mathrm{H}(\theta_1, \theta_2) = \frac{l - \Psi(\theta_1, \theta_2)}{\Psi(\theta_1, \theta_2)}, \tag{6.2.8}$$

$$\Psi(\theta_1, \theta_2) = Q_1Q_2 = [l^2 + 2al(\sin\theta_2 - \sin\theta_1) + 2a^2(1 - \cos(\theta_2 - \theta_1)]^{1/2}. \tag{6.2.9}$$

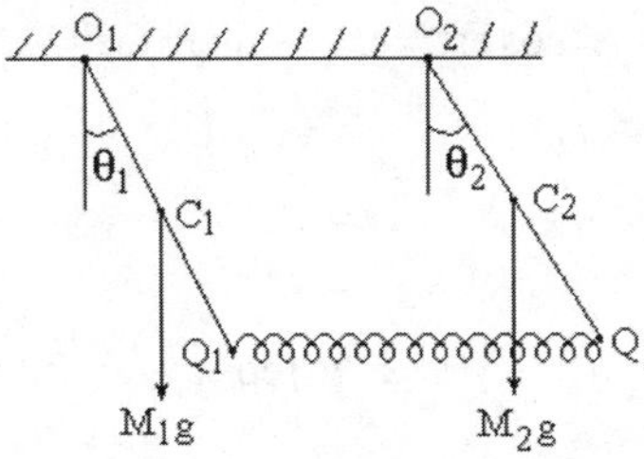

Figure 6.2.1 The coupled pendulum.

Defining the dimensionless variable

$$\tau = t\omega_1 = t\sqrt{\frac{k}{M_1}}, \tag{6.2.10}$$

and notations

$$\frac{M_1^2 g l_1}{I_1 k} = w, \quad \frac{M_1 M_2 g l_2}{I_2 k} = \beta w, \quad \Phi = \frac{\Psi}{l},$$
$$\frac{a}{l} = \xi, \quad \frac{AM_1}{kI_1} = \delta, \quad \frac{alM_1}{I_1} = \alpha, \quad \frac{alM_1}{I_2} = \tilde{\alpha}, \tag{6.2.11}$$

equations (6.2.7) are reduced to the dimensionless equations

$$\begin{cases} \ddot{\theta}_1 + w\sin\theta_1 + \gamma(\theta_1,\theta_2)\alpha[\cos\theta_1 + \xi\sin(\theta_2 - \theta_1)] = \delta\cos\tilde{\omega}\tau, \\ \ddot{\theta}_2 + \beta w\sin\theta_2 - \gamma(\theta_1,\theta_2)\tilde{\alpha}[\cos\theta_2 + \xi\sin(\theta_2 - \theta_1)] = 0, \end{cases} \tag{6.2.12}$$

where $\tilde{\omega} = \omega\sqrt{\frac{M_1}{k}}$, and the dot means the differentiation with respect to τ, and

$$\gamma(\theta_1,\theta_2) = \Phi^{-1/2} - 1,$$

$$\Phi(\theta_1,\theta_2) = 1 + 2\xi(\sin\theta_2 - \sin\theta_1) + 2\xi^2(1 - \cos(\theta_2 - \theta_1)). \tag{6.2.13}$$

The parameters M_2 and l are supposed to be specified, whereas m, ξ, k and δ are considered the control variable parameters, where

$$\frac{M_1}{M_2} = m. \tag{6.2.14}$$

The imposed initial conditions are

$$\theta_1(0) = \theta_1^0, \quad \theta_2(0) = \theta_2^0, \quad \dot{\theta}_1(0) = \theta_{p1}^0, \quad \dot{\theta}_2(0) = \theta_{p2}^0. \tag{6.2.15}$$

Noting

$$\theta_1 = z_1, \quad \theta_2 = z_2, \quad \dot{\theta}_1 = z_3, \quad \dot{\theta}_2 = z_4, \\ \cos\tilde{\omega}\tau = z_5, \quad -\tilde{\omega}\sin\tilde{\omega}\tau = z_6, \tag{6.2.16}$$

the equations (6.2.12) become

$$\begin{aligned} \dot{z}_1 &= z_3, \\ \dot{z}_2 &= z_4, \\ \dot{z}_3 &= -w\sin z_1 - \gamma(z_1,z_2)\alpha[\cos z_1 + \xi\sin(z_2 - z_1)] + \delta z_5, \\ \dot{z}_4 &= -\beta w\sin z_2 + \gamma(z_1,z_2)\tilde{\alpha}[\cos z_2 + \xi\sin(z_2 - z_1)], \\ \dot{z}_5 &= z_6, \\ \dot{z}_6 &= -\tilde{\omega}^2 z_5, \end{aligned} \tag{6.2.17}$$

with initial conditions (6.2.15)

$$z_1(0) = z_1^0, \quad z_2(0) = z_2^0, \quad z_3(0) = z_3^0, \quad z_4(0) = z_4^0, \\ z_5(0) = 1, \quad z_6(0) = 0. \tag{6.2.18}$$

Consider now the case $|z_p| \le \dfrac{\pi}{2}$, and let us approximate the trigonometric functions by series expansions (Abramowitz and Stegun)

$$\sin z = z - \frac{z^3}{3!} + \frac{z^5}{5!} + \varepsilon(z), \quad |\varepsilon(z)| \le 2\cdot 10^{-4}, \\ \cos z = 1 - \frac{z^2}{2!} + \frac{z^4}{4!} + \varepsilon(z), \quad |\varepsilon(z)| \le 9\cdot 10^{-4}. \tag{6.2.19}$$

Substitution of (6.2.19) into (6.2.17) yields to a system of equations we refer to as the *problem E1*

$$\begin{cases} \dot{z}_1 = z_3, \\ \dot{z}_2 = z_4, \\ \dot{z}_3 = -wP(z_1) - \alpha Q_1(z_1,z_2)\gamma(z_1,z_2) + \delta z_5, \\ \dot{z}_4 = -\beta wP(z_2) + \tilde{\alpha}Q_2(z_1,z_2)\gamma(z_1,z_2), \\ \dot{z}_5 = z_6, \\ \dot{z}_6 = -\tilde{\omega}^2 z_5, \end{cases} \tag{6.2.20}$$

with

$$P(z) = z + \tilde{a}z^3 + \tilde{b}z^5, \quad Q_1(z_1,z_2) = R_1(z_1) + R_2(z_1,z_2),$$

$$Q_2(z_1,z_2) = R_1(z_2) + R_2(z_1,z_2), \quad R_1(z) = 1 + \tilde{c}z^2 + \tilde{d}z^4,$$

$$\gamma(z_1,z_2) = -1 + f^{-1/2}(z_1,z_2),$$

$$R_2(z_1,z_2) = -\xi z_1 + \xi z_2 + \tilde{a}\xi z_2^3 - 3\tilde{a}\xi z_2^2 z_1 + \\ +3\tilde{a}\xi z_2 z_1^2 - \tilde{a}\xi z_1^3 + \tilde{b}\xi z_2^5 - 5\tilde{b}\xi z_2^4 z_1 + \\ +10\tilde{b}\xi z_2^3 z_1^2 - 10\tilde{b}\xi z_2^2 z_1^3 + 5\tilde{b}\xi z_2 z_1^4 - \tilde{b}\xi z_1^5,$$

$$f(z_1,z_2) = 1 - 2\xi z_1 + 2\xi z_2 + 4\tilde{c}\xi^2 z_1 z_2 - 2\tilde{c}\xi^2 z_1^2 - \\ -2\tilde{c}\xi^2 z_2^2 - 2\tilde{a}\xi z_1^3 + 2\tilde{a}\xi z_2^3 - 2\tilde{d}\xi^2 z_2^4 - 2\tilde{d}\xi^2 z_1^4 + 8\tilde{d}\xi^2 z_2^3 z_1 - \\ -12\tilde{d}\xi^2 z_2^2 z_1^2 + 8\tilde{d}\xi^2 z_2 z_1^3 + 2\tilde{b}\xi z_2^5 - 2\tilde{b}\xi z_1^5, \qquad (6.2.21)$$

with $\tilde{a} = -\frac{1}{3!}$, $\tilde{b} = \frac{1}{5!}, \tilde{c} = -\frac{1}{2!}, \tilde{d} = \frac{1}{4!}$.

6.3 Problem E2

A way for analyzing a system of equations by following LEM procedure given in section 1.6 is to simplify the exact system of equations (6.2.20) by assuming

$$\gamma(\theta_1,\theta_2) = -\xi(\sin\theta_2 - \sin\theta_1) - \xi^2(1-\cos(\theta_2-\theta_1)), \qquad (6.3.1)$$

under conditions

$$|\, 2\xi(\sin\theta_2 - \sin\theta_1) + 2\xi^2(1-\cos(\theta_2-\theta_1)) \,| < 1. \qquad (6.3.2)$$

Substitution of the function $\gamma(\theta_1,\theta_2)$ given by (6.3.1) into (6.2.17) yields

$$\dot{z}_1 = z_3,$$
$$\dot{z}_2 = z_4,$$
$$\dot{z}_3 = -w\sin z_1 + \alpha\xi^2\cos z_1 + \alpha\xi^3\sin(z_2-z_1) + \alpha\xi\cos z_1\sin z_2 - \\ -\alpha\xi\cos z_1\sin z_1 + \alpha\xi^2\sin z_2\sin(z_2-z_1) - \\ -\alpha\xi^2\sin z_1\sin(z_2-z_1) - \alpha\xi^2\cos z_1\cos(z_2-z_1) - \\ -\alpha\xi^3\sin(z_2-z_1)\cos(z_2-z_1) + \delta z_5,$$

$$\dot{z}_4 = -\beta w\sin z_2 - \tilde{\alpha}\xi^2\cos z_2 - \tilde{\alpha}\xi^3\sin(z_2-z_1) - \tilde{\alpha}\xi\cos z_2\sin z_2 - \\ +\tilde{\alpha}\xi\sin z_1\cos z_2 - \tilde{\alpha}\xi^2\sin z_2\sin(z_2-z_1) + \\ +\tilde{\alpha}\xi^2\sin z_1\sin(z_2-z_1) + \tilde{\alpha}\xi^2\cos z_2\cos(z_2-z_1) + \\ +\tilde{\alpha}\xi^3\sin(z_2-z_1)\cos(z_2-z_1), \qquad (6.3.3)$$
$$\dot{z}_5 = z_6,$$
$$\dot{z}_6 = -\tilde{\omega}^2 z_5.$$

It is important to write the system of equations (6.3.3) under the form

$$\dot{z}_n = A_{np}z_p + g_n(z), \; n, p = 1,\dots,6, \qquad (6.3.4)$$

with

$$
\begin{aligned}
g_n(z) = {} & B_{np}\sin z_p + C_{np}\cos z_p + \\
& + D_{npq}\sin(z_p - z_q) + E_{npq}\cos z_p \sin z_q + \\
& + F_{npqr}\sin z_p \sin(z_q - z_r) + G_{npqr}\cos z_p \cos(z_q - z_r) + \\
& + H_{npq}\sin(z_p - z_q)\cos(z_p - z_q).
\end{aligned}
\tag{6.3.5}
$$

In the above expressions $(n, p, q, r = 1,2,3,4)$, and the summation convection is used over these indices. The nonzero constants of (6.2.5) are given by

$$
\begin{aligned}
& A_{13} = 1, \quad A_{24} = 1, \\
& A_{35} = \delta, \quad A_{56} = 1, \quad A_{65} = -\frac{\omega^2}{\omega_1^2}, \\
& B_{31} = -w, \quad B_{42} = -\beta w, \\
& C_{31} = \alpha\xi^2, \quad C_{42} = -\tilde{\alpha}\xi^2, \\
& D_{321} = \alpha\xi^3, \quad D_{421} = -\tilde{\alpha}\xi^3, \\
& E_{312} = -\alpha\xi, \quad E_{311} = -\alpha\xi, \\
& E_{422} = -\tilde{\alpha}\xi, \quad E_{421} = \tilde{\alpha}\xi, \\
& F_{3221} = \alpha\xi^2, \quad F_{3121} = -\alpha\xi^2, \\
& F_{4221} = -\tilde{\alpha}\xi^2, \quad F_{4121} = \tilde{\alpha}\xi^2, \\
& G_{3121} = -\alpha\xi^2, \quad G_{4221} = \tilde{\alpha}\xi^2, \\
& H_{321} = -\alpha\xi^3, \quad H_{421} = \tilde{\alpha}\xi^3.
\end{aligned}
\tag{6.3.6}
$$

Inserting (6.2.19) into (6.3.5) and neglecting the terms up to the sixth order, we have

$$
g_1(z) = 0,
$$
$$
g_2(z) = 0,
$$

$$
\begin{aligned}
g_3(z) = {} & (-\alpha\xi - w)z_1 + \alpha\xi z_2 - \alpha\xi^2(\tilde{c} - 1)z_1^2 - 2\alpha\xi^2(1 - \tilde{c})z_1 z_2 - \\
& -\alpha\xi^2(\tilde{c} - 1)z_2^2 - (\alpha\xi\tilde{a} + \alpha\xi\tilde{c} + w\tilde{a} - \alpha\xi^3\tilde{c})z_1^3 - \alpha\xi\tilde{c}(3\xi^2 - 1)z_1^2 z_2 + \\
& + 3\alpha\xi^3\tilde{c}z_1 z_2^2 - \alpha\xi(\xi^2\tilde{c} - \tilde{a})z_2^3 - \alpha\xi^2(\tilde{d} + \tilde{c}^2 - 2\tilde{a})z_1^4 - \\
& -\alpha\xi^2(\tilde{a} - 4\tilde{d} - 2\tilde{c}^2)z_1^3 z_2 - \alpha\xi^2(6\tilde{d} + \tilde{c}^2 - 6\tilde{a})z_1^2 z_2^2 - \\
& -\alpha\xi^2(5\tilde{a} - 4\tilde{d})z_1 z_2^3 - \alpha\xi^2(\tilde{d} - 2\tilde{a})z_2^4 - (\tilde{b}w - \alpha\xi^3\tilde{a}\tilde{c} - \alpha\xi^3\tilde{d})z_1^5 - \\
& -\alpha\xi(5\xi^2\tilde{a}\tilde{c} + 5\xi^3\tilde{d} - \tilde{d})z_1^4 z_2 + 10\alpha\xi^3(\tilde{d} + \tilde{a}\tilde{c})z_2^2 z_1^3 - \\
& -\alpha\xi(10\xi^2\tilde{d} + 10\xi^3\tilde{a}\tilde{c} - \tilde{a}\tilde{c})z_1^2 z_2^3 + 5\alpha\xi^3(\tilde{d} + \tilde{a}\tilde{c})z_1 z_2^4 - \\
& -\alpha\xi(\xi^2\tilde{d} + \xi^2\tilde{a}\tilde{c} - \tilde{b})z_2^5,
\end{aligned}
\tag{6.3.7}
$$

$$g_4(z) = \tilde{\alpha}\xi z_1 + (-\tilde{\alpha}\xi - \beta w)z_2 - \tilde{\alpha}\xi^2(1-\tilde{c})z_1^2 - 2\tilde{\alpha}\xi^2(\tilde{c}-1)z_1 z_2 -$$
$$-\tilde{\alpha}\xi^2(1-\tilde{c})z_2^2 - \tilde{\alpha}\xi(\xi^2\tilde{c} - \tilde{a})z_1^3 + 3\tilde{c}\tilde{\alpha}\xi^3 z_1^2 z_2 +$$
$$+\tilde{\alpha}\xi\tilde{c}(3\xi^2 - 1)z_1 z_2^2 - (\tilde{\alpha}\xi\tilde{a} + \tilde{\alpha}\xi\tilde{c} - \alpha\tilde{c}\xi^3 + \beta w\tilde{a})z_2^3 - \tilde{\alpha}\xi^2(2\tilde{a} - \tilde{d})z_1^4 -$$
$$-\tilde{\alpha}\xi^2(4\tilde{d} - 5\tilde{a})z_1^3 z_2 - \tilde{\alpha}\xi^2(6\tilde{a} - 6\tilde{d} - \tilde{c}^2)z_1^2 z_2^2 -$$

$$-\tilde{\alpha}\xi^2(-5\tilde{a} + 4\tilde{d} + 2\tilde{c}^2)z_1 z_2^3 + \tilde{\alpha}\xi^2(\tilde{d} + \tilde{c}^2)z_2^4 - \tilde{\alpha}\xi(-\tilde{b} + \xi^2\tilde{d} + \xi^2\tilde{a}\tilde{c})z_1^5 +$$
$$+5\tilde{\alpha}\xi^3(\tilde{b} + \tilde{a}\tilde{c})z_1^4 z_2 - \tilde{\alpha}\xi(10\tilde{d}\xi^2 + 10\tilde{a}\tilde{c}\xi^2 - \tilde{a}\tilde{c})z_2^2 z_1^3 +$$
$$+10\tilde{\alpha}\xi^3(\tilde{d} + \tilde{a}\tilde{c})z_1^2 z_2^3 - 5\tilde{\alpha}\xi^3(\tilde{d} + \tilde{a}\tilde{c})z_1 z_2^4 + \beta w\tilde{b}z_2^5,$$

$$g_5(z) = 0,\ g_6(z) = 0.$$

Using (6.2.7), the system of equations (6.2.4) yields to a simplified system of Bolotin type (Bolotin), we refer to as the *problem E2*

$$\dot{z} = Az + \sum_{i=1}^{4} F_i(z), \qquad (6.3.8)$$

where

$$Az = \sum_{p=1}^{6} a_{np} z_p,$$

$$F_1(z) = \sum_{p,q=1}^{6} b_{npq} z_p z_q,$$

$$F_2(z) = \sum_{p,q,r=1}^{6} c_{npqr} z_p z_q z_r, \qquad (6.3.9)$$

$$F_3(z) = \sum_{p,q,r,l=1}^{6} d_{npqr} z_p z_q z_r z_l,$$

$$F_4(z) = \sum_{p,q,r,l,m=1}^{6} e_{npqrlm} z_p z_q z_r z_l z_m,$$

with $n = 1,2,3,...,6$. The matrix $A = a_{np}$ is

$$A = \begin{bmatrix} 0 & 0 & 1 & 0 & 0 & 0 \\ 0 & 0 & 0 & 1 & 0 & 0 \\ -\alpha\xi - w & \alpha\xi & 0 & 0 & \delta & 0 \\ \tilde{\alpha}\xi & -\tilde{\alpha}\xi - \beta w & 0 & 0 & 0 & 0 \\ 0 & 0 & 0 & 0 & 0 & 1 \\ 0 & 0 & 0 & 0 & -\tilde{\omega}^2 & 0 \end{bmatrix}. \qquad (6.3.10)$$

The constants of (6.3.9) are defined as follows

$$\tilde{a} = -0.16605, \quad \tilde{b} = 0.00761, \quad \tilde{c} = -0.49670, \quad \tilde{d} = 0.03705,$$

$$c_{3111} = -\alpha\xi\tilde{a} - \alpha\xi\tilde{c} + \alpha\xi^3\tilde{c} - w\tilde{a},$$

$$c_{4122} = -\tilde{\alpha}\xi\tilde{c}(3\xi^2 - 1), \quad c_{4112} = 3\tilde{\alpha}\xi^3\tilde{c}, \quad c_{3222} = -\alpha\xi(\xi^2\tilde{c} - \tilde{a})$$

$$d_{31111} = -\alpha\xi^2(\tilde{d} + \tilde{c}^2 - 2\tilde{a}), \quad d_{31112} = -\alpha\xi^2(\tilde{a} - 4\tilde{d} - 2\tilde{c}^2),$$

$$d_{31122} = -\alpha\xi^2(6\tilde{d} + \tilde{c}^2 - 6\tilde{a}), \quad d_{31222} = -\alpha\xi^2(5\tilde{a} - 4\tilde{d}),$$

$$d_{32222} = -\alpha\xi^2(\tilde{d} - 2\tilde{a}), \quad d_{41111} = -\alpha\xi^2(2\tilde{a} - \tilde{d}),$$

$$d_{41112} = -\tilde{\alpha}\xi^2(4\tilde{d} - 5\tilde{a}), \quad d_{41122} = -\tilde{\alpha}\xi^2(6\tilde{a} - 6\tilde{d} - \tilde{c}^2),$$

$$d_{41222} = -\tilde{\alpha}\xi^2(-5\tilde{a} + 4\tilde{d} + 2\tilde{c}^2), \quad d_{42222} = \tilde{\alpha}\xi^2(\tilde{d} + \tilde{c}^2),$$

$$e_{311111} = -\tilde{b}w + \alpha\xi^3\tilde{a}\tilde{c} + \alpha\xi^3\tilde{d},$$

$$e_{311122} = 10\alpha\xi^3(\tilde{a}\tilde{c} + \tilde{d}), \quad e_{411222} = 10\tilde{\alpha}\xi^3(\tilde{a}\tilde{c} + \tilde{d}),$$

$$e_{411122} = -\tilde{\alpha}\xi(10\xi^2\tilde{a}\tilde{c} + 10\xi^2\tilde{d} - \tilde{a}\tilde{c}), \quad e_{312222} = 5\alpha\xi^3(\tilde{a}\tilde{c} + \tilde{d}),$$

$$e_{322222} = -\alpha\xi(\xi^2\tilde{a}\tilde{c} + \xi^2\tilde{d} - \tilde{b}),$$

$$e_{411111} = -\tilde{\alpha}\xi(\xi^2\tilde{a}\tilde{c} + \xi^2\tilde{d} - \tilde{b}), \quad e_{422222} = -\beta w\tilde{b}.$$

6.4 LEM solutions of the system E2

In this section we derive the LEM solutions of the system E2 (Toma, Munteanu *et al.* 2002b). For this, we apply to equations of the system E2 an exponential transform depending on four parameters of the form (1.6.2)

$$v(t,\sigma_1,\sigma_2,\sigma_3,\sigma_4) = \exp(\sigma_1 z_1 + \sigma_2 z_2 + \sigma_2 z_2 + \sigma_4 z_4), \quad \sigma_i \in R, \tag{6.4.1}$$

that yields a linear first order partial differential equation

$$\frac{\partial v}{\partial t} = \sum_{n=1}^{4}\Big(\sum_{p=1}^{4}\sigma_n a_{np}\frac{\partial v}{\partial\sigma_p} + \sum_{p,q=1}^{4}\sigma_n b_{npq}\frac{\partial^2 v}{\partial\sigma_p\partial\sigma_q} + \sum_{p,q,r=1}^{4}\sigma_n c_{npqr}\frac{\partial^3 v}{\partial\sigma_p\partial\sigma_q\partial\sigma_r} + \\ + \sum_{p,q,r,l=1}^{4}\sigma_n d_{npqrl}\frac{\partial^4 v}{\partial\sigma_p\partial\sigma_q\partial\sigma_r\partial\sigma_l} + \sum_{p,q,r,l,m=1}^{4}\sigma_n e_{npqrlm}\frac{\partial^5 v}{\partial\sigma_p\partial\sigma_q\partial\sigma_r\partial\sigma_l\partial\sigma_m}\Big), \tag{6.4.2}$$

with initial conditions

$$v(0,\sigma_1,\sigma_2,\sigma_3,\sigma_4) = \exp(\sigma_1 z_1^0 + \sigma_2 z_2^0 + \sigma_3 z_3^0 + \sigma_4 z_4^0), \tag{6.4.3a}$$

and bounding condition at infinity

$$|v(t,\sigma_1,\sigma_2,\sigma_3,\sigma_4)| \le v_0 \quad \text{as} \quad t \to \infty , \tag{6.4.3b}$$

where $v_0 = \sup\{|v(t,\sigma)| : t \ge 0, \ \sigma \in R\}$.

An important step in solving the equation (6.4.2) is the observation that a series of the form $\sum_{m=0} (\mu t)^m \sigma^m$ verifies the equation. By looking on the explicit form of this particular solution we can derive, without any information and in a purely deductive way, other series of the form $\sum_{m,n=0} (\mu t)^m (\mu t)^n \sigma^m \sigma^n$, which verify the equation. So, there must be a family of series that can be taken into consideration. In fact, this observation yields to a solution of (6.4.2) and (6.4.3) under the form

$$\begin{aligned} v = 1 &+ \sum_{k=1}^{4} \sum_{i=1} A_k^{\,i} \frac{\sigma_{,k}^{i}}{i!} + \sum_{\substack{k,l=1\\k\ne l}}^{4} \sum_{i,j=1} A_k^{\,i} A_l^{\,j} \frac{\sigma_{,k}^{i}\sigma_{,l}^{j}}{i!\,j!} + \\ &+ \sum_{\substack{k,l,m=1\\k\ne l\ne m}}^{4} \sum_{i,j,r=1} A_k^{\,i} A_l^{\,j} A_m^{\,r} \frac{\sigma_{,k}^{i}\sigma_{,l}^{j}\sigma_{,m}^{r}}{i!\,j!\,r!} + \\ &+ \sum_{\substack{k,l,m,n=1\\k\ne l\ne m\ne n}}^{4} \sum_{i,j,r,s=1} A_k^{\,i} A_l^{\,j} A_m^{\,r} A_n^{\,s} \frac{\sigma_{,k}^{i}\sigma_{,l}^{j}\sigma_{,m}^{r}\sigma_{,n}^{s}}{i!\,j!\,r!\,s!}, \end{aligned} \tag{6.4.4}$$

where $A_n(t)$, $n = 1,2,3,4$, are defined as

$$A_n(t) = \sum_{k,\eta-0} \{(\mu t)^{k+1} \tilde{A}_{nk}(\eta) \Phi_k(\mu t,\eta) + (\mu t)^k \tilde{B}_{nk}(\eta) \psi_k(\mu t,\eta)\}. \tag{6.4.5}$$

Indices take integers values $k = 0,1,2...$, $\eta = 0,1,2,...$. In (6.4.5) the functions $\Phi_k(\mu t,\eta)$ and $\psi_k(\mu t,\eta)$ and $\mu = \mu(\eta)$ are defined as

$$\Phi_k(\mu t,\eta) = \sum_{m=k+1} (\mu t)^{m-k-1} A_m^{(k)}(\eta) , \quad \psi_k(\mu t,\eta) = \sum_{m=k+1} (\mu t)^{m-k-1} B_m^{(k)}(\eta) , \tag{6.4.6}$$

$$\mu = \sum_{j=1}^{4} \alpha_j \lambda_j , \tag{6.4.7}$$

with $\sum_{l=1}^{4} \alpha_j = \eta + 1$, $\alpha_j \ge 0$, $j = 1,2,3,4$. In (6.4.7) p_j, $j = 1,2,3,4$, are the roots

$$\lambda_1 = p_1, \ \lambda_2 = -p_1, \ \lambda_3 = p_2, \ \lambda_4 = -p_2 , \tag{6.4.8}$$

of the characteristic equation

$$(\lambda^4 + p\lambda^2 + \Delta) = 0 , \tag{6.4.9}$$

with

$$p = \tilde{\alpha}\xi + \alpha\xi + \beta w + w\,, \quad \Delta = \alpha\xi\beta w + \tilde{\alpha}\xi w + w^2\beta\,. \tag{6.4.10}$$

The unknowns $\tilde{A}_{nk}$ and $\tilde{B}_{nk}$ depend on η in the following way

$$\tilde{A}_{nk}(\eta) = C_k(\eta)B_{nk}(\eta),\ \tilde{B}_{nk}(\eta) = C_k(\eta)C_{nk}(\eta). \tag{6.4.11}$$

The constants $C_k(\eta)$ verify the recurrence relation

$$C_k(\eta) = \frac{\sqrt{k^2+\eta^2}}{k(2k+1)}C_{k-1}(\eta)\,, \tag{6.4.12}$$

with

$$C_0(\eta) = \frac{p\,|\,\Gamma(1+i\eta)\,|}{2!}\,, \tag{6.4.13}$$

where Γ is the Gamma function, and

$$|\frac{\Gamma(1+i\eta)}{\Gamma(2)}|^2 = \prod_{n=0}[1+\frac{\eta^2}{(2+n)^2}]^{-1}\,. \tag{6.4.14}$$

Constants $A_m^{(k)}$, $B_m^{(k)}$ are related between themselves by

$$B_m^{(k)}(\eta) = m\Delta A_m^{(k)}(\eta)\,, \tag{6.4.15}$$

where $A_m^{(k)}(\eta)$ is defined by

$$\begin{gathered} A_{k+1}^{(k)} = 1, \quad A_{k+2}^{(k)} = \frac{\eta}{k+1}, \\ (m+k)(m-k-1)A_m^{(k)} = 2\eta A_{m-1}^{(k)} - A_{m-2}^{(k)}, \quad m > k+2. \end{gathered} \tag{6.4.16}$$

We mention that $B_{nk}(\eta)$ and $C_{nk}(\eta)$ depend on the initial constants and on the constants a,b,c,d,e of (3.1.1). Substituting $\tilde{B}_m^{(k)}(\eta)$ given by (6.4.15) into (6.3.6)$_2$, we obtain

$$\psi_k(\mu t,\eta) = \sum_{m=k+1}(\mu t)^{m-k-1}B_m^{(k)}(\eta)\,. \tag{6.4.17}$$

We point out that $\psi(\mu t,\eta)$ can be calculated from $\Phi(\mu t,\eta)$ by formulae

$$\psi(\mu t,\eta) = \mu\frac{\mathrm{d}}{\mathrm{d}t}\Phi(\mu t,\eta) = \mu\dot{\Phi}(\mu t,\eta)\,. \tag{6.4.18}$$

Denoting

$$F_k(\mu t,\eta) = C_k(\eta)\quad(\mu t)^{k+1}\Phi_k(\mu t,\eta)\,, \tag{6.4.19}$$

we obtain for (6.4.5)

$$A_n(t) = \sum_{\eta,k=0}\{B_{nk}(\eta)F_k(\mu t,\eta) + C_{nk}(\eta)\dot{F}_k(\mu t,\eta)\}\,. \tag{6.4.20}$$

It is easy to observe that

$$F_0(\mu t, 0) = \sin \mu t,\ \dot{F}_0(\mu t, 0) = \cos \mu t. \tag{6.4.21}$$

Now, return to (6.4.4), and after a little manipulation, we have

$$v = (1 + \sum_i A_1^i \frac{\sigma_1^i}{i!})(1 + \sum_j A_2^j \frac{\sigma_2^j}{j!})(1 + \sum_k A_3^k \frac{\sigma_3^k}{k!})(1 + \sum_l A_4^l \frac{\sigma_4^l}{l!}), \tag{6.4.22}$$

and

$$v(x, \sigma_1, \sigma_2, \sigma_3, \sigma_4) = \exp(\sigma_1 A_1 + \sigma_2 A_2 + \sigma_3 A_3 + \sigma_4 A_4). \tag{6.4.23}$$

Finally, the solution of (6.3.8) results from (6.4.27)

$$z_n(t) = A_n(t) = \sum_{k,\eta=0} \{B_{nk}(\eta) F_k(\mu t, \eta) + C_{nk}(\eta) \dot{F}_k(\mu t, \eta)\}. \tag{6.4.24}$$

We refer to (6.4.24) as the LEM representations of the solutions of problem E2. These solutions are bounded at $t \to \infty$. We name the functions $F_k(\mu t, \eta)$ the incomplete Coulomb functions of vibration, or Coulomb functions of vibrations, since they are similar to Coulomb wave functions (Abramowitz and Stegun). So, the LEM solutions are describable as a linear superposition of Coulomb vibrations (Donescu 2003). The first terms in (6.4.24) represent the linear part of the solutions of the problem E2 ($n = 1, 2,\ k = 0, 1$)

$$z_1 = C_{10} \cos p_1 t + B_{10} \sin p_1 t + C_{11} \cos p_2 t + B_{11} \sin p_2 t,$$

$$z_2 = C_{20} \cos p_1 t + B_{20} \sin p_1 t + C_{21} \cos p_2 t + B_{21} \sin p_2 t, \tag{6.4.25}$$

where the constants are defined as $\eta = 0$ ($n = 1, 2$, $k = 0, 1, ..., 9$),

$$C_{10} = \frac{z_2^0 - v_2 z_1^o}{D_1},\ C_{11} = \frac{-z_2^0 + v_1 z_1^o}{D_1},$$

$$B_{10} = \frac{z_3^0 \frac{p_2}{a_{13}} N_2 - z_4^o \frac{p_2}{a_{13}}}{D_2},\ B_{11} = \frac{-z_3^0 \frac{p_1}{a_{13}} N_1 + z_4^o \frac{p_1}{a_{13}}}{D_2},$$

$$C_{20} = C_{10} v_1,\ C_{21} = C_{11} v_2,\ B_{20} = B_{10} v_1,\ B_{21} = B_{11} v_2,$$

$$D_1 = \frac{p_2^2 - p_1^2}{a_{32} a_{13}},\ D_2 = \frac{p_1 p_2 (N_2 - N_1)}{a_{13}^2},$$

$$v_1 = \frac{1}{a_{32}}(-\frac{p_1^2}{a_{13}} - a_{31}),\ v_2 = \frac{1}{a_{32}}(-\frac{p_2^2}{a_{13}} - a_{31}),\ N_1 = \frac{a_{42}}{p_1 a_{32}}(\frac{p_1^2}{a_{13}} + a_{31}) - \frac{a_{41}}{p_1}$$

$$N_2 = \frac{a_{42}}{p_2 a_{32}}\left(\frac{p_2^2}{a_{13}} + a_{31}\right) - \frac{a_{41}}{p_2}, \; C_{12} = \frac{z_1^0\left(z_2^0 \frac{p_2 w_2}{b_{412}} - z_1^0 \frac{p_2}{b_{312}}\right) - z_2^0 z_1^0 \frac{p_1 p_2}{b_{322}}}{D_3},$$

$$C_{13} = \frac{-z_1^0\left(z_2^0 \frac{p_1}{b_{312}} + w_1 z_1^o \frac{p_1}{b_{412}}\right) - z_2^0 z_1^0 \frac{p_1 p_2}{b_{422}}}{D_3},$$

$$B_{12} = \frac{-z_3^0\left(z_3^0 \frac{p_1 w_2}{b_{422}} + z_4^o \frac{p_1}{b_{322}}\right) + z_3^0 z_4^0 \frac{p_1 p_2}{b_{312}}}{D_4},$$

$$B_{13} = \frac{z_4^0\left(z_3^0 \frac{p_2}{b_{322}} - w_1 z_4^o \frac{p_2}{b_{422}}\right) + z_3^0 z_4^0 \frac{p_1 p_2}{b_{412}}}{D_4},$$

$$C_{22} = C_{12} w_1 \,, \; C_{23} = C_{13} w_2 \,, \; B_{22} = B_{12} w_1 \,, \; B_{23} = B_{13} w_2 \,,$$

$$D_3 = \frac{b_{312} p_2^2 p_1 - b_{422} p_1^2 p_2}{b_{322} b_{412}}, \; D_4 = \frac{p_1 p_2 (b_{312} p_2 w_1 - b_{422} p_1 w_2)}{b_{312} b_{422}},$$

$$w_1 = \frac{3}{b_{312}}\left(-\frac{p_1^2}{b_{322}} - b_{412}\right), \; w_2 = \frac{3}{b_{312}}\left(-\frac{p_2^2}{b_{322}} - b_{412}\right),$$

$$C_{14} = \frac{-z_1^0\left(z_2^{02} \frac{p_2 w_4}{c_{4111} c_{3222}} + z_1^{02} \frac{p_2}{c_{4111} c_{3222}}\right) + s_1 z_2^0 z_1^{02} - s_2 z_1^{03} + s_3 z_2^{03}}{D_5},$$

$$C_{15} = \frac{z_2^0\left(z_1^{02} \frac{p_2 w_3}{c_{3111} c_{4222}} - z_2^{02} \frac{p_2}{c_{3111} c_{4222}}\right) - s_4 z_2^{02} z_1^0 + s_5 z_1^{03} - s_6 z_2^{03}}{D_5},$$

$$B_{14} = \frac{z_3^0\left(z_3^{02} \frac{p_1 w_4}{c_{3222} c_{4111}} - z_4^{02} \frac{p_1}{c_{3222} c_{4111}}\right) + (s_1 z_3^0 z_4^{02} - s_2 z_3^{03} + s_3 z_4^{03})\alpha_1}{D_6},$$

$$B_{15} = \frac{-z_4^0\left(z_3^{02} \frac{p_2 w_3}{c_{3222} c_{4111}} + z_4^{02} \frac{p_2}{c_{3222} c_{4111}}\right) + (-s_4 z_3^{02} z_4^0 + s_5 z_3^{03} - s_6 z_4^{03})\alpha_2}{D_6},$$

$$C_{24} = C_{14} w_3 \,, \; C_{25} = C_{15} w_4 \,, \; B_{24} = B_{14} w_3 \,, \; B_{25} = B_{15} w_4 \,,$$

$$D_5 = \frac{1}{16}\frac{c_{3112}c_{4122}p_2^2 p_1 - c_{3122}c_{4112}p_1^2 p_2}{c_{3111}b_{412} + c_{4222}b_{322}}.$$

The following theorem holds:

THEOREM 6.4.1 *The system of equations E2*

$$\frac{dz_n}{dt} = \sum_{p=1}^{N} a_{np} z_p + \sum_{p,q=1}^{N} b_{npq} z_p z_q + \sum_{p,q,r=1}^{N} c_{npqr} z_p z_q z_r +$$

$$+ \sum_{p,q,r,l=1}^{N} d_{npqrl} z_p z_q z_r z_l + \sum_{p,q,r,l,m=1}^{N} e_{npqrlm} z_p z_q z_r z_l z_m, \quad n = 1,2,3,\dots,N,$$

with initial conditions

$$z_n(0) = z_n^0, \; z_n^0 \in R,$$

admit for $\xi \le 0.3$, *bound-stated solutions at* $t \to \infty$, *of the form*

$$z_n(t) = \sum_{k,\eta=0} \{B_{nk}(\eta) F_k(\mu t, \eta) + C_{nk}(\eta) \dot{F}_k(\mu t, \eta)\},$$

where

$$F_k(\mu t, \eta) = C_k(\eta)(\mu t)^{k+1} \Phi_k(\mu t, \eta),$$

are Coulomb functions of vibration, and

$$\Phi_k(\mu t, \eta) = \sum_{m=k+1} (\mu t)^{m-k-1} A_m^{(k)}(\eta),$$

$$C_k(\eta) = \frac{\sqrt{k^2 + \eta^2}}{k(2k+1)} C_{k-1}(\eta), \quad C_0(\eta) = \frac{p\,|\Gamma(1 + i\eta)|}{2!},$$

$$A_{k+1}^{(k)} = 1, \quad A_{k+2}^{(k)} = \frac{\eta}{k+1},$$

$$(m+k)(m-k-1) A_m^{(k)} = 2\eta A_{m-1}^{(k)} - A_{m-2}^{(k)}, \quad m > k+2.$$

6.5 Cnoidal solutions

The aim of this section is to apply the cnoidal method to both problems E1 and E2. First we analyze a particular case which can reduce the equation of motion to a Weierstrass equation of the type (1.4.14) that admits an analytical solution represented by a sum of a linear superposition and a nonlinear superposition of cnoidal vibrations. We consider the uncoupled case $\alpha = \tilde{\alpha}$ ($I_1 = I_2 = I$)

$$\begin{cases} \ddot{\theta}_1 + w \sin\theta_1 + \alpha\cos\theta_1 = 0, \\ \ddot{\theta}_2 + \beta w \sin\theta_2 - \alpha\cos\theta_2 = 0, \end{cases} \tag{6.5.1}$$

where

$$\frac{M_1^2 g l_1}{Ik} = w, \quad \frac{M_1 M_2 g l_2}{Ik} = \beta w,$$
$$\frac{a}{l} = \xi, \quad \frac{a l M_1}{I} = \alpha, \quad \frac{M_1}{M_2} = m,$$

and initial conditions

$$\theta_1(0) = \theta_1^0, \quad \theta_2(0) = \theta_2^0 \,, \ \dot{\theta}_1(0) = \theta_{p1}^0, \quad \dot{\theta}_2(0) = \theta_{p2}^0. \tag{6.5.2}$$

Multiplying the first equation by $2\dot{\theta}_1$, and the second one by $2\dot{\theta}_2$, and integrating we obtain

$$\dot{\theta}_1^2 = 2w\cos\theta_1 - 2\alpha\sin\theta_1 + C_1, \quad \dot{\theta}_2^2 = 2w\beta\cos\theta_2 + 2\alpha\sin\theta_2 + C_2, \tag{6.5.3}$$

where C_i, $i = 1,2$, are integration constants. Approximating the trigonometric functions by series of five-order, (6.5.3) yields

$$\dot{\theta}_i^2 = P_i(\theta_i), \ i = 1,2, \tag{6.5.4}$$

where $P_i(\theta_i)$ are polynomials of fifth-order in θ_i

$$P_i(\theta_i) = a_{0i} + a_{1i}\theta_i + a_{2i}\theta_i^2 + a_{3i}\theta_i^3 + a_{4i}\theta_i^4 + a_{5i}\theta_i^5, \ i = 1,2, \tag{6.5.5}$$

with

$$a_{01} = 2w + C_1, \quad a_{02} = 2\beta w + C_2, \quad a_{11} = -2\alpha,$$

$$\begin{gathered} aA_{12} = 2\alpha, \quad a_{21} = 2w\tilde{c}, \quad a_{22} = 2\beta w\tilde{c}, \\ a_{31} = -2\alpha\tilde{a}, \quad a_{32} = 2\alpha\tilde{a}, \end{gathered} \tag{6.5.6}$$

$$\begin{gathered} a_{41} = 2w\tilde{d}, \quad a_{42} = 2w\tilde{d}\beta, \\ a_{51} = -2\alpha\tilde{b}, \quad a_{52} = 2\alpha\tilde{b}, \end{gathered}$$

where, for sake of simplicity, we take

$$-2w = C_1, \quad -2\beta w = C_2, \ a_{11} = -a_{12} = 2\alpha \neq 0.$$

We recognize in (6.5.4) the Weierstrass equations of the form (1.4.14), for $n = 5$

$$\dot{\theta}^2 = A_1\theta + A_2\theta^2 + A_3\theta^3 + A_4\theta^4 + A_5\theta^5, \tag{6.5.7}$$

where, by dropping the second index for constants a

$$A_1 = \frac{1}{2}a_1, \quad A_2 = a_2, \quad A_3 = \frac{3}{2}a_3, \quad A_4 = 2a_4, \quad A_5 = \frac{5}{2}a_5.$$

The assumed initial conditions for (6.5.7) are

$$\theta(0) = \theta_0, \quad \dot{\theta}(0) = \theta_{p0}. \tag{6.5.8}$$

We know that the equation (6.5.7) admits a particular solution expressed as an elliptic Weierstrass function that is reduced, in this case, to the cnoidal function cn

$$\wp(t+\delta'; g_2, g_3) = e_2 - (e_2 - e_3)\mathrm{cn}^2(\sqrt{e_1 - e_3}t + \delta'),$$

where δ' is an arbitrary constant, e_1, e_2, e_3, are the real roots of the equation $4y^3 - g_2 y - g_3 = 0$ with $e_1 > e_2 > e_3$, and g_2, g_3 are expressed in terms of the constants A_i, $i = 1, 2, ..., 5$, and satisfy the condition $g_2^3 - 27g_3^2 > 0$.

The general solution of (6.5.7) is expressed as a linear superposition of cnoidal vibrations as given by (1.4.13)

$$\theta_{lin} = \sum_{l=1}^{n} \alpha_l \mathrm{cn}^2[\omega_l t; m_l],$$

where $0 \le m_l \le 1$, and the angular frequencies ω_k, amplitudes α_k depend on the initial conditions. The Weierstrass equations admit also solutions expressed as a nonlinear superposition of cnoidal vibrations. To derive these solutions, we adopt the Krishnan solution (Krishnan)

$$\theta_{\mathrm{int}}(t) = \frac{\lambda \wp(t)}{1 + \mu \wp(t)}, \tag{6.5.9}$$

where $\wp(t)$ is the Weierstrass elliptic function given by (6.5.8), and λ, μ, are arbitrary constants. Substituting (6.5.9) into (6.5.7) we obtain four equations in λ, μ, g_2 and g_3

$$-2\lambda\mu^2 = A_1\mu^4 + A_2\lambda\mu^3 + A_3\lambda^2\mu^2 + A_4\lambda^3\mu + A_5\lambda^4, \tag{6.5.10a}$$

$$4\lambda\mu = 4A_1\mu^3 + 3A_2\lambda\mu^2 + 2A_3\lambda^2\mu + A_4\lambda^3, \tag{6.5.10b}$$

$$6\lambda + \frac{3}{2}\lambda\mu^2 g_2 = 6A_1\mu^2 + 3A_2\lambda\mu + A_3\lambda^2, \tag{6.5.10c}$$

$$\lambda\mu g_2 + 2\lambda\mu^2 g_3 = A_1\mu + A_2\lambda. \tag{6.5.10d}$$

From (6.5.10a), (6.5.10b) we get

$$6A_1\mu^4 + 5A_2\lambda\mu^3 + 4A_3\lambda^2\mu^2 + 3A_4\lambda^3\mu + 2A_5\lambda^4 = 0. \tag{6.5.11}$$

Let us consider the special case where (6.5.11) is reducible to

$$(R\lambda + S\mu)^4 = 0, \tag{6.5.12}$$

$$\mu = -\left(\frac{A_5}{3A_1}\right)^{1/4}\lambda, \tag{6.5.13}$$

with

$$R = (2A_5)^{1/4}\,,\ S = (6A_1)^{1/4}\,,\ A_2 = \frac{4}{5}RS^3\,,\ A_3 = \frac{3}{2}R^2S^2\,,\ A_4 = \frac{4}{3}R^3S\,. \tag{6.5.14}$$

We observe that both quantities R and S in (6.5.12) are both real or imaginary. In the last case this equation leads to $\mathrm{i}(R'\lambda + S'\mu)^4 = 0$ with R' and S' real quantities.

We calculate $A_1 = \frac{1}{2}a_1 = -\alpha \neq 0$ for the first equation (6.5.5), and $A_1 = \frac{1}{2}a_1 = \alpha \neq 0$, for the second equation (6.5.5), with $\alpha > 0$. In both cases we have $\frac{A_5}{3A_1} = \frac{5}{3}\tilde{b} > 0$.

Then

$$\mu = -\left(\frac{5\tilde{b}}{3}\right)^{1/4}\lambda\,. \tag{6.5.15}$$

Using (6.5.13) we have a unique constant λ from (6.5.10a) and (6.5.10b)

$$\lambda = -30(3A_1A_5)^{-3/2}\,. \tag{6.5.16}$$

The definition of the constant A_5 yields

$$A_5 = \frac{5}{2}a_5 = -5\alpha\tilde{b}\,,\quad \text{or}\quad A_5 = \frac{5}{2}a_5 = 5\alpha\tilde{b}\,.$$

From (6.5.16) we have for both situations

$$\lambda = -30(15\alpha^2\tilde{b})^{-3/2}\,,\ \alpha^2\tilde{b} > 0\,. \tag{6.5.17}$$

From (6.5.15) and (6.5.16) we obtain

$$\mu = 30\left(\frac{5\tilde{b}}{3}\right)^{1/4}(15\alpha^2\tilde{b})^{-3/2}\,. \tag{6.5.18}$$

The unknowns g_2 and g_3 are computable from (6.5.10c) and (6.5.10d). It results that λ, μ, g_2 and g_3 are always real. The expression (6.5.9) becomes

$$\theta_{nonlin}(t) = \frac{\lambda[e_2 - (e_2 - e_3)\mathrm{cn}^2(\sqrt{e_1 - e_3}\,t)]}{1 + \mu[e_2 - (e_2 - e_3)\mathrm{cn}^2(\sqrt{e_1 - e_3}\,t)]}\,, \tag{6.5.19}$$

where λ and μ are given by (6.5.17) and (6.5.18).

A general form for nonlinear part of the solution of (6.5.7) can be written as

$$\theta_{\mathrm{int}}(x,t) = \frac{\sum_{k=0}^{n}\beta_k \mathrm{cn}^2[\omega_k t; m_k]}{1 + \sum_{k=0}^{n}\lambda_k \mathrm{cn}^2[\omega_k t; m_k]}\,. \tag{6.5.20}$$

So, we can conclude that the solution of the equation (6.4.7) and the arbitrary initial conditions (6.4.8) consist of a linear superposition of cnoidal vibrations and a nonlinear interaction between them.

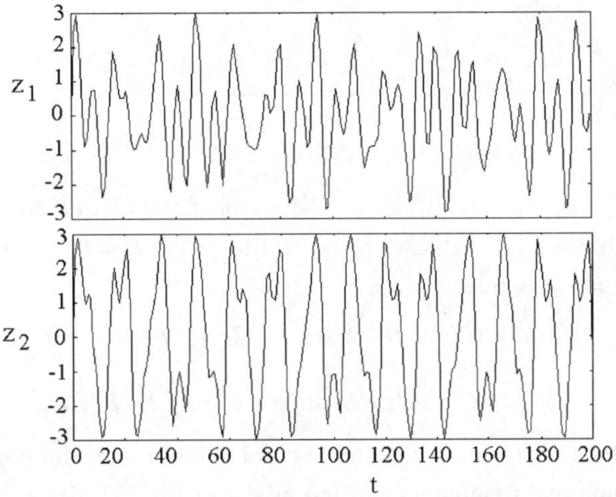

Figure 6.5.1 Solutions z_1 and z_2 calculated by cnoidal and LEM methods.

Next, we show that the cnoidal method is suitable for solving the system of equations E1. Suppose the representations of solutions $z_k(t)$, $k = 1,2,3,4$, of the form

$$z_k(t) = 2\frac{\partial^2}{\partial t^2}\log\Theta_n^{(k)}(\eta_1,\eta_2,...,\eta_n), \tag{6.5.21}$$

where the Θ function is defined by

$$\Theta_n^{(k)}(\eta_1,\eta_2,...,\eta_n) = \sum_{\substack{M_i^{(k)}=-\infty \\ 1\le i\le n}}^{\infty} \exp\left(\sum_{j=1}^{n} \mathrm{i}M_j^{(k)}\eta_j + \frac{1}{2}\sum_{i,j=1}^{n} M_i^{(k)}B_{ij}^{(k)}M_j^{(k)}\right), \tag{6.5.22}$$

with

$$\eta_j = -\omega_j t + \beta_j,\ 1 \le j \le n. \tag{6.5.23}$$

Here, k_j are the wave numbers, ω_j are the frequencies, β_j are the phases, and n the number of degrees of freedom for a particular solution.

The following result holds:

THEOREM 6.5.1 *The bound-state solutions* $z_k(t)$, $k = 1,2,3,4$, *of the system E1 given by (6.2.20), (6.2.21) can be written as*

$$z_k(t) = z_{lin}^{(k)}(\eta) + z_{\mathrm{int}}^{(k)}(\eta),\ \eta = [\eta_1\eta_2...\eta_n], \tag{6.5.24}$$

where $z_{lin}^{(k)}$ *represents a linear superposition of cnoidal vibrations*

$$z_{lin}^{(k)} = 2\sum_{l=0}^{n} \alpha_{l(k)}\mathrm{cn}^2[\omega_{l(k)}t; m_{l(k)}], \tag{6.5.25}$$

and $z_{\rm int}^{(k)}$ *represents a nonlinear interaction among the cnoidal vibrations*

$$z_{\rm int}^{(k)}(x,t) = \frac{\sum_{l=0}^{n} \beta_{l(k)} \mathrm{cn}^2[\omega_{l(k)}t; m_{l(k)}]}{1+\sum_{l=0}^{n} \lambda_{l(k)} \mathrm{cn}^2[\omega_{l(k)}t; m_{l(k)}]}, \tag{6.5.26}$$

where $0 \le m \le 1$, ω_k *and* α_k, β_k, γ_k *are determined from initial conditions (6.2.18).*

In the small oscillations case $M_1 = M_2$, $l_1 = l_2$, $I_1 = I_2 = I$ ($\alpha = \tilde{\alpha}$, $\beta = 1$) the solutions for $m = 0$ are given by

$$z_1 = C_{10}\mathrm{cn}p_1t + B_{10}\mathrm{sn}p_1t + C_{11}\mathrm{cn}p_2t + B_{11}\mathrm{sn}p_2t,$$

$$z_2 = C_{20}\mathrm{cn}p_1t + B_{20}\mathrm{sn}p_1t + C_{21}\mathrm{cn}p_2t + B_{21}\mathrm{sn}p_2t.$$

The equivalence between LEM and cnoidal solutions for the problem E2 ($\xi \le 0.3$) is shown by numerical treatment of differential equations (Collatz, Halanay, Scalerandi *et al.*).

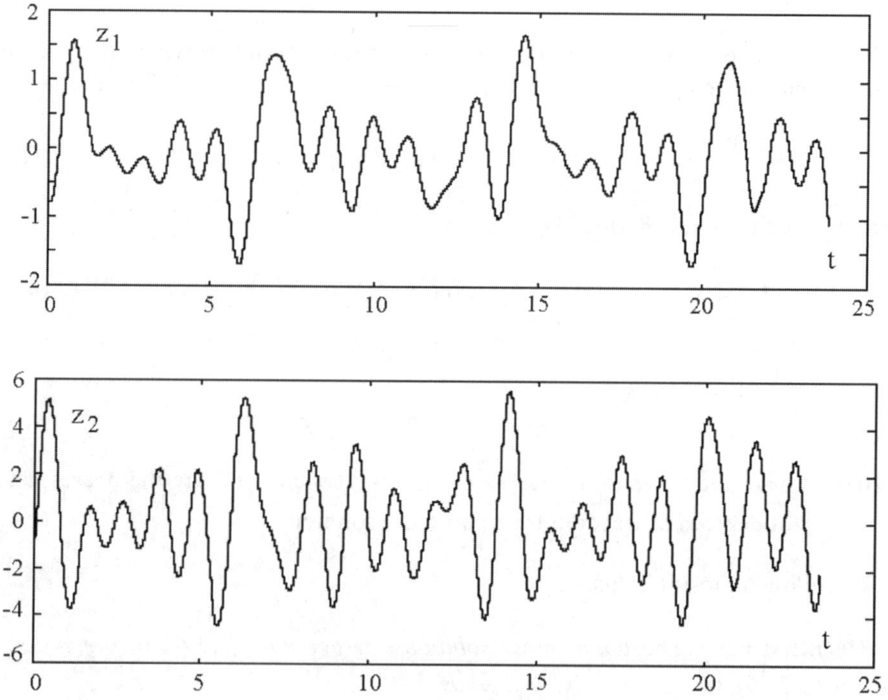

Figure 6.5.2 The transient solutions z_1 and z_2 of the pendulum.

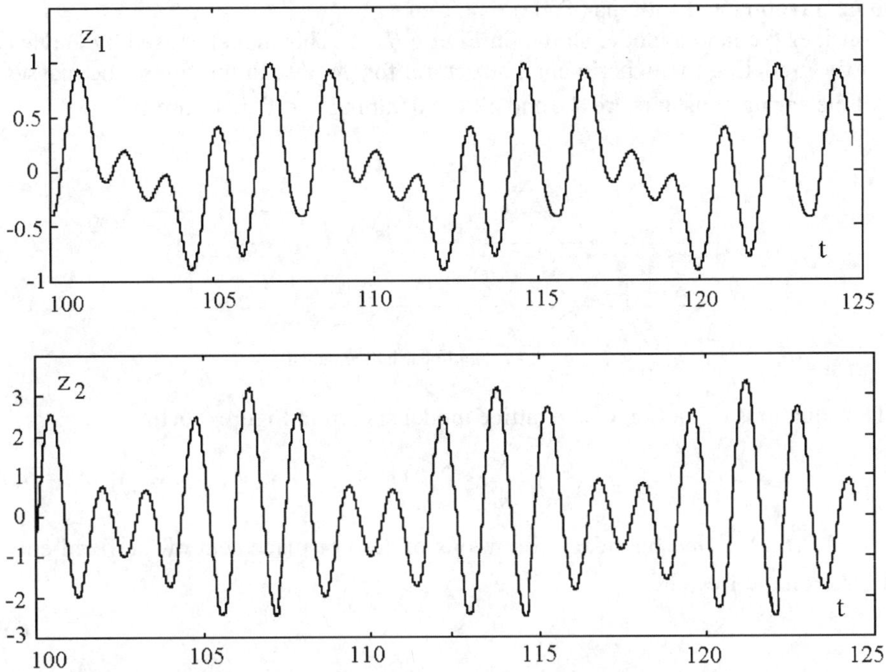

Figure 6.5.3 Stabilized solutions z_1 and z_2 of the pendulum.

Consider an example $g = 10\,\mathrm{m/s}^2$, $l_1 = l_2 = \dfrac{a}{2}$, $M_2 = 10\,\mathrm{kg}$, for which $\xi = 0.25$, $m = 1.5$, $k = 70$. In Figure 6.5.1 are represented these solutions. As a result the solutions calculated by LEM and cnoidal methods coincide in a proportion of 97.7%. No visible differences are observed.

Figure 6.5.2 displays the transient evolution of solutions $z_1(t)$ and $z_2(t)$ for $m = 1$, $\xi = 0.9$, $k = 1.6 \times 10^4$, and the initial conditions $z_1 = z_2 = -0.8$, $z_3 = z_4 = 0.02$, calculated by the cnoidal method. Figure 6.5.3 shows the stabilized solutions $z_1(t)$ and $z_2(t)$ for $t > 100$.

6.6 Modal interaction in periodic structures

Davies and Moon have studied in 2001 an experimental structure consisting of nine harmonic oscillators coupled through buckling sensitive elastica. The structure was modeled by a modified Toda lattice, and analytical results confirm the soliton-like nature of waves observed in the structural motions. These systems exhibit complex nonlinear wave propagation including solitons. The modal interactions and the modal trading of energy were studied in the early 1950s by Fermi, Pasta and Ulam. Elastic models of crystal lattices are developed by Born and Huang, Teodosiu. The Toda analytic solutions for a lattice with exponential interactions take the form of both periodic traveling and solitons (Toda).

Consider the lattice model shown in Figure 6.6.1. This model is used by Davies and Moon for modeling the experimental structure, for $N = 9$. In the figure the masses are m, all the spring constants are k, and all the damping coefficients are c.

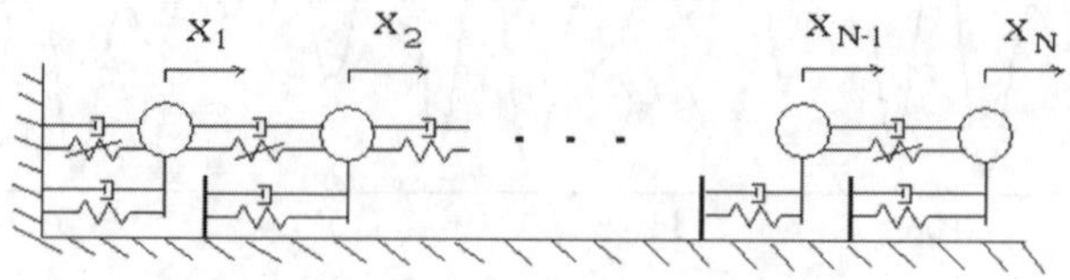

Figure 6.6.1 Nonlinear lattice model.

The equations of motion of the lattice model shown in Figure 6.6.1 are

$$\ddot{x}_j = -\frac{c}{m}(3\dot{x}_j - \dot{x}_{j+1} - \dot{x}_{j-1}) - \frac{k}{m}x_j + \frac{1}{m}[F(x_{j+1} - x_j) - F(x_j - x_{j-1})], \tag{6.6.1}$$

with $j = 1, 2, ..., N$, and boundary conditions of zero displacement of the fixed end and zero force on the free end

$$x_0 = 0, \; x_{N+1} = x_N, \tag{6.6.2}$$

where m is the block mass, k is the cantilevered beam stiffness, c the damping coefficient, F represents the nonlinear force-displacement relationship for the buckling elements, and N the number of masses. The damping term $3\dot{x}_j$ in the motion equation comes about because the dampers between masses and the ones attached between the masses and ground are taken to have identical coefficients. The expression for the force-displacement behavior of the buckling elements is the same as that used by Toda to model the interaction forces in an atomic lattice (Toda and Wadati, Toda)

$$F(x_{j+1} - x_j) = \frac{\gamma}{b}[\exp(b(x_{j+1} - x_j)) - 1], \tag{6.6.3}$$

where γ is the element's linearised stiffness, and b determines the strength of the nonlinearity. For the atoms interaction in a lattice of a solid the Morse interaction force is valid in quantum mechanics of an electron motion, α being a constant

$$F(x_{j+1} - x_j) = F_0[\exp(-2\alpha(x_{j+1} - x_j)) - 2\exp(-\alpha(x_{j+1} - x_j))].$$

Morse force of interaction governs precisely the nearest-neighbor interaction in an anharmonic lattice of the atoms in a diatomic molecule of solids in a continuum limit. But, no exact solutions are found for elasticity problems governed by Morse force. For Toda interaction force there are some exact explicit solutions for the model. With k and c set to zero, equations (6.6.1) and (6.6.3) reduce to a Toda lattice, and the functional form of the nonlinear soliton solution is given by

$$x_{j+1} - x_j = \frac{1}{b}\ln[\frac{m\beta^2}{\gamma}\operatorname{sech}^2(kj - \beta t) + 1]. \tag{6.6.4}$$

Substitution of (6.6.4) into (6.6.1) and (6.6.3) with k and c set to zero, shows that it solves these equations if we have

$$\beta = \sqrt{\frac{\lambda}{m}}\sinh k. \tag{6.6.5}$$

If $\gamma > 0$ and $b < 0$, $x_{j+1}(t) - x_j(t) < 0$ for all values of j and t. This corresponds to the compressive atomic lattice waves studied by Toda. If $\gamma > 0$ and $b > 0$, then $x_{j+1}(t) - x_j(t) > 0$, we have the tensile waves, which propagate to the speed $c = \frac{\beta a}{k}$, where a is the distance between masses. The speed of this soliton is dependent upon the amplitude.

For the case of small displacement and zero damping, (6.6.1) and (6.6.3) reduce to the linear conservative system defined by

$$\ddot{x}_j + \frac{k}{m}x_j + \frac{\gamma}{m}(x_{j+1} - 2x_j + x_{j-1}) = 0, \quad j = 1, 2, ..., N. \tag{6.6.6}$$

Substituting a periodic traveling wave solution of the form $x_j(t) = A\exp[\mathrm{i}(kaj - \omega t)]$, into (6.6.6) we obtain the dispersion relation

$$\omega = \pm\sqrt{\frac{4\gamma}{m}\sin^2(\frac{ka}{2}) + \frac{k}{m}}, \tag{6.6.7}$$

where ω is the wave frequency, k is the wave number and a is the spacing between adjacent masses. From (6.6.7) we see that the system allows only in the Brillouin zone of frequencies to propagate without attenuation. The equation (6.6.6) has the solution

$$x_j(t) = \sum_{k=1}^{N} A_k(t)\sin\frac{(2k-1)j\pi}{2N+1}, \quad j = 1, 2, ..., N, \tag{6.6.8}$$

where A_k are time-dependent coefficients. The energy of the k-mode of vibration is (Davies and Moon)

$$E_k = \frac{2N+1}{4}\left(\frac{m}{2}\dot{A}_k^2 + A_k^2(\frac{k}{2} + 2\gamma\sin^2\frac{(2k-1)\pi}{2(2N+1)})\right). \tag{6.6.9}$$

If initial conditions are small the nonlinear model (6.6.1) will have approximately linear behavior, and the modal energies given by (6.6.9) will be nearly constant in time. For large initial conditions, the nonlinearities may lead to complex modal interactions (Duncan *et al.*).

The experimental results of Davies and Moon in studying a periodically reinforced structure with strong buckling nonlinearity, have shown that the modal interaction is characterized by highly localized waves. These waves were shown to be similar to Toda solitons.

Let us analyze the modal interaction by using the Toda interacting equations. We know that the N-soliton solution of a nonlinear equation can be regarded as a system of N-interacting solitons, each of which becomes a single soliton as $t \to \pm\infty$. When we have N single solitons initially apart in space they interact with each other and become apart without exchanging their identities.

To study the interaction modal solitons, we consider (6.6.1) and (6.6.3) under the form of the original Toda equation

$$\mathrm{d}\frac{\mathrm{d}V(n)}{1+V(n)} = V(n+1)+V(n-1)-2V(n), \tag{6.6.10}$$

or

$$\mathrm{d}^2 \ln[1+V(n)] - \Delta^2 V = 0, \tag{6.6.11}$$

$$\Delta^2 V(n) = V(n+1)+V(n-1)-2V(n). \tag{6.6.12}$$

Here, the operator d is $\mathrm{d} = \partial/\partial t$ and $n = -\infty,\ldots,-2,-1,0,1,2,\ldots,\infty$.

Yoneyama proposes the following form of interacting Toda equations, which represent a natural extension of the Toda equation (6.6.10)

$$\mathrm{d}\frac{\mathrm{d}V_i(n)}{1+V(n)} = V_i(n+1)+V_i(n-1)-2V_i(n),\ i=1,2,\ldots,N, \tag{6.6.13}$$

with the total wave $V(n)$ is

$$V(n) = \sum_{i-1}^{N} V_i(n). \tag{6.6.14}$$

The equation (6.6.13) is the Toda–Yoneyama equation. In the following we present the explicit form of the solution $V_i(n)$ given by Yoneyama. We see that summing up (6.6.13) and using (6.6.14) we obtain (6.6.10). The solution $V(n)$ has the form

$$V(n) = \mathrm{d}^2 \ln f. \tag{6.6.15}$$

The function f is given by

$$f(\gamma_1,\gamma_2,\ldots,\gamma_N) = \det[I + B(n+1)], \tag{6.6.16}$$

where I and B are $N\times N$ matrices

$$I_{kl} = \delta_{kl},\ B_{kl}(n) = \frac{1}{1-z_k z_l}\phi_k(n)\phi_l(n),\ \phi_i(n) = C_i \exp\gamma_i(n),\ \gamma_i(n) = \beta_i t - \alpha_i n, \tag{6.6.17}$$

with arbitrary real constants α and C, $1 \le k$, $l \le N$, $i=1,2,\ldots,N$, and

$$z_i = \pm\exp(-\alpha_i),\ \beta_i = \frac{z_i^{-1} - z_i}{2}.$$

Let us introduce N-independent time variables t_i, $i=1,2,\ldots,N$, and define as

$$\Phi_i(n) = C_i \exp\Gamma_i(n),\ \Gamma_i(n) = \beta_i t_i - \alpha_i n, \tag{6.6.18}$$

$$F(\Gamma_1, \Gamma_2, ..., \Gamma_N) = A(\Gamma_i)\det[I + B(n+1)], \tag{6.6.19}$$

where B is a $N \times N$ matrix with elements

$$B_{kl}(n) = \frac{1}{1 - z_k z_l}\Phi_k(n)\Phi_l(n),\ 1 \le k,\ l \le N, \tag{6.6.20}$$

$$A(\Gamma_i) = \exp[\sum_{i=1}^{n} \alpha_i(n)\Gamma_i + \beta(n)], \tag{6.6.21}$$

with arbitrary functions $\alpha_i(n)$ and $\beta(n)$. Introduce the operators

$$\hat{\mathrm{d}}_i F(\Gamma_1, \Gamma_2, ..., \Gamma_N) = (\partial / \partial t_i) F(\Gamma_1, \Gamma_2, ..., \Gamma_N)|_{t_1 = t_2 = ... = t_N = t}. \tag{6.6.22}$$

The solution of (6.6.13) is given by (6.6.15), where $V(n)$ is

$$V(n) = \mathrm{d}\sum_k \hat{\mathrm{d}}_\mathrm{k} \ln F = \sum_k \mathrm{d}\sum_k \hat{\mathrm{d}}_\mathrm{k} \ln F. \tag{6.6.23}$$

Here $\mathrm{d}f = \sum_{k=1}^{N} \hat{\mathrm{d}}_k F$ and $\mathrm{d}\hat{\mathrm{d}}_\mathrm{i} = \hat{\mathrm{d}}_\mathrm{i} \sum_{k=1} \mathrm{d}_k$. Substituting (6.6.23) into (6.6.12) we have

$$\sum_i \hat{\mathrm{d}}_\mathrm{i} G = 0,\ G = \sum_k \mathrm{d}_\mathrm{k} \ln[1 + (\sum_l \hat{\mathrm{d}}_\mathrm{l})^2 F] - \Delta^2 \sum_m \mathrm{d}_\mathrm{m} \ln F. \tag{6.6.24}$$

This equation is satisfied if G is a constant and if $\hat{\mathrm{d}}_\mathrm{i} G = 0$ for each i. So, we must have

$$\hat{\mathrm{d}}_\mathrm{i} \sum_k \mathrm{d}_\mathrm{k} \ln[1 + (\sum_l \hat{\mathrm{d}}_\mathrm{l})^2 F] = \hat{\mathrm{d}}_\mathrm{i} \Delta^2 \sum_m \mathrm{d}_\mathrm{m} \ln F, \tag{6.6.25}$$

or

$$\mathrm{d}\frac{\mathrm{d}(\mathrm{d}\hat{\mathrm{d}}_\mathrm{i} \ln F)}{1 + V(n)} = \Delta^2(\mathrm{d}\hat{\mathrm{d}}_\mathrm{i} \ln F). \tag{6.6.26}$$

Therefore, the solution of (6.6.13) is

$$V(n) = \mathrm{d}\sum_{i=1}^{N} \hat{\mathrm{d}}_\mathrm{i} \ln F = \sum_{i=1}^{N} V_i(n). \tag{6.6.27}$$

The explicit form of $V_i(n)$ is obtained by defining a symmetric $N \times N$ matrix $\psi(n,m)$ and three diagonal $N \times N$ matrices β, λ and Z of elements

$$\Psi_{ij}(n,m) = \Psi_{ji}(n,m) = \frac{1}{2}[\phi_i(n)\phi_j(m) + \phi_i(m)\phi_j(n)], \tag{6.6.28}$$

$$\beta_{ij} = \delta_{ij}\beta_j,\ \lambda_{ij} = \delta_{ij}\lambda_j,\ Z_{ij} = \delta_{ij} z_j,\ \lambda_i = \frac{z_i^{-1} + z_i}{2}. \tag{6.6.29}$$

Also define the N-component column vector $\phi(n)$ of elements $\phi_i(n) = C_i \exp\gamma_i(n)$. In these conditions, the identities hold

$$\mathrm{d}B(n+1) = \beta B(n+1) + B(n+1)\beta\,,\ \mathrm{d}B(n+1) = \Psi(n,n+1)\,. \tag{6.6.30}$$

Also, we have

$$\Psi(n,n+1) \neq \phi(n)\phi^T(n+1)\,, \tag{6.6.31}$$

where the transpose of $\phi(n)$ is $\phi^T(n) = (\phi_1(n), \phi_2(n), ..., \phi_N(n))$, and

$$\mathrm{tr}[P(n+1)\Psi(n,n+1)P(n+1)] = \mathrm{tr}[P(n+1)\phi(n)\phi^T(n+1)P(n+1)]\,, \tag{6.6.32}$$

$$P(n) = \frac{I}{I + B(n)}\,. \tag{6.6.33}$$

THEOREM 6.6.1 (Yoneyama) *The explicit form of $V_i(n)$ is given by*

$$\begin{aligned} V_i(n) &= \mathrm{d}\hat{\mathrm{d}}_i \ln F = 2\beta_i[P(n+1)\mathrm{d}B(n+1)P(n+1)]_{ii} = \\ &= 2\beta_i[P(n+1))\phi(n)\phi^T(n+1)P(n+1)]_{ii} = \\ &= 2[\beta P(n+1)\phi(n)\phi^T(n+1)P^T(n+1)]_{ii} = 2\beta_i P_i(n+1)\phi_i(n)P_i(n+1)\phi_i(n+1). \end{aligned} \tag{6.6.34}$$

Proof. From (6.6.33) and taking account that for any matrix A,

$$\mathrm{d}A^{-1} = -A^{-1}(\mathrm{d}A)A^{-1}\,, \tag{6.6.35}$$

it results

$$-\mathrm{d}\ln\det P(n+1) = \mathrm{tr}\,\{\mathrm{d}\ln[I + B(n+1)]\} = \mathrm{tr}[\frac{I}{I + B(n+1)}\mathrm{d}B(n+1)]\,. \tag{6.6.36}$$

From (6.6.30) and (6.6.33) it results

$$\mathrm{tr}[\frac{I}{I + B(n+1)}\mathrm{d}B(n+1)] = 2\sum_m \beta_m[\frac{B(n+1)}{I + B(n+1)}]_{mm} = 2\sum_m \beta_m[I - P(n+1)]_{mm}\,. \tag{6.6.37}$$

The equations (6.6.15), (6.6.16), (6.6.33), (6.6.36) and (6.6.30) give

$$V(n) = \mathrm{d}\sum_m 2\beta_m[I - P(n+1)]_{mm} = 2\sum_m \beta_m P_{mk}(n+1)\phi_k(n)\phi_l(n+1)P_{lm}(n+1)\,. \tag{6.6.38}$$

From (6.6.38) we obtain

$$\begin{aligned} V(n) &= 2tr[\beta P(n+1)\phi(n)\phi^T(n+1)P^T(n+1)] = \\ &= 2\sum_i \beta_i P_i(n+1)\phi_i(n)P_i(n+1)\phi_i(n+1). \end{aligned} \tag{6.6.39}$$

Finally, (6.6.40), (6.6. 20) and (6.6.22) yield

$$\mathrm{d}_i B_{kl}(n+1) = \beta_i(\delta_{ik} + \delta_{il})B_{kl}(n+1)\,,\ \hat{\mathrm{d}}_i \ln\det P(n+1) = 2\beta_i[I - P(n+1)]_{ii}\,. \tag{6.6.40}$$

Chapter 7

DYNAMICS OF THE LEFT VENTRICLE

7.1 Scope of the chapter

The heart consists of two pumps (the right and the left) connected in a series that pump blood through the circulatory system. The left ventricle generates the highest pressures in the heart, about 16 kPa, which is four times the pressure developed by the right ventricle (Taber). The left ventricle receives the most attention in the literature because most infarcts occur in this chamber. The left ventricle is a thick-walled body composed of myocardium between a thin outer membrane (epicardium) and an inner membrane (endocardium). The dynamics of the left ventricle is the result of the contractile motion of the muscle cells in the left ventricular wall. Heart muscle is a mixture of muscle and collagen fibers, coronary vessels, coronary blood and the interstitial fluid. The fibers wind around the ventricle, and their orientation, relative to the circumferential direction, changes continuously from about 60° at the endocardium to -60° at the epicardium.

This anisotropy influences the transmural distribution of wall stress. Huyghe and his coworkers, Van Campen, Arts and Heethaar, developed a theory of myocardial deformation and intramyocardial coronary flow. In this theory the tissue is considered as a two-phase mixture (a solid phase and a fluid phase representing the different coronary microcirculatory components).

A central problem in modeling the dynamics of the heart is in identifying functional forms and parameters of the constitutive equations, which describe the material properties of the resting and active, normal and diseased myocardium. Recent models capture some important properties including: the nonlinear interactions between the responses to different loading patterns; the influence of the laminar myofiber sheet architecture; the effects of transverse stresses developed by the myocytes; and the relationship between collagen fiber architecture and mechanical properties in healing scar tissue after myocardial infarction.

We consider in this chapter the cardiac tissue as a mixture of an incompressible solid and an incompressible fluid. Following studies by Van Campen and his coworkers, Huyghe, Bovendeerd and Arts, we construct a model in which the constitutive laws are specified within the broad framework of the intrinsic assumptions of the theory. The cnoidal method is applied to solve the set of nonlinear dynamic equations of the left ventricle. By using the theta-function representation of the solutions and a genetic algorithm, the ventricular motion is describable as a linear superposition of cnoidal pulses and additional terms, which include nonlinear interactions among them.

In the last section, the evolution equations governing the human cardiovascular system and distortions of waves along this system for perturbed initial conditions, which are responsible to the energy influx conditions, are analyzed. The formation of asymmetric solitons from initial data under and above threshold is the principal topic of this analysis.

We refer the readers to the works by Van Campen *et al.* (1994), Munteanu and Donescu (2002), Munteanu *et al.* (2002a, c), and Chiroiu V. *et al.* (2000).

7.2 The mathematical model

The underformed heart, in a stress-free reference state, is modeled as a super ellipsoid surface S defined by the implicit equations (Bardinet *et al.* 1994)

$$\left[\left(\frac{x}{a_1}\right)^{\frac{2}{c_2}}+\left(\frac{y}{a_2}\right)^{\frac{2}{c_2}}\right]^{\frac{c_2}{c_1}}+\left(\frac{z}{a_3}\right)^{\frac{2}{c_1}}=1, \tag{7.2.1}$$

where the constants a_i, $i=1,2,3$ and c_i, $i=1,2$, are given by

$$c_1=c_2=0.773,\ a_1=a_2=0.892R,\quad a_3=R, \tag{7.2.2}$$

with $R=0.0619\,\mathrm{m}$ for a particular heart considered in this paper. For a sphere we have $c_1=c_2=1$ and $a_1=a_2=a_3=R$.

Representation of the cylindrical coordinates is given in Figure 7.2.1. The z-axis corresponds to the z-axis of inertia of the super ellipsoid model. The muscle fibers in the ventricular wall are assumed to be parallel to the endocardial and epicardial surfaces.

The governing equations of the dynamics of the left ventricle are presented by following the above-mentioned papers. Cardiac muscle is considered to be a mixture of two phases, a solid phase and a fluid phase.

The equations are derived from the general equations of the continuum theory of mixtures (Truesdell).

Nomenclature:

V,	actual volume of the heart,
$x=(r,\theta,z)$,	spatial cylindrical (Eulerian) coordinates, centred in O, $\begin{pmatrix}x\\y\\z\end{pmatrix}\to\begin{pmatrix}r\\\theta\\z\end{pmatrix}$,
$X_i, i=1,2,3$,	material cylindrical coordinates (Lagrangian coordinates), corresponding to a reference state which may be subject to an initial finite deformation,
t,	time coordinate,

σ^s, effective Cauchy stress in the solid representing the stress induced by the deformation in the absence of fluid and measured per unit bulk surface,

p, intramyocardial pressure representing the stress in the liquid component of the bi-phases mixture,

$\sigma = \sigma^s - pI$, total Cauchy stress tensor in the mixture, $\sigma = \sigma^T$,

u, displacement vector

$$\begin{pmatrix} u_1 \\ u_2 \\ u_3 \end{pmatrix} \to \begin{pmatrix} u_r \\ u_\theta \\ u_z \end{pmatrix},$$

q, Eulerian spatial fluid flow vector,

K^0, permeability tensor of the underformed tissue,

N^b, averaged porosity of the underformed tissue,

c^c, volumetric modulus of the empty solid matrix,

$H = \nabla u$, displacement gradient,

F, deformation gradient tensor

$$F = 1 + H,$$

$J = \det F > 0$, Jacobean of the deformation,

C, isotropic energy function defined as

$$C = \frac{c^c}{2}(J-1)^2,$$

which is zero in the underformed state and positive elsewhere,

W, strain energy function, zero in the underformed state and positive elsewhere,

K, permeability tensor given by

$$K = (\frac{J-1}{N^b} + 1)^2 K^0,$$

E, Green–Lagrange strain tensor defined as

$$E = \frac{1}{2}(F^T F - I),$$

$S(E,t)$, effective second Piola–Kirchhoff stress tensor

$$S = JF^{-1}\sigma^s (F^{-1})^T, \; S = S^T,$$

split into two components $S = S^a + S^p$,

$S^a(E,t)$, active stress tensor, $S^p(E,t)$, the passive stress tensor, split into two components $S^p = S^c + S^s$,

$S^c(E)$, component of the passive stress tensor resulting from elastic

	volume change of the myocardial tissue, zero in the underformed state,
$S^s(E,t)$,	component of the passive stress tensor resulting from viscoelastic shape change of the myocardial tissue,
$S^e(E)$,	anisotropic (orthotropic) elastic response of the tissue. S^e is zero in the underformed state,
$G(t)$,	a scalar relaxation function,
T^a,	first Piola–Kirchhoff active stress (not symmetric) related to the second Piola–Kirchhoff active stress by $S^a = F^{-1}T^a$,
l,	current sarcomere length,
v,	velocity of shortening of the sarcomeres $v = \dfrac{dl}{dt}$,
ϕ_{helix},	angle between the muscle fiber direction and the local circumferential direction, varying from 60° at the endocardium through 0° in the midwall layers to -60° at the epicardium, while ϕ_{trans} is kept zero (Figure 7.2.1),
ϕ_{trans},	angle between the local circumferential direction and the projection of the fiber on the plane perpendicular to the local longitudinal direction, varying from 13.5° at the base through 0° at the equator to -13.5° at the apex,
$\nabla = (\dfrac{\partial}{\partial r}, \dfrac{1}{r}\dfrac{\partial}{\partial \theta}, \dfrac{\partial}{\partial z})$	gradient operator with respect to the current configuration.

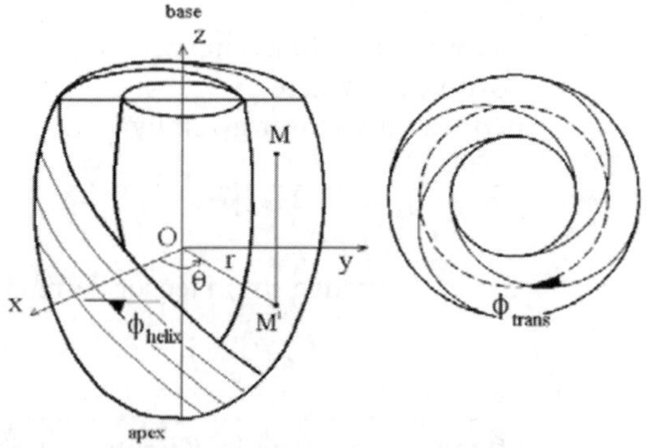

Figure 7.2.1 Cylindrical coordinates and definition of the angle ϕ_{helix}, ϕ_{trans} (Van Campen *et al.*)

The equations of the beating left ventricle are composed from:

1. The equilibrium equation of the deformed myocardium (by neglecting the inertia forces)

$$\nabla\sigma^s - \nabla p = 0. \tag{7.2.3}$$

2. Darcy's law in Eulerian form (by neglecting the transmural pressure differences across blood vessel walls)

$$q = -K\nabla p\,, \tag{7.2.4}$$

with $K = (\frac{J-1}{N^b}+1)^2 K^0$, the parameters K^0 and N^b being specified.

3. Continuity equation (conservation of mass)

$$\nabla\dot{u} + \nabla q = 0\,. \tag{7.2.5}$$

4. Passive constitutive laws

$$S^c = \frac{\partial C}{\partial E}\,, \tag{7.2.6}$$

where

$$S = S^a + S^p\,,\ S^p = S^c + S^s\,, \tag{7.2.7}$$

$$C = \frac{c^c}{2}(J-1)^2\,,\ J = \det F > 0\,, \tag{7.2.8}$$

$$S(E,t) = JF^{-1}\sigma^s (F^{-1})^T\,,\ S = S^T\,, \tag{7.2.9}$$

$$E = \frac{1}{2}(F^T F - I)\,,\ F = 1 + H\,,\ H = \nabla u\,. \tag{7.2.10}$$

C is the isotropic energy function, E is the Green–Lagrange strain tensor, and $S(E,t)$ the effective second Piola–Kirchhoff stress tensor, split into an active stress $S^a(E,t)$ and a passive stress $S^p(E,t)$. The passive stress tensor is split into a component resulting from elastic volume change of the myocardial tissue $S^c(E)$, and a component resulting from viscoelastic shape $S^s(E,t)$ described in the form of quasi-linear viscoelasticity (Fung) as

$$S^s = \int_{-\infty}^{t} G(t-\tau)\frac{\mathrm{d}}{\mathrm{d}\tau}S^e \mathrm{d}\tau\,, \tag{7.2.11}$$

$$S^e = \frac{\partial W}{\partial E}\,, \tag{7.2.12}$$

where S^e is the anisotropic elastic response of the material, $G(t)$ is a scalar function (reduced relaxation function) derived from a continuous relaxation spectrum, W is the potential energy of deformation per unit volume (or elastic potential).

The form of the strain energy C (equation 7.2.4) and W are chosen so that C and W are zero in the unstrained state and positive elsewhere, and $S^c(E)$ and S^e are zero in the underformed state. The expression (7.2.4) satisfies those conditions.

Some expressions for W satisfying the above conditions were proposed in literature: the full orthotropic behavior, the transversely isotropic behavior with respect to the fiber orientation (Bovendeerd *et al.*), W is an exponential function of E_{ij} (Huyghe, Fung).

Because passive ventricular myocardium is nearly incompressible, biaxial tissue testing is a valuable method for characterizing its material properties.

Demer and Yin were the first to report the results of such measurements on passive myocardium. They demonstrated that biaxial loaded passive myocardium behaves like a nonlinear, anisotropic, viscoelastic material that can be approximated as pseudoelastic.

Huyghe and coworkers concluded that quasi-linear viscoelasticity does not adequately describe the material properties of passive ventricular myocardium. Therefore, resting myocardium is frequently modeled as a finite elastic material using a hyperelastic pseudo-strain function.

In this work we have investigated four approaches for the functional form of W :

1. The polynomial function for a subclass of transverse isotropy (Humphrey *et al.*, Huyghe *et al.* 1991, Bovendeerd *et al.*)

$$W = c_1(\alpha-1)^2 + c_2(\alpha-1)^3 + c_3(l_1-3) + c_4(l_1-3)(\alpha-1) + c_5(l_1-3)^2, \quad (7.2.13)$$

where l_1 is the first principal strain invariant and the transversely isotropic invariant α is the extension ratio in the fiber direction

$$l_1 = 2\mathrm{tr}E + 3, \ \alpha = \sqrt{2E_{jj}+1}, \quad (7.2.14)$$

where E_{ij} is the Lagrangian Green's strain.

2. The transversely isotropic 3D strain energy function (Guccione *et al.*, Guccione and McCulloch, Fung, Huyghe)

$$W = \frac{C}{2}(\exp Q - 1), \quad (7.2.15)$$

where

$$Q = b_1E_{ff}^2 + b_2(E_{cc}^2 + E_{rr}^2 + 2E_{cr}E_{rc}) + 2b_3(E_{fc}E_{cf} + E_{fr}E_{rf}), \quad (7.2.16)$$

with E_{ij} representing strain components referred to a system of local fiber (*f*), cross-fiber in-plane (*c*) and radial (*r*) coordinates.

3. The form developed by Costa *et al.*, Holmes *et al.*,

$$W = \frac{C}{2}(\exp Q - 1), \quad (7.2.17)$$

where

$$Q = c_1E_{ff}^2 + c_2E_{ss}^2 + c_3E_{nn}^2 + 2c_4E_{fs}E_{sf} + 2c_5E_{fn}E_{nf} + 2c_6E_{sn}E_{ns}. \quad (7.2.18)$$

The material parameters c_1, c_2 and c_3 represent the stiffness along the fiber axis, the sheet axis and the sheet normal axis, respectively. The ratio of $\frac{c_2}{c_3}$ governs anisotropy in the plane normal to the local fiber axis, with unity indicating transverse isotropy. The parameter c_4 represents the shear modulus in the sheet plane, and c_5 and c_6 represent shear stiffness between adjacent sheets.

4. The expression of the ion-core (Born–Mayer) repulsive energy (Delsanto *et al.*, Jankowski and Tsakalakos)

$$W = \frac{0.5}{V}\int_V \alpha(\phi)\exp[-\beta(\phi)R]\mathrm{d}V\,, \tag{7.2.19}$$

where $\alpha(\phi(x))$ is the repulsive energy function, $\beta(\phi(x))$ the repulsive range function and V is the heart volume, and

$$R = \sqrt{(x-x_0)^2+(y-y_0)^2+(z-z_0)^2}\,,$$

with (x_0, y_0, z_0) an arbitrary point. We suppose that $\alpha(\phi)$ and $\beta(\phi)$ depend on the angles angle ϕ_{helix} ϕ_{trans} in the form

$$\alpha(\phi) = \alpha_1(\phi_{helix})\alpha_2(\phi_{trans})\,,$$

$$\beta(\phi) = \beta_1(\phi_{helix})\beta_2(\phi_{trans})\,. \tag{7.2.20}$$

We mention that the values of the angles depend on the position x.

A common problem with estimating parameters of the strain energy function, whether directly from isolated tissue testing or indirectly from strains measured in the intact heart, is that the functions are nonlinear.

The kinematic response terms, whether principal invariants (7.2.13) or strain components (7.2.15), (7.2.16), usually co-vary under any real loading condition. This is probably the main factor responsible for the very wide variation in parameter estimates between individual mechanical tests that is typically reported. Although the first three equations are well motivated, a difficulty with any constitutive model based on biaxial tissue tests is uncertainty as to how the biaxial properties of isolated tissue slices are related to the properties of the intact ventricular wall.

Though the energy function (7.2.19) is referring to metallic bilayers and noble metals, this form might be more recommendable from a practical point of view. We refer to the fact that the parameter estimation could be greatly improved. By separating the volume change from fiber extension modeled by the repulsive energy function, and shearing distortions modeled by the repulsive range function $\beta(\phi(x))$, the resulting set of response terms should provide an improved foundation for myocardial constitutive modeling. Also, it is an easier identification of only two functions by using a genetic algorithm based on the inversion of the experimental data.

In this chapter we adopt the pseudopotential energy approach and consider for W the expression of the ion-core (Born–Mayer) repulsive energy.

5. Active constitutive laws (Van Campen *et al.*, Arts *et al.*)

$$T^a = T^{a0}A(t,l,v)\,, \tag{7.2.21}$$

where T^a is the first order Piola–Kirchhoff non-symmetric active stress tensor, related to the second Piola–Kirchhoff active stress by $S^a = F^{-1}T^a$, and T^{a0} is a constant associated with the load of maximum isometric stress.

The stress tensor T^a is convenient for some purposes; it is measured relative to the initial underformed configuration and can be determined experimentally. The cardiac muscle is striated across the fiber direction.

The sarcomere length l (the distance between the striations) is used as a measure of fiber length. The experiments show that the active stress generated by cardiac muscle depends on time t, sarcomere length l and velocity of shortening of the sarcomeres $v = \frac{\mathrm{d}l}{\mathrm{d}t}$. The active stress generated by the sarcomeres is directed parallel to the fiber orientation. The function $A(t,l,v)$ represent the dependency on t, l and v.

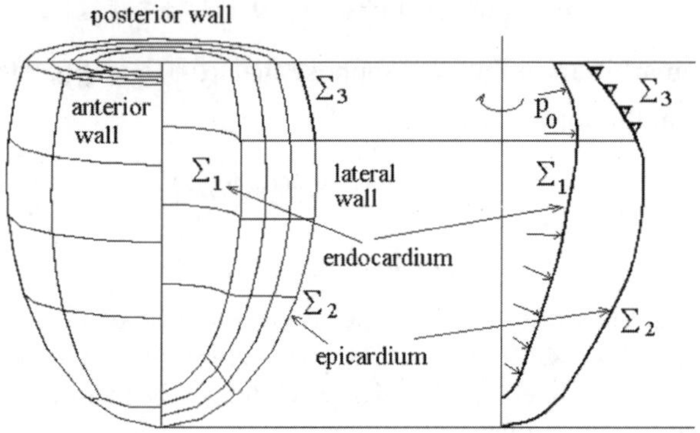

Figure 7.2.2 Representations of surfaces Σ_1 - endocardium, Σ_2 - epicardium and $\Sigma_3 \subset \Sigma_2$ - the portion where only radial displacement is allowed (adapted from Van Campen *et al.*)

We suppose that $A(t,l,v)$ has the form

$$A(t,l,v) = f(t)g(l)h(v) \,. \tag{7.2.22}$$

Functions $A(t,l,v)$, $\alpha(\phi)$ and $\beta(\phi)$ are determined from experimental data (Bardinet *et al.* 1994, 1995) by using an optimization approach (Popescu and Chiroiu). A genetic algorithm is considered in section 7.4.

The equations (7.2.3)–(7.2.22) represent a coupled set of four nonlinear equations for the displacements $u_k(x,t)$, $k = 1,2,3$, and the intramyocardial pressure $p(x,t)$ that can be written in the form

$$\nabla \dot{u} + \nabla(-K\nabla p) = 0 \,,$$

$$\nabla (J^{-1}F[S^a + S^c + S^s]F^T) - \nabla p = 0 \,, \tag{7.2.23}$$

with

$$K = (\frac{J-1}{N^b} + 1)^2 K^0 \,,$$

$$S^a = F^{-1}T^{a0}A(t,l^s,v^s) \,,$$

$$S^c = \frac{\partial C}{\partial E},$$

$$S^s = \int_{-\infty}^{t} G(t-\tau)\frac{\mathrm{d}}{\mathrm{d}\tau}\frac{\partial W}{\partial E}\mathrm{d}\tau, \tag{7.2.24}$$

$$C = \frac{c^c}{2}(J-1)^2, \; E = \frac{1}{2}(F^T F - I),$$

$$W = \frac{0.5}{V}\int_V \alpha(\phi)\exp[-\beta(\phi)R]\mathrm{d}V.$$

The boundary conditions are

$$p(x,0) = p_0(x), \; x \in \Sigma_1, \; k = 1,2,3, \; t \in [0,T],$$

$$u_k(x,0) = u_k^0, \; x \in \Sigma_1, \; k = 1,2,3, \; t \in [0,T], \tag{7.2.25}$$

$$u_k(x,0) = 0, \; x \in \Sigma_3 \subset \Sigma_1, \; k = 2,3, \; t \in [0,T],$$

where Σ_1 is the epicardial surface, and $[0,T]$ the time interval during a cardiac cycle.

The cardiac cycle is composed from a systole (contraction of the ventricle) and a diastole (relaxation of the ventricle) phases.

We have supposed that at the endocardial surface Σ_1 a uniform intraventricular pressure p_0 is applied as an external load (Figure 7.2.2). The loads exerted by the papillary muscles and by the pericardium are neglected.

The surface Σ_3 represents the upper end of the annulus fibrosis and is a non-contracting surface with a circumferential fiber orientation. At $\Sigma_3 \subset \Sigma_1$ only radial displacement u_1 is allowed.

To compute $\frac{\partial W}{\partial E}$ we use the formula

$$\frac{\partial}{\partial E_{ij}} = \frac{1}{2}\left(X_i\frac{\partial}{\partial x_j} + X_j\frac{\partial}{\partial x_i}\right), \tag{7.2.26}$$

where X_i ($X_1 \equiv X$, $X_2 \equiv Y$, $X_3 \equiv Z$) are the Lagrange coordinates corresponding to a reference state which may be subject to an initial finite deformation, and x_i, the final Eulerian coordinates ($x_1 \equiv x$, $x_2 \equiv y$, $x_3 \equiv z$) differing from X_i by an infinitesimal deformation. It is important to note that, in applying (7.2.26) the Lagrangian coordinates must be considered as constants since they refer to a predefined reference state.

For specified form for $A(t,l,v)$, $\alpha(\phi)$ and $\beta(\phi)$ the analytical solutions of (7.2.23) – (7.2.26) are determined in the next section.

7.3 Cnoidal solutions

Let us consider the solutions

$$z_i(x,t) = \{u_k(x,t),\ k = 1,2,3,\ p(x,t)\},\ i = 1,2,3,4,$$

of the system of nonlinear equations (7.2.23)–(7.2.26) under the form

$$z(x,t) = z_{cn}(x,t) + z_{\text{int}}(x,t). \tag{7.3.1}$$

The first term z_{cn} represents a linear superposition of cnoidal waves and the second term z_{int}, a nonlinear superposition of the cnoidal waves

$$z_{cn}(x,t) = 2\sum_{m=1}^{n} \alpha_m \text{cn}^2(k_{mj}x_j - C_m t),\ j = 1,2,3, \tag{7.3.2}$$

$$z_{\text{int}}(x,t) = \frac{\sum_{m=1}^{n} \gamma_m \text{cn}^2(k_{mj}x_j - C_m t)}{1 + \sum_{m=1}^{n} \lambda_m \text{cn}^2(k_{mj}x_j - C_m t)},\ j = 1,2,3. \tag{7.3.3}$$

We employ the usual summation convention over the repeated indices. We take the boundary conditions $(7.2.25)_1$ under the form

$$p_0(x,t) = a_0 \text{cn}^2(k_{0j}x_j - C_0 t), \tag{7.3.4}$$

with k_{0j}, C_0, a_0, $j = 1,2,3$, specified.

Consider that $n = 2$. Experimental calculations have shown us that solutions do not earn any improvements for $n > 2$. For specified form for $A(t,l,v)$, $\alpha(\phi)$ and $\beta(\phi)$ the analytical solutions $z_i(x,t)$, $i = 1,2,3,4$, are given by (7.3.2) and (7.3.3)

$$z_i(x,t) = 2\sum_{m=1}^{2} \alpha_{mi} \text{cn}^2(k_{mj}x_j - C_m t) + \frac{\sum_{m=1}^{2} \gamma_{mi} \text{cn}^2(k_{mj}x_j - C_m t)}{1 + \sum_{m=1}^{2} \lambda_{mi} \text{cn}^2(k_{mj}x_j - C_m t)}, \tag{7.3.5}$$

where α_{mi}, k_{mj}, γ_{mi}, λ_{mi} and C_m, $i = 1,2,3,4$, $m = 1,2$, $j = 1,2,3$, are unknown.

By introducing (7.3.2) into (7.2.23)–(7.2.25) the unknowns α_{mi}, k_{mj}, γ_{mi}, λ_{mi} and C_m, are easily determined by an identification procedure in terms of the controlling functions

$$\mathfrak{M} = \{\alpha_1(\phi_{helix}), \alpha_2(\phi_{trans}), \beta_1(\phi_{helix}), \beta_2(\phi_{trans}), f(t), g(l), h(v)\}, \tag{7.3.6}$$

that determine the constitutive laws, the initial data and the constants that appear in governing equations.

The analytic expressions (7.3.2)–(7.3.3) of the solutions are available once the controlling functions $\mathfrak{M}_i$, $i = 1,2,...,7$, are specified. The problem to be addressed here is the inverse of the forward problem. The aim is to use the difference between experimental and predicted by theory parameters to provide a procedure, which iteratively corrects the controlling parameters towards values leading to the least discord between predictions and experimental observations. In the following we consider that the functions $\mathfrak{M}_i$, $i = 1,2,...,7$, are approximated by polynomials of five degree $\mathfrak{M}_i(b_{6i-5},....b_{6i})$, $i = 1,2,...7$, characterized by coefficients b_j, $j = 1,2,...,42$.

To extract the functions $\mathfrak{M}_i(b_{6i-5},....b_{6i})$, $i = 1,2,...7$, from the experimental data, an objective function $\mathfrak{I}$ must be chosen that measures the agreement between theoretical and experimental data

$$\mathfrak{I}(P) = 42^{-1}\sum_{j=1}^{42} 4^{-1}\sum_{i=1}^{4}\sum_{m=1}^{M} [z_{im}(b_j) - z_{im}^{\exp}(b_j)]^2 , \qquad (7.3.7)$$

where $z_{im}(b_j)$ are the predicted values of the solutions $z_i(b_j)$, $i = 1,2,3,4$, $j = 1,2,...,42$, given by the forward problem, calculated at M points belonging to the volume between the inner and outer wall of the left ventricle. The functions $z_{im}^{\exp}$ are the experimental values of solutions z_i measured at the same points. We have extracted these values from the analysis of the volumetric deformation of the left ventricle of the heart, developed by Bardinet *et al.*

The model gives a compact representation of a set of points in a 3D image. Experimental results are shown in time sequences of two kinds of medical images, Nuclear Medicine and X-Ray Computed Tomography. Controlling functions $\mathfrak{M}_i$, $i = 1,2,...,7$, are determined by using a genetic algorithm (GA). GA assures an iteration scheme that guarantees a closer correspondence of predicted and experimental values of controlling parameters at each iteration.

We use a binary vector with 42 genes representing the real values of the parameters b_j, $j - 1,2,...,42$ (Chiroiu et al. 2000). The length of the vector depends on the required precision, which in this case is six places after the decimal point. The domain of parameters $b_j \in [-a_j, a_j]$, with length $2a_j$ is divided into at least 15000 equal size ranges. That means that each parameter b_j, $j = 1,2,...,42$, is represented by a gene (string) of 22 bits ($2^{21} < 3000000 \le 2^{22}$). One individual consists of a row of 42 genes, that is, a binary vector with 22 × 42 components.

$$(b_{21}^{(1)} b_{20}^{(1)} ... b_0^{(1)} b_{21}^{(2)} b_{20}^{(2)} ... b_0^{(2)} ... b_{21}^{(42)} b_{20}^{(42)} ... b_0^{(42)}).$$

The mapping from this binary string into 42 real numbers from the range $[-a_j, a_j]$ is completed in two steps:

– convert each string ($b_{21}^{(j)} b_{20}^{(j)} ... b_0^{(j)}$) from the base 2 to base 10

$$(b_{21}^{(j)} b_{20}^{(j)} ... b_0^{(j)})_2 = b_j' \ , \ j = 1,2,...,42 ,$$

– find a corresponding real number b_j, $j = 1,2,...,42$.

GA is linked to the problem that is to be solved through the fitness function, which measures how well an individual satisfies the real data. From one generation to the next

GA usually decreases the objective function of the best model and the average fitness of the population. The starting population (with K individuals) is usually randomly generated. Then, new descendant populations are iteratively created, with the goal of an overall objective function decrease from generation to generation. Each new generation is created from the current one by the main operations: selection, crossover and reproduction, mutation and fluctuation. By selection two individuals of the current population are randomly selected (parent 1 and parent 2) with a probability that is proportional to their fitness.

This ensures that individuals with good fitness have a better chance to advance to the next generation. In the crossover and reproduction operation some crossover sites are chosen randomly and exchanging some genes between parents reproduces two individuals.

In the newly produced individuals, a randomly selected gene is changed with a random generated integer number by the mutation operation. In the fluctuation operation we exchange a discretized value of an unknown parameter in a random direction, by extending the search in the neighborhood of a current solution.

The fitness function is evaluated for each individual that corresponds to the gene representation. The alternation of generations stops when the convergence is detected. Otherwise, the process stops when a maximum number of generations are reached.

The procedure consists of the following steps:

- The initial population is generated by random selection.
- The crossover operator reproduces two new individuals.
- New individuals are obtained by the mutation operator.

The fluctuation operator extends the search in the neighborhood of a current solution.

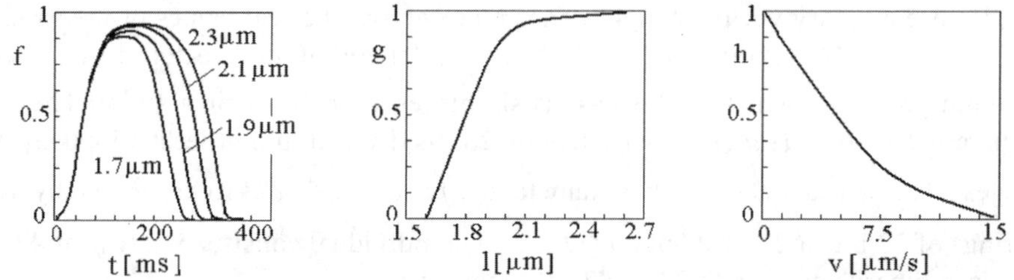

Figure 7.3.1 Active material behavior as obtained by the genetic algorithm (time dependence of active stress for sarcomere lengths of 1.7, 1.9, 2.1 and 2.3 μm, length dependence of active stress, and velocity dependence of active stress).

The fitness value is evaluated for each individual and in the total population only individuals with a higher fitness remain at the next generation.

The alternation of generations is stopped when convergence is detected. If there is no convergence the iteration process continues until the specified maximum number of generations is reached.

Next, we report the results of the genetic algorithm. Figure 7.3.1 shows the active material behavior as obtained by GA after 312 iterations. There are shown the time dependence of active stress for sarcomere lengths of 1.7, 1.9, 2.1 and 2.3 μm, the length dependence of active stress, and the velocity dependence of active stress. The diagrams are very similar to the active material behavior assumed by Van Campen *et al.*

We consider that the curves are correctly predicted by the genetic algorithm, the results being qualitatively and quantitatively consistent with experimental data given by Bovendeerd *et al.*, Arts *et al.*

Figure 7.3.2 shows the dependence of $\alpha(\phi)$ and $\beta(\phi)$ on angles ϕ_{helix}, ϕ_{trans} given by GA after 247 iterations. The distribution of ϕ_{helix} and ϕ_{trans} from the endocardium to the epicardium is also shown.

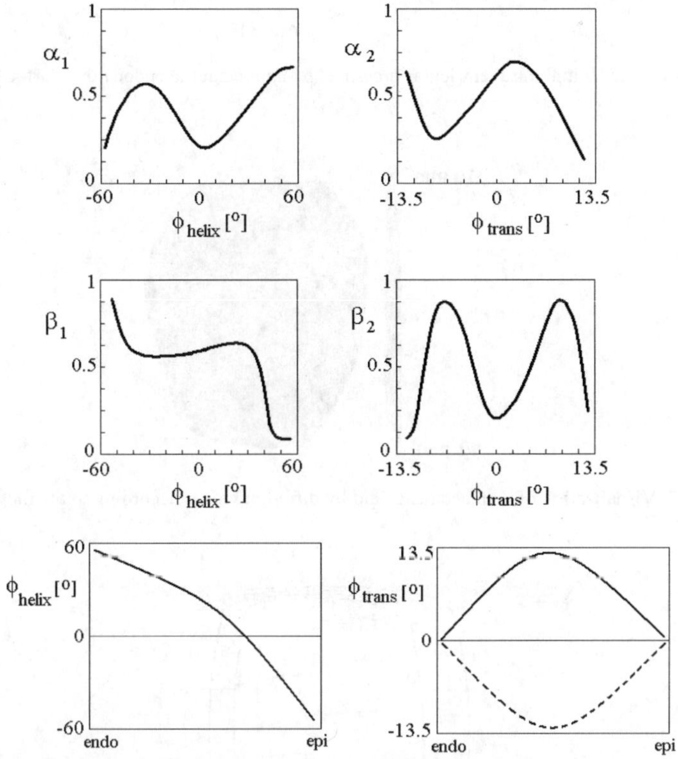

Figure 7.3.2 Angle dependence of $\alpha(\phi)=\alpha_1(\phi_{helix})\alpha_2(\phi_{trans})$ and $\beta(\phi)=\beta_1(\phi_{helix})\beta_2(\phi_{trans})$ as given by the genetic algorithm.

7.4 Numerical results

The solutions of the equations (7.2.23)–(7.2.26) are represented analytically by using the cnoidal method. The unknown parameters from these representations are determined from a genetic algorithm. The analytical solutions $u_i(x,t)$, $i=1,2,3$, and $p(x,t)$ are given by (7.3.5) being describable as a linear superposition of two cnoidal pulses and additional terms, which include nonlinear interactions among them.

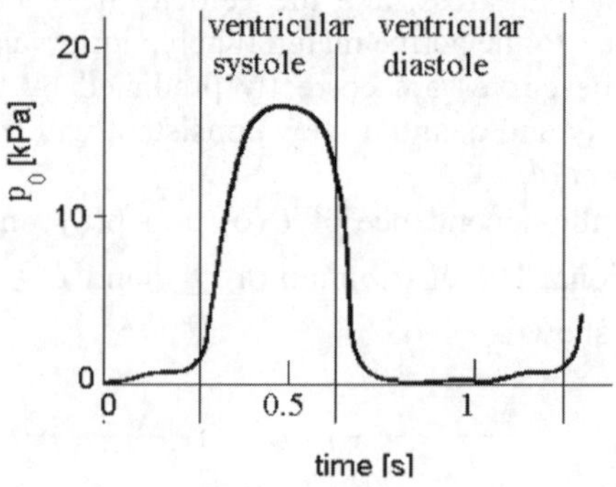

Figure 7.4.1 Initial intraventricular pressure p_0 applied at the endocardial surface Σ_1.

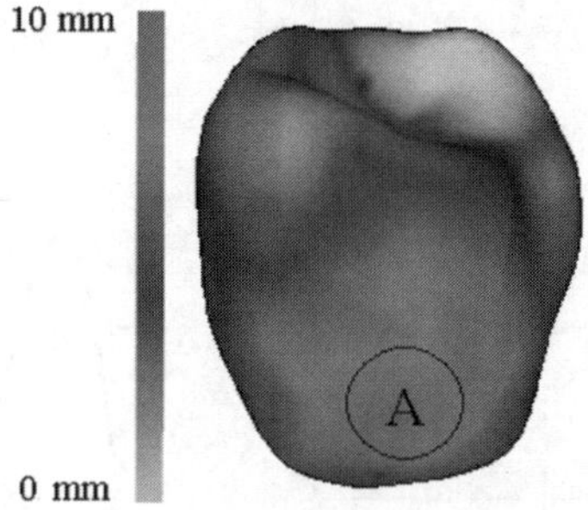

Figure 7.4.2 Visualization of displacement field by different values according to the range 0–10 mm.

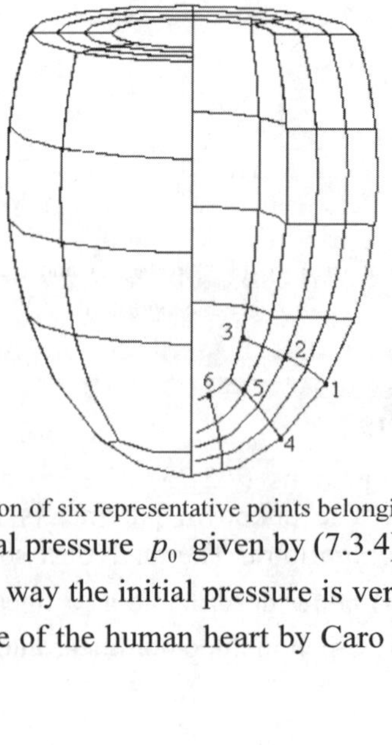

Figure 7.4.3 Location of six representative points belonging to region A

Figure 7.4.1 shows the initial pressure p_0 given by (7.3.4) applied at the endocardial surface Σ_1 ($a_0 = 0.34$). In this way the initial pressure is very similar to the diagram of the pressure in the left ventricle of the human heart by Caro *et al.* The visualization of the displacement field by different values according to the range 0–10 mm is shown in Figure 7.4.2.

We can see clearly areas on the ventricle where the displacements are high (for example the area A). In Figure 7.4.3 the location of some representative points belonging to the region A are displayed.

Figures 7.4.4–7.4.6 represent the time variation of displacements $u_k(x,t)$, $k=1,2,3$, during a cardiac cycle calculated in six points, displayed in Figure 7.4.3. Figure 7.4.7 shows the variation of the intramyocardial pressure during a cardiac cycle in the same points.

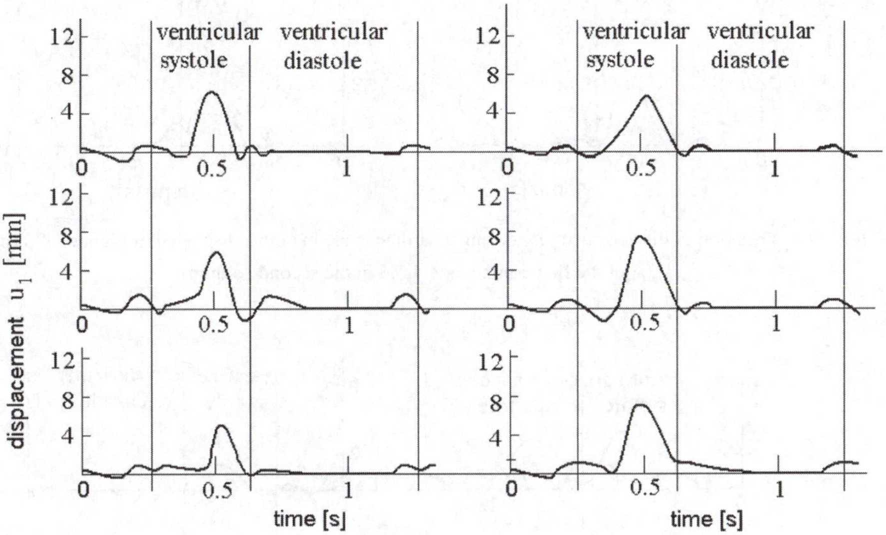

Figure 7.4.4 Variation of displacement u_1 during a cardiac cycle in the points displayed in Figure 7.4.3 (points 1,2,3 in the first column and 4,5,6 in the second column)

Comparison of these analytical results to the results obtained by Van Campen *et al.* through the finite element method, shows a good consistency.

The analytical solutions allow the possibility of investigating in detail the field of displacements and the field of pressure in each point of the left ventricle. This helps to predict the important features of the left ventricle motion.

In the hyperactive region A there are points that move out of phase with the other points (for example point 5). The physician, to help localize pathologies, such as infarcted regions, could use the visualization of these fields. In conclusion, the equations that govern the motion of the left ventricle have the remarkable property whereby the solutions can be represented by a sum of a linear and a nonlinear superposition of cnoidal vibrations.

So, we can say that the real virtue of the cnoidal method is to give the elegant and compact expressions for the solutions in the spirit of this property.

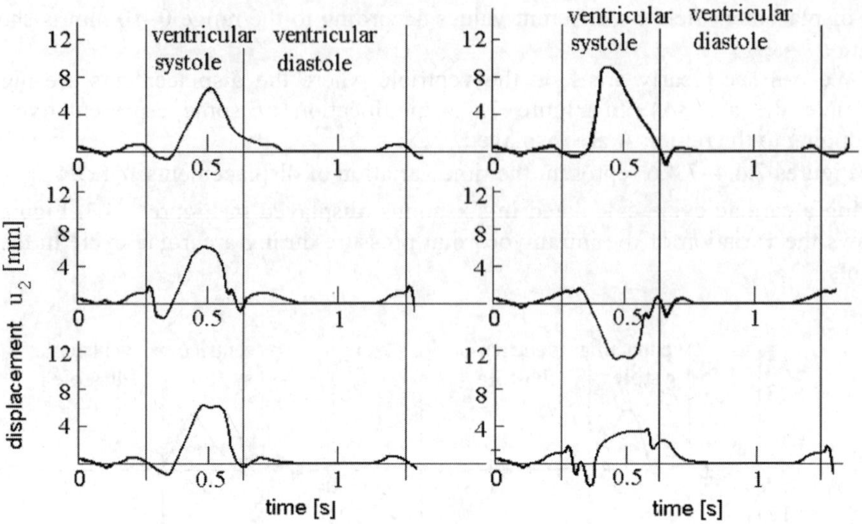

Figure 7.4.5 Variation of displacement u_2 during a cardiac cycle in points displayed in Figure 7.4.3 (points 1,2,3 in the first column and 4,5,6 in the second column)

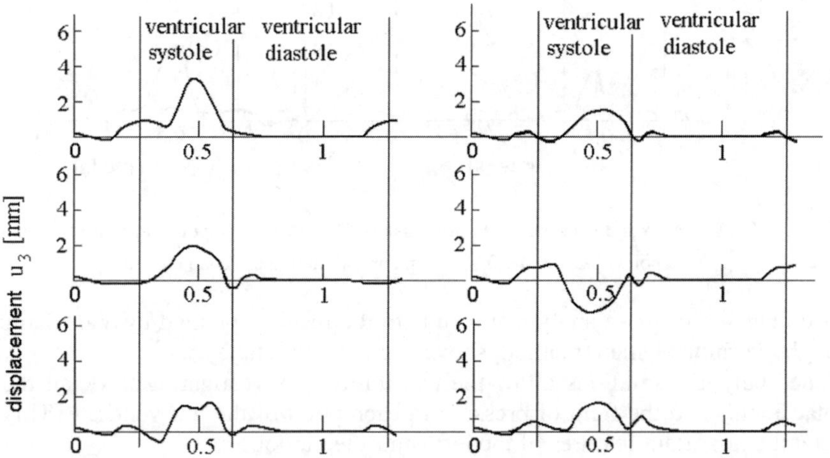

Figure 7.4.6 Variation of displacement u_3 during a cardiac cycle in points displayed in Figure 7.4.3 (points 1,2,3 in the first column and 4,5,6 in the second column).

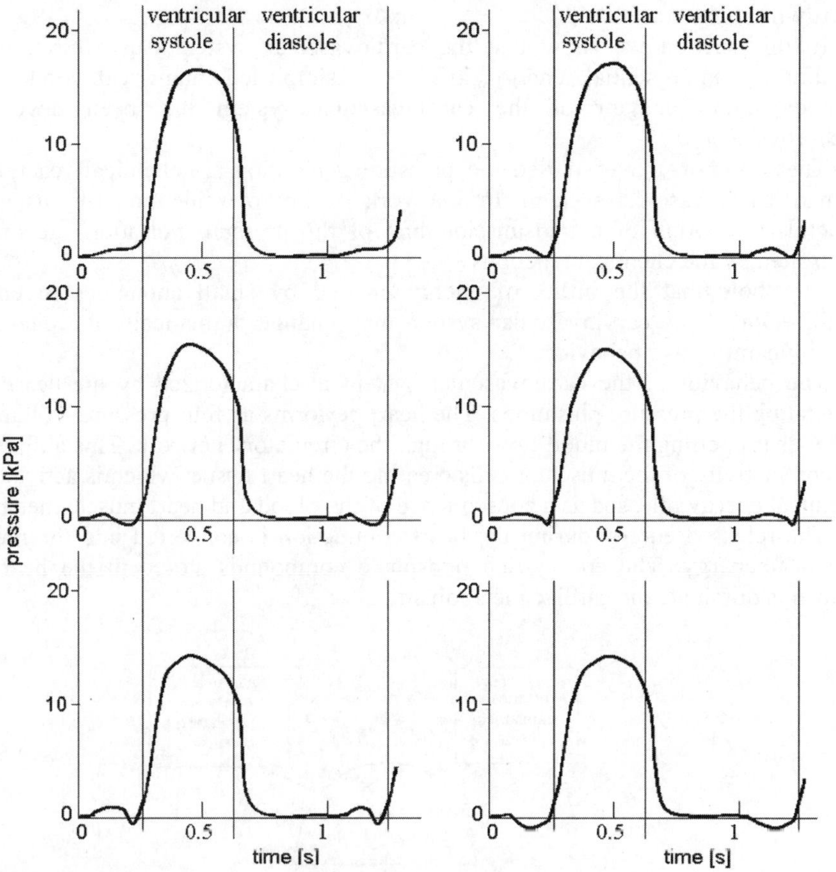

Figure 7.4.7 Variation of intramyocardial pressure p during a cardiac cycle in points displayed in Figure 7.4.3 (points 1,2,3 in the first column and 4,5,6 in the second column).

7.5 A nonlinear system with essential energy influx

In nature, energy is usually unbalanced and there exists either outflux or influx of energy. The energy influx is not so well understood or modeled (Engelbrecht).

The weak energy influx systems that may be called perturbed systems are based on the dilaton concept, which means negative density fluctuations with loosened bonds between the structural elements. The dilatons are able either to absorb energy from the surrounding medium or to give it away (Zhurkov).

The mechanism of energy influx yields to the amplification and/or attenuation phenomena due to the energy pumping from one subprocess to wave motion. In mathematical terms the energy influx gives rise to source-like terms in governing equations. For example, a burning candle is a classic case where the velocity of the flame is a nonlinear wave, which depends on the rate of heat release (Engelbrecht).

The electromagnetic nerve pulse transmission, modeled by Engelbrecht, takes its energy from ion currents.

In this section we show that the cardiovascular system can exhibit, in certain conditions, an essential energy influx. A simplified numerical model for the hemodynamic behavior of the cardiovascular system has been developed by Převorovská *et al.*

These authors have studied the pressure generation by chemical reaction in the human cardiovascular system. In this work we improve this model by using a more general description of a transmission line of the pressure pulsations generated and controlled by the chemical reactions.

We show that the influx of energy created by small anomalies in energetical equilibrium of the cardiovascular system may change dramatically the picture of the hemodynamic wave behavior.

The behavior of the cardiovascular system is characterized by the heart qualities generating the pressure pulsations. The heart performs as four pressure–volume pumps in series propelling the blood flow through the circulatory network. This ability is given by contractivity of the muscular cells creating the heart tissue, which is activated by the chemical energy released as a consequence of the blood and heart muscle metabolism.

The released energy during the heart contraction is converted into the mechanical and heat energy. The energy-rich phosphate compounds stored in the heart muscle filaments dominate the cardiac metabolism.

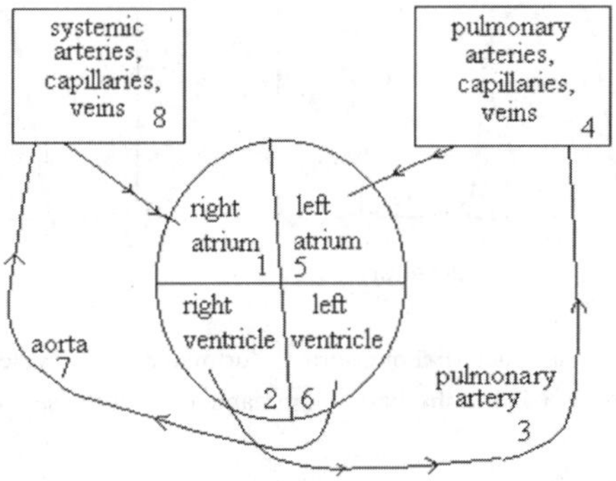

Figure 7.5.1 The cardiovascular system model.

The key source of the chemical energy is ATP (adenosine triphosphate) reacting both aerobic and anaerobic pathways.

The cardiovascular system is modeled by eight elastic segments connected by elastic tubes in a serial circuit depicting both pulmonary (low-pressure) and systemic (high-pressure) blood circulation (Figure 7.5.1).

Deoxygenated blood returns from the body through the vens cavae and fills the right atrium (1), which contracts, sending the blood into the right ventricle (2). The right ventricle then contracts, forcing the blood into the pulmonary artery, which carries it to the lungs to pick up oxygen.

Next, the oxygenated blood flows through the pulmonary vein into the left atrium (5), which delivers it to the left ventricle (6). Finally, the left ventricle contracts, ejecting the blood into the aorta and out to the systemic circulation.

During the cardiac cycle, each chamber fills during a period of relaxation, or diastole, and ejects the blood during a period of active contraction, or systole.

The heart segments passing through the passive and active states during the cardiac cycle have been considered as anisotropic and viscoelastic incompressible material. They act as chemomechanical pumps converting the chemical energy released by the anaerobic hydrolysis of ATP into the mechanical energy and heat.

The behavior of the cardiovascular system has been described by the mechanical variables (pressure, volume, flow) characterized by the cardiovascular parameters (compliance, resistance and inductance) and by the physicochemical variables (ATP consumption, chemical work, molar enthalpy).

The governing equations of the cardiovascular system are:

1. The pressure–volume relation

$$\dot{p}_i = \alpha_{1i}(F_i + \gamma_i p_i + \beta_{1i}\dot{V}_i + \delta_i V_i),\ i = 1,2,5,6, \tag{7.5.1}$$

$$p_i = \alpha_{2i} V_i + \beta_{2i},\ i = 3,4,7,8, \tag{7.5.2}$$

where $p(t)$ is the pressure, $V(t)$ the volume, the dot means the differentiation with respect to time. The function $F(t)$ and the constants α_i, β_i, γ_i, δ_i are defined as

$$F_i(t) = \frac{k_i}{E} w_i(t) - 1,\ \alpha_{1i} = \frac{2h_i E}{r_i \tau_i},\ \gamma_i = -\frac{r_i}{2h_i E},$$

$$\beta_{1i} = \frac{\tau_i}{V_{0i}},\ \delta_i = \frac{1}{V_{0i}},\ \alpha_{2i} = \frac{1}{c_i},\ \beta_{2i} = -\frac{V_{ri}}{c_i}, \tag{7.5.3}$$

where E is the Young's elastic modulus, h thickness of the myocardial wall, k parameter of the chemical energy release, r radius of the atrium, V_0 initial atrial or ventricle volume, w chemical reaction rate, V_r the residual volume, c the compliance and τ relaxation time characterizing the muscle plasticity.

The index i denotes the corresponding heart segments. The function $w(t)$ has an exponential form and describes all four phases of the cardiac cycle

$$w(t) = w_0(1 - \exp(-t'/\tau_3))\exp(-t''/\tau_2)\exp(-t'''/\tau_1),$$

with w_0 an initial given constant.

The four phases of the cardiac cycle are given by:

A. stretching systole-isometric contraction

$$t \in (t_1, t_2),\ t' = t,\ t'' = 0,\ t''' = 0,$$

B. emptying systole-auxotonic contraction

$$t \in (t_2, t_3),\ t' = t_2,\ t'' = t,\ t''' = 0,$$

C. relaxing diastole-isometric contraction

$$t \in (t_3, t_4)\,, t' = t_2,\ t'' = t_3,\ t''' = t\,,$$

D. filling diastole-noncontractive state

$$t \in (t_4, t_1)\,, t' = 0,\ t'' = 0,\ t''' = 0\,.$$

We have noted by τ_1, τ_2, τ_3 and τ_4 the relaxation time of the *C* phase, of the *B* phase, of the *A* phase, and respectively the relaxation time of the *D* phase.

2. The influence of the pressure pulsations

$$\dot{G}_{ij} = \frac{1}{l_i}(p_i - p_j - r_{ij}G_{ij} - H_i G_{ij}^2)\,, \tag{7.5.4}$$

where $G_{ij}(t)$ is the blood flux entering from the *i*-th to *j*-th segment, respectively, the blood flux getting out *j*-th segment to *k*-th segment, r_{ij} the hydrodynamic resistance, l_i the blood inertia, $H_i(t) = \dfrac{\varsigma^2 \rho}{2A_i^2(t)}$ with $\varsigma = 0.001$ the loss coefficient, $\rho = 1.062 \times 10^3\,\text{kg/m}^3$ the blood density and $A_i(t)$ the flow area. The flow index rule is

$$(i, j) \in \{(1,2), (2,3), (3,4), (4,5), (5,6), (6,7), (7,8), (8,1)\}\,.$$

The flow area is given by

$$A_i(t) = \frac{aG_{ij}}{\sqrt{2g(p_i - p_j)}}\,, \tag{7.5.5}$$

with $a = \dfrac{60}{cTf}$, c an empirical constant, f the heart frequency (beats/min), T the time duration of the cardiac cycle, and $g = 9.81\text{m/s}^2$ the gravitational acceleration.

3. The continuity equation

$$\dot{V}_i = G_{ij} - G_{jk}\,. \tag{7.5.6}$$

We consider a set of small-perturbed initial conditions (pressure–time, pressure–volume and chemical reaction rate–time) during one cardiac cycle in the left ventricle, represented in Figures 7.5.2–7.5.4. The values of the used parameters are given in Table 7.5.1 and Table 7.5.2.

We have considered that $V_{ri} = 5 \times 10^{-5}\,\text{m}^3$ for all *i*. This set of perturbed data has been obtained numerically starting from the clinical published stable data (Taber, Převorovská *et al.*, Chiroiu V. *et al.* 2000).

We analyze only the evolution of the pressure wave profile with respect to time, due to the perturbed initial conditions in the left ventricle (Figures 7.5.2–7.5.4), during one cardiac cycle. The perturbed data are accepted from the physical and clinical point of view. We present the results of both analytical and numerical integration in order to get information about the pressure wave profiles. The standard Runge–Kutta method of the fourth order is used in the numerical calculation.

The transient pressure wave has the form of an asymmetric soliton represented in Figure 7.5.5 (the curve *a* represents an asymmetric soliton for unperturbed initial data).

We observe that, for perturbed initial conditions, this wave is amplified from its initial form to the certain profile, the amplitude of which reaches the asymptotic value.

An intriguing question is the amplification, which is characterized by a fast change in amplitude for small-perturbed initial data.

We have analyzed the transient waves taking for the related perturbed initial conditions (Figure 7.5.5, curve *b* – for Figure 7.5.2, curve *c* – for Figure 7.5.3 and curve *d* – for Figure 7.5.4).

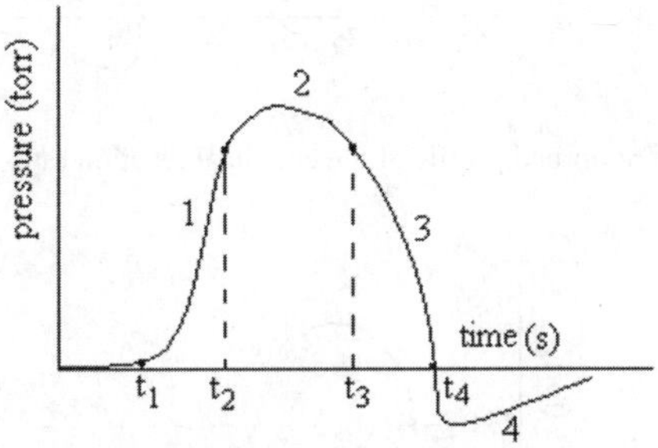

Figure 7.5.2 The perturbed profile of the pressure–time diagram during one cardiac cycle.

The result is that the maximum amplification happens for initial excitations given by the chemical reaction rate during the one cardiac cycle. The changes are both in the amplitude and in the width of the pulse.

The broken line denotes the asymptotic value of the pressure amplitude after a large interval of time. The aperiodic time-dependent phenomenon which appears here is in correspondence with the causality principle which states that the transport or propagation processes in the cardiovascular system are due to (causal) chains of interactions (cause–effect) between a source of perturbation (emission of signal) and the response (reception) and this can be realized (due to the inertia of the interacting human body) only after a certain time delay (relaxation), so that the transport occurs at a finite velocity (Kranys).

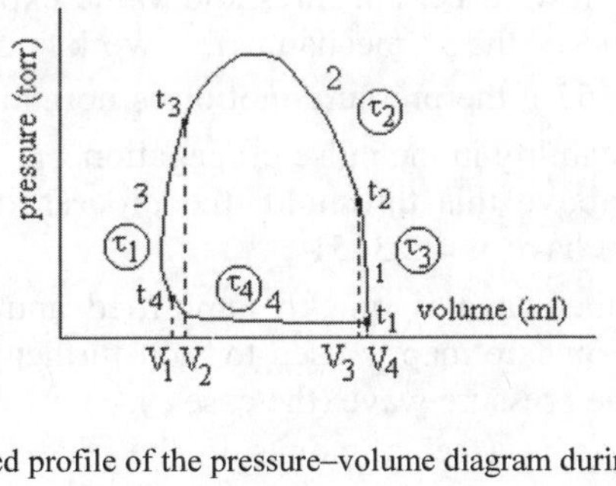

Figure 7.5.3 The perturbed profile of the pressure–volume diagram during one cardiac cycle.

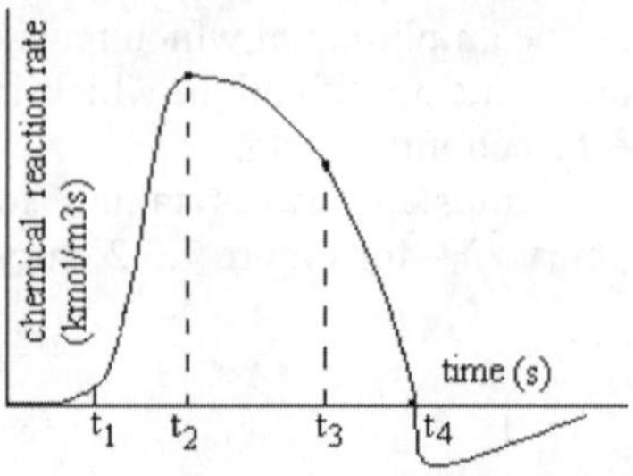

Figure 7.5.4 The perturbed profile of the chemical reaction rate–time during one cardiac cycle.

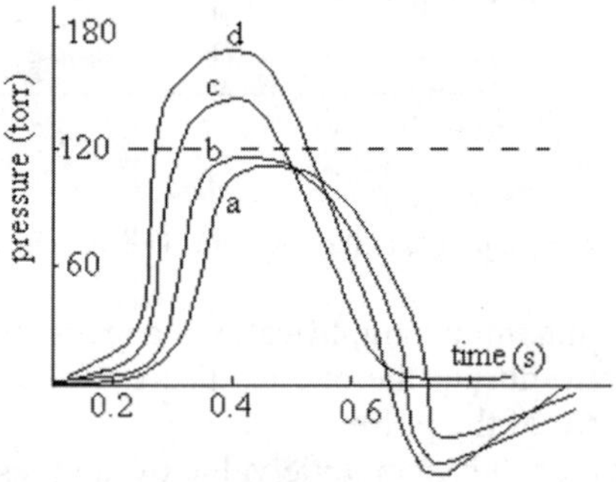

Figure 7.5.5 The transient profiles of pressure for the perturbed initial data (Figures 7.5.2–7.5.4) in the left ventricle.

As a principal conclusion attention has to be paid to the energetic aspects of the cardiac activity. The influx of energy created by small anomalies in energetical equilibrium of the cardiovascular system changes dramatically the picture of the hemodynamic waves behavior. The perturbation of the initial chemical reaction data acts as a stimulus.

If the stimulus is below a certain threshold value expressed as ratio between the chemical work and the mechanical work in the left ventricle ($\mu = W_{chem} / W_{mec} \leq 2.11557$), the pressure motion is normal and the stable state returns quickly without any instability in the pulse propagation.

If the stimulus is above this threshold, the process turns out to be much more dramatic. In our case we have $\mu = 2.3451$.

The amplitude of the wave is quickly amplified and the wave exhibits a clear tendency to chaos. A Poincaré map is used to gain further insight into the structure of the chaotic motion of the pressure wave (the case *d*).

Here a Poincaré map is a set of points in the phase plane $p(t), \dot{p}(t)$ plotted at discrete intervals of time, one each cardiac cycle, after 1000 cardiac cycles (Figure 7.5.6). The map reveals the properties of a strange attractor, namely stretching and folding of the sheet of pressure trajectories (Guckenheimer and Holmes).

Table 7.5.1 The quantities used in the governing equations (Převorovská *et al.*).

	k [Js/kmol]	V_0 [m^3]	τ [s]	w_0 [kmol/m^{3}s]	τ_1 [s]	τ_2 [s]	τ_3 [s]
1-A	14	3×10^{-5}	0.06	6.44×10^{-5}	0.06	0.1	0.006
1-B	400	6×10^{-5}	0.01	6.44×10^{-5}	0.01	0.03	0.01
1-C	3×10^{5}	7×10^{-5}	0.01	5.085×10^{-5}	0.06	0.01	0.06
1-D	1×10^{-9}	7×10^{-5}	0.1	0	0.06	0.3	0.06
2-A	5.4×10^{5}	1.2×10^{-4}	0.4	1.65×10^{-4}	0.06	0.3	0.006
2-B	5.57×10^{4}	1×10^{-5}	0.14	2.75×10^{-4}	0.06	0.3	0.6
2-C	1.6×10^{3}	8×10^{-5}	0.03	2.89×10^{-4}	0.06	0.01	0.06
2-D	1×10^{-9}	1.8×10^{-4}	0.3	0	0.06	0.1	0.06
5-A	1.4×10^{4}	2×10^{-5}	0.1	7.02×10^{-5}	0.06	0.003	0.006
5-B	1×10^{4}	3×10^{-5}	0.1	1.026×10^{-4}	0.06	0.1	0.06
5-C	50	7×10^{-5}	0.1	8.1×10^{-5}	0.06	0.03	1.6
5-D	1×10^{-9}	7×10^{-5}	0.01	0	0.01	0.03	0.01
6-A	4.4×10^{5}	5×10^{-5}	1.25	8.855×10^{-4}	0.06	0.29	0.006
6-B	2.55×10^{6}	3×10^{-4}	0.2	2.818×10^{-3}	0.06	0.1	0.6
6-C	1×10^{5}	1.2×10^{-4}	0.063	1.449×10^{-3}	0.06	0.02	1.1
6-D	1×10^{-9}	1.2×10^{-4}	0.1	0	0.01	0.3	0.01

Table 7.5.2 The constants characterizing the mechanical properties of the CVS (Převorovská *et al.*).

i	j	r_{ij} [Pas/m^3]	l_j [Pas^2/m^3]	c_j [m^3/Pa]	E_j [Pa]	p_i [Pa]	G_{ij} [m^3/s]
1	2	9.5×10^{4}	1×10^{2}	-	2.2×10^{-3}	7.53×10^{4}	2×10^{-7}
2	3	1.2×10^{5}	1×10^{2}	-	3.75×10^{-5}	1.33×10^{4}	1×10^{-7}
3	4	4×10^{6}	5.5×10^{3}	5.2×10^{-8}	-	-	2×10^{-7}
4	5	5.9×10^{5}	1×10^{4}	5.9×10^{-8}	-	-	1×10^{-7}
5	6	$1{,}5\times10^{5}$	1.1×10^{2}	-	1.5×10^{-3}	-	3×10^{-7}
6	7	5×10^{5}	3×10^{2}	-	1.2×10^{-5}	-	1×10^{-7}
7	8	4.5×10^{7}	5×10^{5}	6.8×10^{-9}	-	9.33×10^{4}	2×10^{-7}
8	1	1.3×10^{7}	9×10^{4}	6.9×10^{-9}	-	3.99×10^{4}	1×10^{-7}

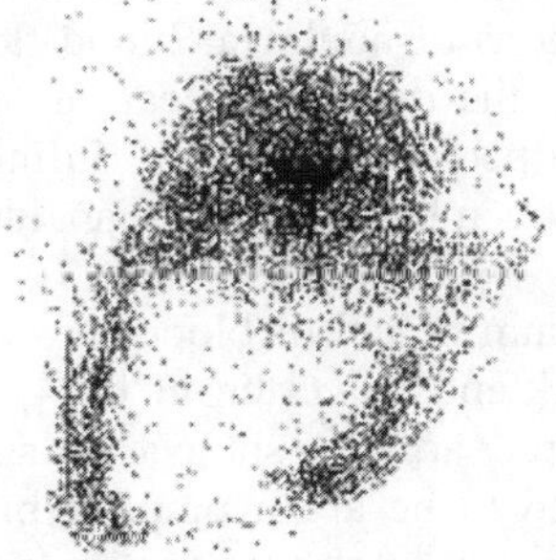

Figure 7.5.6 Poincaré map $p(t)$, $\dot{p}(t)$ after 1000 cardiac cycles.

Chapter 8

THE FLOW OF BLOOD IN ARTERIES

8.1 Scope of the chapter

Arteries conduct blood from the heart to the tissues and peripheral organs. The left ventricle pumps blood into the aorta, and the right ventricle pumps blood into the pulmonary artery. These main conduits branch into smaller vessels, and narrow arterioles vary their dimensions to regulate blood flow. Pulsatile flow of blood in large arteries has attracted much attention in blood dynamics. Computational fluid dynamics has emerged as a powerful alternative tool to study the hemodynamics at arterial tubes. Compared with experimental methods, fluid dynamics can easily accommodate changes in blood flow theory. Experimental studies of blood pulses revealed that they propagate with a solitonic characteristic pattern as they propagate away from the heart (McDonald, 1974).

Euler in 1775 obtained the one-dimensional nonlinear equations of blood motion through arteries, for the first time. Rudinger (1966), Skalak (1966), Ariman *et al.* (1974), Yomosa (1987), Moodie and Swaters (1989) have developed further this nonlinear theory. In 1958 Lambert used the method of characteristics to analyze the motion equations of blood. The finite difference method and the finite elements method are used also for the computations of nonlinear blood flow. In this chapter the soliton theory is employed to describe the dynamical features of the pulsatile blood flow in large arteries. The theory is performed for an infinitely long, straight, circular, homogeneous thin-walled elastic tube filled with an ideal fluid, in the spirit of the Yomosa theory (1987).

The theory of microcontinuum model of blood developed by Eringen in 1966 and Ariman and his coworkers, Turk and Sylvester, in 1974, is applied next to describe the transient flow of blood in large arteries subject to an arbitrary two-soliton blood pressure. The blood is assumed to be an incompressible micropolar fluid. The flow velocity, the micro-gyration and the cross-sectional area are calculated as functions of the two-soliton blood pressure pulse. The effects of increasing hematocrit on the amplitudes of the flow velocity and of the microgyration are analyzed.

The main bibliographies of this chapter are the works of Ariman *et al.* (1974), Yomosa (1987), Eringen (1966, 1970), Munteanu *et al.* (1998).

8.2 A nonlinear model of blood flow in arteries

McDonald in 1974 analyzed the form of the flow and pressure waves at certain locations from the ascending aorta to the saphenous artery in the dog (Figure 8.2.1). The propagation of the pressure pulse is accompanied by an increase in amplitude and a decrease in pulse-width which have been noted as *peaking* and *steepening*. The increase in amplitude is in accordance with an increase of the pulse–wave velocity and is combined with the generation of a dicrotic wave (Hashizume).

Sections 8.2 and 8.3 present the Yomosa theory, performed for an infinitely long, straight, circular, homogeneous thin-walled elastic tube embedded in the tissue. The blood is assumed an incompressible and nonviscous fluid.

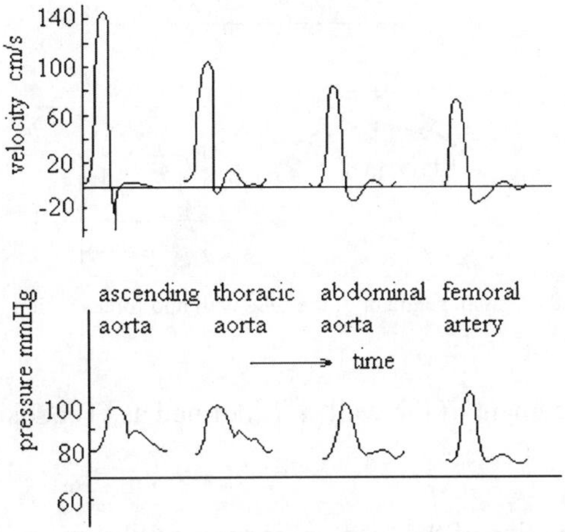

Figure 8.2.1 The behavior of the flow velocity and pressure pulses from the ascending aorta to the femoral artery (McDonald).

Consider the one-dimensional fields of longitudinal flow velocity $v_z(z,t)$, the blood pressure $p(z,t)$, and radial displacement of the arterial wall $u_r(z,t)$, expressed in cylindrical coordinates (r,θ,z), where r is the radial coordinate and z, axial coordinate. Assume rotational symmetry and the hypothesis of uniform distributions of flow velocity $v_z(z,t)$ and the fluid pressure $p(z,t)$ over the cross-section of the vessel.

– Navier–Stokes equation of motion for longitudinal flow

$$\frac{\partial v_z}{\partial t} + v_z \frac{\partial v_z}{\partial z} + \frac{1}{\rho}\frac{\partial p}{\partial z} = 0, \tag{8.2.1}$$

where ρ is the density of blood.

– Continuity equation which expresses the incompressibility of blood

$$\frac{\partial A}{\partial t} + \frac{\partial (v_z A)}{\partial z} = 0, \tag{8.2.2}$$

where $A(z,t)$ is the cross-sectional area of the tube.

The third equation describes the radial motion of the arterial wall. This equation was derived by Yomosa. We present his derivation of the equation. Consider the small segment of the tube wall surrounded by inner and outer surfaces of the wall and parallel two cross-sectional surfaces perpendicular to z -axis defined by constant coordinates of z and $z+\mathrm{d}z$, and two surfaces defined by constant angles of θ and $\theta+\mathrm{d}\theta$ (Figure 8.2.2).

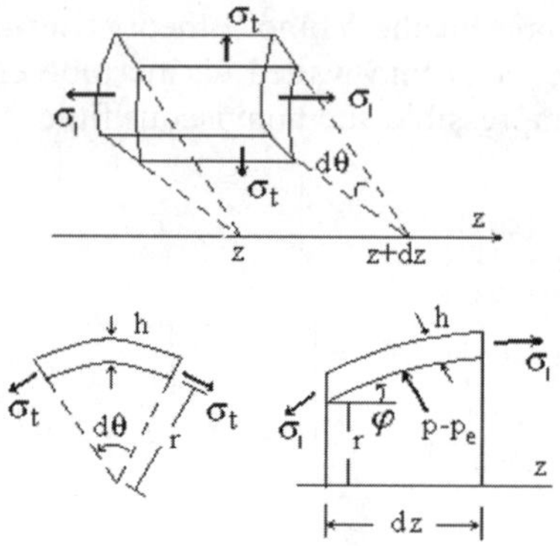

Figure 8.2.2 A small segment of the tube wall and forces acting on it (Yomosa).

The radial displacement of the wall u_r is defined as (Teodosiu, Solomon)

$$r = r_0 + u_r\,, \tag{8.2.3}$$

where $r(z,t)$ is the radius of the tube, r_0 is the equilibrium radius of the tube when the wall is static and $p = p_e$, where p_e is the pressure outside the tube which can be equal to the atmospheric pressure. The constitutive law of the elastic arterial wall can be written as

$$\sigma_t = E\varepsilon_r(1 + a\varepsilon_r)\,, \tag{8.2.4}$$

where $\varepsilon_r = \dfrac{r - r_0}{r_0}$ is the radial strain, E the Young's elasticity modulus, and a, a nonlinear coefficient of elasticity. The equation of radial motion of the segment considered in Figure 8.2.2 is

$$\begin{aligned}\rho_0(r\,\mathrm{d}\theta\,h\,\mathrm{d}z)\frac{\partial^2 r}{\partial t^2} &= (p - p_e)\cos\varphi(r\,\mathrm{d}\theta\mathrm{d}l) - 2\sigma_t\sin(\mathrm{d}\theta/2)h\mathrm{d}z + \\ &+ \frac{\partial}{\partial z}(\sigma_l \sin\varphi)\,\mathrm{d}z\,h\,r\,\mathrm{d}\theta),\end{aligned} \tag{8.2.5}$$

where ρ_0 is the density of the wall material, h the thickness of the wall in radial direction, σ_l the longitudinal extending stress in the direction of the meridian line on the wall surface, σ_t the extending stress in the tangential direction, and $\varphi(z,t)$ the angle between a tangent to a meridian line on the wall surface and the z-axis, and $\mathrm{d}l$ the elementary length of the meridian line in the segment, defined as $\mathrm{d}l\cos\varphi = \mathrm{d}z$.

For a small angle φ, we have the approximations

$$\cos\varphi = [1+(\partial r/\partial z)^2]^{1/2} \cong 1 \, , \; \sin\varphi \cong \tan\varphi = \partial r/\partial z \, . \tag{8.2.6}$$

These approximations are related to the fact that the term $(\partial r/\partial z)^2$ contributes higher order small quantity in the later perturbation expansions, proportional to ε^3, whereas the present theory is carried out by taking into account up to the terms proportional to ε^2, where ε denotes a small parameter $0<\varepsilon<<1$. Thus, the term $\partial(\sigma_l \sin\varphi)/\partial z$ can be approximated as

$$\begin{aligned}\partial(\sigma_l \sin\varphi)/\partial z &= E_1(\partial/\partial z)[((\mathrm{d}l-\mathrm{d}z)/\mathrm{d}z)(\partial r/\partial z)] = \\ &= (3E_1/2)(\partial r/\partial z)^2(\partial^2 r/\partial z^2),\end{aligned} \tag{8.2.7}$$

where E_1 is Young's longitudinal elasticity modulus of the wall. This term is proportional to ε^5, and therefore can be neglected.

Taking $\sin(\mathrm{d}\theta/2) \cong \mathrm{d}\theta/2$, for small angles $\mathrm{d}\theta$, we obtain the radial motion equation

$$\rho_0 h \frac{\partial^2 r}{\partial t^2} = (p-p_e) - \frac{h\sigma_t}{r} \, . \tag{8.2.8}$$

We must mention that the real thickness of the wall $\bar{h}$ in the direction perpendicular to the wall surface is given by $\bar{h} = h\cos\varphi \cong h$. Also, the inertial effects on the radial motion of the wall caused by the tissue must be considered. So, the effective inertial thickness is $H = h + h'$, where h is the thickness of the wall in which the material participates in the elastic deformation and h' is the additional effective inertial thickness of the tissue in which the material does not participate in the elastic deformation. With these assumptions and by considering that the densities of the materials of the wall and the tissue are the same, the equation of radial motion becomes

$$\rho_0 H \frac{\partial^2 r}{\partial t^2} = (p-p_e) - \frac{h\sigma_t}{r} \, . \tag{8.2.9}$$

For $p = p_e$, the relation (8.2.9) yields $\sigma_t = 0$, and then from (8.2.4) we have $u_r = 0$, and $r = r_0$. Denoting by h_0, H_0 the equilibrium values of the thickness of the wall and of the effective inertial thickness, the conditions for the conservation of mass of the wall and the tissue are given by

$$\rho_0 r \mathrm{d}\theta H \,\mathrm{d}z = \rho_0 r_0 \mathrm{d}\theta H_0 \mathrm{d}z \, , \; \rho_0 r \mathrm{d}\theta h \,\mathrm{d}z = \rho_0 r_0 \mathrm{d}\theta h_0 \mathrm{d}z \, , \tag{8.2.10}$$

or

$$rH = r_0 H_0 \, , \; rh = r_0 h_0 \, . \tag{8.2.11}$$

Taking into account that

$$A = \pi(r_0 + u_r)^2, \tag{8.2.12}$$

the equation (8.2.2) becomes

$$\frac{\partial u_r}{\partial t} + \frac{r_0}{2}(1 + \frac{u_r}{r_0})\frac{\partial v}{\partial z} + v\frac{\partial u_r}{\partial z} = 0. \tag{8.2.13}$$

Substituting (8.2.4) and (8.2.11) into (8.2.9) the equation of motion of the wall becomes

$$\rho_0 \frac{\partial^2 u_r}{\partial t^2} = \frac{p - p_e}{H_0}(1 + \frac{u_r}{r_0}) - Eu_r \frac{h_0(1 + a\frac{u_r}{r_0})}{H_0 r_0^2 (1 + \frac{u_r}{r_0})}. \tag{8.2.14}$$

Let us introduce the dimensionless independent variables

$$z = L_0 z', \; t = T_0 t', \tag{8.2.15}$$

with

$$L_0 = \sqrt{\frac{r_0 H_0 \rho_0}{2\rho}}, \; T_0 = \sqrt{\frac{r_0^2 H_0 \rho_0}{h_0 E}}, \tag{8.2.16}$$

and dimensionless dependent variables $\tilde{v}$, $\tilde{p}$ and $\tilde{\gamma}$

$$v_z = c_0 \tilde{v}, \; p - p_e = p_0 \tilde{p}, \; u_z = r_0 \tilde{\gamma}, \tag{8.2.17}$$

where c_0 and p_0 are given by

$$c_0 = \frac{L_0}{T_0} = \sqrt{\frac{h_0 E}{2\rho r_0}}, \; p_0 = \frac{h_0 E}{2 r_0}. \tag{8.2.18}$$

Therefore, the basic equations (8.2.1), (8.2.13) and (8.2.14) become

$$\frac{\partial \tilde{v}}{\partial t'} + \tilde{v}\frac{\partial \tilde{v}}{\partial z'} + \frac{\partial \tilde{p}}{\partial z'} = 0, \tag{8.2.19}$$

$$\frac{\partial \tilde{\gamma}}{\partial t'} + \frac{1}{2}(1 + \tilde{\gamma})\frac{\partial \tilde{v}}{\partial z'} + \tilde{v}\frac{\partial \tilde{\gamma}}{\partial z'} = 0, \tag{8.2.20}$$

$$\frac{\partial^2 \tilde{\gamma}}{\partial t'^2} = \frac{1}{2}\tilde{p}(1 + \tilde{\gamma}) - \frac{\tilde{\gamma}(1 + a\tilde{\gamma})}{(1 + \tilde{\gamma})}. \tag{8.2.21}$$

At the end of the diastole period, the flow velocity v_z and the acceleration of tube wall vanish. The diastolic pressure equals the lowest blood pressure. The quantities $\tilde{\gamma}_0$, $\tilde{p}_0$ can be related by (8.2.21), if we write $p_0 - p_e = p_0 \tilde{p}_0$ and $(u_r)_0 = r_0 \tilde{\gamma}_0$, that is

$$\tilde{p}_0 = \frac{2\tilde{\gamma}_0(1 + a\tilde{\gamma}_0)}{(1 + \tilde{\gamma}_0)^2}. \tag{8.2.22}$$

Now, let us study the asymptotic behavior of the governing equations (8.2.19)–(8.2.21).

To describe the nonlinear asymptotic behavior, Gardner and Morikawa introduced a scale transformation to combine it with a perturbation expansion of the dependent variable. Employing this method, we firstly linearise the equations in the form

$$\frac{\partial \tilde{v}}{\partial t'} + \frac{\partial \tilde{\tilde{p}}}{\partial z'} = 0\,, \tag{8.2.23}$$

$$\frac{\partial \tilde{\tilde{\gamma}}}{\partial t'} + \frac{1}{2}(1+\tilde{\gamma}_0)\frac{\partial \tilde{v}}{\partial z'} = 0\,, \tag{8.2.24}$$

$$\frac{\partial^2 \tilde{\tilde{\gamma}}}{\partial t'^2} = \frac{1}{2}\tilde{\tilde{p}}(1+\tilde{\gamma}_0) - \frac{\tilde{\tilde{\gamma}}(1+(2a-1)\tilde{\gamma}_0)}{(1+\tilde{\gamma}_0)^2}\,, \tag{8.2.25}$$

where $\tilde{\tilde{p}}$ and $\tilde{\tilde{\gamma}}$ are defined as

$$\tilde{p} = \tilde{p}_0 + \tilde{\tilde{p}}\,, \quad \tilde{\gamma} = \tilde{\gamma}_0 + \tilde{\tilde{\gamma}}\,. \tag{8.2.26}$$

Suppose we have harmonic solutions $\tilde{v}$, $\tilde{\tilde{p}}$, $\tilde{\tilde{\gamma}}$, with exponential factor of the form

$$\exp[\mathrm{i}(kz' - \omega t')]\,. \tag{8.2.27}$$

In this case, we obtain a system of homogeneous equations in amplitudes $\tilde{v}$, $\tilde{\tilde{p}}$, $\tilde{\tilde{\gamma}}$. The dispersion relations are obtained by requiring that the determinant of the system vanishes

$$\omega = \frac{gk}{\sqrt{1+k^2}}\,, \quad g = \frac{\sqrt{1+(2a-1)\tilde{\gamma}_0}}{(1+\gamma_0)}\,. \tag{8.2.28}$$

We expand $\omega(k)$ into a Taylor series about the origin for small k, and retain terms up to the third power of k. So, we have

$$\omega(t) \approx gk\left(1 - \frac{k^2}{2}\right), \tag{8.2.29}$$

and the exponential (8.2.27) becomes

$$\exp\left[\mathrm{i}\left\{k(z' - gt') + \frac{k^3 gt'}{2}\right\}\right]. \tag{8.2.30}$$

The similarity of asymptotic behavior holds for a coordinate transformation which satisfies

$$\frac{z' - gt'}{(gt')^{1/3}} = \text{const.} \tag{8.2.31}$$

So, the scale transformation, for which the invariance of (8.2.31) hold, can be defined as

$$\xi = \sqrt{\varepsilon}(z' - gt'),\ \tau = \varepsilon\sqrt{\varepsilon}\, gt', \tag{8.2.32}$$

with ε a small parameter. The perturbation expansions of $\tilde{v}$, $\tilde{\tilde{p}}$, $\tilde{\tilde{\gamma}}$ with respect to ε, are given by

$$\tilde{v} = \sum_{n=1}^{\infty} \varepsilon^n v_n(\xi,\tau),\ \tilde{\tilde{p}} = \sum_{n=1}^{\infty} \varepsilon^n p_n(\xi,\tau),\ \tilde{\tilde{\gamma}} = \sum_{n=1}^{\infty} \varepsilon^n \gamma_n(\xi,\tau). \tag{8.2.33}$$

The dependent variables $\tilde{v}$, $\tilde{\tilde{p}}$, $\tilde{\tilde{\gamma}}$ depend on ξ and τ, and the nonlinear equations (8.2.19)–(8.2.21), written with respect to ξ and τ, are

$$g(-\frac{\partial}{\partial\xi} + \varepsilon\frac{\partial}{\partial\tau})\tilde{v} + \tilde{v}\frac{\partial\tilde{v}}{\partial\xi} + \frac{\partial\tilde{\tilde{p}}}{\partial\xi} = 0, \tag{8.2.34}$$

$$g(-\frac{\partial}{\partial\xi} + \varepsilon\frac{\partial}{\partial\tau})\tilde{\tilde{\gamma}} + \frac{1}{2}(1 + \tilde{\gamma}_0 + \tilde{\tilde{\gamma}})\frac{\partial\tilde{v}}{\partial\xi} + \tilde{v}\frac{\partial\tilde{\tilde{\gamma}}}{\partial\xi} = 0, \tag{8.2.35}$$

$$\begin{aligned} g^2(\varepsilon\frac{\partial^2}{\partial\xi^2} - 2\varepsilon^2\frac{\partial^2}{\partial\xi\partial\tau} + \varepsilon^3\frac{\partial^2}{\partial\tau^2})\tilde{\tilde{\gamma}} - \frac{1}{2}(\tilde{p}_0 + \tilde{\tilde{p}})(1 + \tilde{\gamma}_0 + \tilde{\tilde{\gamma}}) + \\ + \frac{(\tilde{\gamma}_0 + \tilde{\tilde{\gamma}})[1 + a(\tilde{\gamma}_0 + \tilde{\tilde{\gamma}})]}{(1 + \tilde{\gamma}_0 + \tilde{\tilde{\gamma}})} = 0. \end{aligned} \tag{8.2.36}$$

Substitute (8.2.33) into (8.2.34)–(8.2.36) and equate the coefficients of like powers of ε. The terms that are proportional to ε^0, give equation (8.2.22), and the terms proportional to ε give

$$\begin{aligned} &-g\frac{\partial v_1}{\partial\xi} + \frac{\partial p_1}{\partial\xi} = 0, \\ &-g\frac{\partial\gamma_1}{\partial\xi} + \frac{1}{2}(1 + \tilde{\gamma}_0)\frac{\partial v_1}{\partial\xi} = 0, \\ &-\frac{1}{2}\tilde{p}_0\gamma_1 - \frac{1}{2}(1 + \tilde{\gamma}_0)p_1 + \frac{(1 + 2a\tilde{\gamma}_0 + a\tilde{\gamma}_0^2)\gamma_1}{(1 + \tilde{\gamma}_0)^2} = 0. \end{aligned} \tag{8.2.37}$$

Also, the terms proportional to ε^2 give

$$\begin{aligned} &-g\frac{\partial v_2}{\partial\xi} + g\frac{\partial v_1}{\partial\tau} + v_1\frac{\partial v_1}{\partial\xi} + \frac{\partial p_2}{\partial\xi} = 0, \\ &-g\frac{\partial\gamma_2}{\partial\xi} + g\frac{\partial\gamma_1}{\partial\tau} + \frac{1}{2}(1 + \tilde{\gamma}_0)\frac{\partial v_2}{\partial\xi} + \frac{1}{2}\gamma_1\frac{\partial v_1}{\partial\xi} + v_1\frac{\partial\gamma_1}{\partial\xi} = 0, \end{aligned} \tag{8.2.38}$$

$$g^2\frac{\partial^2\gamma_1}{\partial\xi^2}-\frac{1}{2}\tilde{p}_0\gamma_2-\frac{1}{2}(1+\tilde{\gamma}_0)p_2-\frac{1}{2}\tilde{p}_1\gamma_1+$$
$$+\frac{(1+2a\tilde{\gamma}_0+a\tilde{\gamma}_0^2)\gamma_2}{(1+\tilde{\gamma}_0)^2}+\frac{(a-1)\gamma_1^2}{(1+\tilde{\gamma}_0)^3}=0.$$

From (8.2.37) we obtain by integrating

$$v_1=\frac{1}{g}p_1+f(\tau), \tag{8.2.39}$$

$$v_1=\frac{2g}{1+\tilde{\gamma}_0}\gamma_1+\varphi(\tau), \tag{8.2.40}$$

$$p_1=\frac{2g^2}{1+\tilde{\gamma}_0}\gamma_1, \tag{8.2.41}$$

where the functions $f(\tau)$ and $\varphi(\tau)$ can be determined from initial conditions. We can have in particular

$$f(\tau)=\varphi(\tau)=0, \tag{8.2.42}$$

for

$$p=p_0,\ \tilde{\gamma}=\tilde{\gamma}_0,\ \text{for } v_z=0, \tag{8.2.43}$$

or

$$p_1=0,\ \gamma_1=0,\ \text{for } v_1=0. \tag{8.2.44}$$

In these conditions, (8.2.39) and (8.2.40) give again (8.2.41). The quantities v_1, p_1 and γ_1 are related between themselves by

$$v_1=\frac{p_1}{g}=\frac{2g}{1+\tilde{\gamma}_0}\gamma_1. \tag{8.2.45}$$

Eliminating v_2, p_2 and γ_2 from (8.2.38) and (8.2.45), we obtain the KdV equations

$$\frac{\partial v_1}{\partial\tau}+Kv_1\frac{\partial v_1}{\partial\xi}+\frac{1}{2}\frac{\partial^3 v_1}{\partial\xi^3}=0, \tag{8.2.46}$$

$$\frac{\partial p_1}{\partial\tau}+Lp_1\frac{\partial p_1}{\partial\xi}+\frac{1}{2}\frac{\partial^3 p_1}{\partial\xi^3}=0, \tag{8.2.47}$$

$$\frac{\partial\gamma_1}{\partial\tau}+M\gamma_1\frac{\partial\gamma_1}{\partial\xi}+\frac{1}{2}\frac{\partial^3\gamma_1}{\partial\xi^3}=0, \tag{8.2.48}$$

where the constants K, L and M are given by

$$K = \frac{(1+\tilde{\gamma}_0)[(1+2a)+3(2a-1)\tilde{\gamma}_0]}{4[1+(2a-1)\tilde{\gamma}_0]^{3/2}}, \tag{8.2.49}$$

$$L = \frac{K}{g}, \quad M = \frac{2g}{1+\tilde{\gamma}_0}K. \tag{8.2.50}$$

8.3 Two-soliton solutions

The flow velocity equation (8.2.46) may be written as

$$U_\tau - 6UU_x + U_{xxx} = 0, \tag{8.3.1}$$

by the transformations

$$v_1 = -\frac{3}{K}U, \quad \xi = X, \quad \tau = 2T. \tag{8.3.2}$$

The equation (8.3.1) admits a soliton solution given by

$$U = 2k^2 \operatorname{sech}^2[k(X-4k^2T)-\delta], \tag{8.3.3}$$

with $\delta = \text{const}$. The solution for flow velocity is obtained from (8.2.17) and (8.2.33)

$$v_z(z,t) = \frac{6}{K}c_0\tilde{k}^2 \operatorname{sech}^2[\frac{\tilde{k}}{L_0}(z-Vt)-\delta], \tag{8.3.4}$$

where

$$\varepsilon k^2 = \tilde{k}^2, \quad V = gc_0(1+2\tilde{k}^2). \tag{8.3.5}$$

The equations (8.2.47), (8.2.48) can be solved in a similar way, and the solutions for fluid pressure and radial displacement are given by

$$p(z,t) = p_0 + \frac{6}{L}p_0\tilde{k}^2 \operatorname{sech}^2[\frac{\tilde{k}}{L_0}(z-Vt)-\delta], \tag{8.3.6}$$

$$u_r(z,t) = r_0\tilde{\gamma}_0 + \frac{6}{M}r_0\tilde{k}^2 \operatorname{sech}^2[\frac{\tilde{k}}{L_0}(z-Vt)-\delta]. \tag{8.3.7}$$

The soliton solutions describe the pulsatile pulses in which the amplitude and velocity are related. The pulses with large amplitude are narrow in width and move rapidly. This feature can explain that the *steepening* in the arterial pulse occurs in accordance with the increase in the velocity pulse.

For arbitrary initial conditions the solutions $U(X,T)$ of (8.3.1) are obtained exactly by applying the inverse scattering transform. As $T \to \infty$, the solution approaches asymptotically to N soliton solutions. The amplitudes and the velocities of these solutions are determined by the eigenvalues $-k_n^2$ of N bound states of the Schrödinger equation for which the potential is the initial condition $U(X,0)$.

Consider now the arbitrary initial conditions for the KdV equation (8.3.1) which can support two bound states at $-k_1^2$ and $-k_2^2$

$$U(X,0) = -\frac{K}{3}v_1(\xi,0) = -\frac{K}{3\varepsilon}v_z(z,0) = -\frac{K}{3\varepsilon c_0}\tilde{v}(z,0),$$

$$X = \xi = \sqrt{\varepsilon}z' = \sqrt{\varepsilon}\frac{1}{L_0}z\ . \tag{8.3.8}$$

The solution is obtained with no reflection, $R(k,0)=0$, and yields as $T \to \infty$ to a two-soliton solution given by

$$\begin{aligned} U(X,T) = -2k_1^2\operatorname{sech}^2[k_1(X-4k_1^2T)-\delta_1]- \\ -2k_2^2\operatorname{sech}^2[k_2(X-4k_2^2T)-\delta_2]. \end{aligned} \tag{8.3.9}$$

The constants δ_1, δ_2 are arbitrary. The flow velocity is described as a two-soliton solution and can be regarded as being composed from a main pulse and an associated dicrotic wave. From (8.3.9) it results the solution for the flow velocity of blood

$$\begin{aligned} v_z(z,t) = \frac{6}{K}c_0\tilde{k}_1^2\operatorname{sech}^2[\frac{\tilde{k}_1}{L_0}(z-4k_1^2t)-\delta_1]+ \\ +\frac{6}{K}c_0\tilde{k}_2^2\operatorname{sech}^2[\frac{\tilde{k}_2}{L_0}(z-4k_2^2t)-\delta_2]. \end{aligned} \tag{8.3.10}$$

In a similar way we obtain two-soliton solutions for blood pressure and for radial displacement

$$\begin{aligned} p(z,t) = p_0 + \frac{6}{L}p_0\tilde{k}_1^2\operatorname{sech}^2[\frac{\tilde{k}_1}{L_0}(z-4k_1^2t)-\delta_1]+ \\ +\frac{6}{L}p_0\tilde{k}_2^2\operatorname{sech}^2[\frac{\tilde{k}_2}{L_0}(z-4k_2^2t)-\delta_2], \end{aligned} \tag{8.3.11}$$

$$\begin{aligned} u_r(z,t) = u_0\tilde{\gamma}_0 + \frac{6}{M}r_0\tilde{k}_1^2\operatorname{sech}^2[\frac{\tilde{k}_1}{L_0}(z-4k_1^2t)-\delta_1]+ \\ +\frac{6}{M}r_0\tilde{k}_2^2\operatorname{sech}^2[\frac{\tilde{k}_2}{L_0}(z-4k_2^2t)-\delta_2]. \end{aligned} \tag{8.3.12}$$

The time t_0 required for a pulse starting from the heart to cause the *steepening* in the aorta may be estimated by considering the equation (8.2.46) in which the third term is discarded

$$\frac{\partial v_1}{\partial \tau} + K v_1 \frac{\partial v_1}{\partial \xi} = 0 . \tag{8.3.13}$$

The solution of (8.3.13) can be regarded as a functional relation

$$v_1 = f(\xi - K v_1 \tau) . \tag{8.3.14}$$

Therefore, the scaled time t_0, which causes the steepening, may be connected with a scaled initial pulse of width ξ_0, on the ξ -axis, by an approximate relation

$$\tau_0 K (v_1)_m \cong \frac{1}{2} \xi_0 , \tag{8.3.15}$$

where $(v_1)_m$ is the maximum value of v_1. Inserting the transform relations obtained from (8.2.15), (8.2.17), (8.2.32) and (8.2.33)

$$\xi_0 = \sqrt{\varepsilon} z' = \sqrt{\varepsilon} L_0^{-1} d_0 ,$$

$$\tau_0 = \varepsilon \sqrt{\varepsilon} \ t_0' = \varepsilon \sqrt{\varepsilon} T_0^{-1} t_0 , \tag{8.3.16}$$

$$v_{1m} = \varepsilon^{-1} (\tilde{v})_m = \varepsilon^{-1} c_0^{-1} (v_z)_m ,$$

into (8.3.15) we have

$$t_0 \cong \frac{d_0}{2Kg(v_z)_m} , \tag{8.3.17}$$

where d_0 is the width of the initial pulse on the z -axis.

Taking K =1, g=1 and $(v_z)_m = 0.5$ m/s, $d_0 = 0.3$ m, from the experimental data, the time t_0 is estimated to be about 0.3 s. For $V = 5$ m/s, the distance that the wave needs to travel until the soliton is formed is estimated to be about 1.5 m. But the distance from the heart to the abdominal aorta is about 0.5 m. Therefore, in the abdominal aorta the soliton is not yet formed. The steepening phenomenon can be interpreted as the generating and growing of solitons in large arteries.

The condition for an equilibrium state of the system is given by

$$p = p_0 , \ v_z = 0 , \text{ for all } z , \tag{8.3.18}$$

and the boundary conditions of the system are

$$p = p_0 , \ v_z = 0 , \text{ as } z \pm \infty . \tag{8.3.19}$$

From (8.2.22), (8.2.28) and (8.3.6) we have

$$c \cong c_0 \{1 + (\frac{a}{2} - \frac{3}{4}) \frac{p_0 - p_e}{p_0} \} \frac{1}{\sqrt{1+k^2}} , \tag{8.3.20}$$

$$V \cong c_0 \{1 + (\frac{a}{2} - \frac{3}{4}) \frac{p_0 - p_e}{p_0} \} \{1 + \frac{L}{3} \frac{p_m - p_0}{p_0} \} . \tag{8.3.21}$$

The velocity of sound wave with small amplitude near the equilibrium state is given by (8.2.28), that is $gc_0(1+k^2)^{-1/2}$, and the phase velocity of a soliton with large amplitude, which satisfy the boundary condition (8.3.19), is given as $gc_0(1+2\tilde{k}^2)$. These velocities depend on g and $\tilde{\gamma}_0$. But $\tilde{\gamma}_0$ and $\tilde{p}_0$ are related by (8.2.22), and thus theses velocities change with the lowest pressure p_0. The phase velocity includes a term of $\tilde{k}^2$ which is proportional to the amplitude of the pulse. Then the efficiency of blood circulation depends on the lowest blood pressure p_0 and on the difference between the highest and the lowest pressure of blood $p_m - p_0$.

In this section, the results obtained from this theory are compared with the experimental measurements given by McDonald for the dog. These experimental results are summarized as follow (Yomosa):

– for the thoracic aorta

$$r_0 = 5\times 10^{-3}\,\text{m},\quad \frac{h_0}{r_0} = 0.12,\quad \rho = 1.05\times 10^{-9}\,\text{kg/m}^3,$$
$$\rho_0 = 1.06\times 10^{-9}\,\text{kg/m}^3,\quad E = 5.49\times 10^5\,\text{Pa}, \tag{8.3.22}$$
$$v_m = 0.55\,\text{m/s},\quad p_m - p_e = 102\,\text{mmHg} = 102\times 133.3\,\text{Pa},$$
$$p_0 - p_e = 81\,\text{mmHg} = 81\times 133.3\,\text{Pa},$$
$$p_m = 21\,\text{mmHg} = 21\times 133.3\,\text{Pa},\quad u_{rm} = 0.05 r_0,$$
$$v = 5.5\,\text{m/s},\quad f = 3.65\,\text{cycles/s},$$

where f is frequency.

– for the femoral artery

$$r_0 = 1.5\times 10^{-3}\,\text{m},\quad \frac{h_0}{r_0} = 0.12,\quad \rho = 1.05\times 10^{-9}\,\text{kg/m}^3,$$
$$\rho_0 = 1.06\times 10^{-9}\,\text{kg/m}^3,\quad E = 14.1\times 10^5\,\text{Pa}, \tag{8.3.23}$$
$$v_m = 0.4\,\text{m/s},\quad p_m - p_e = 110\,\text{mmHg} = 110\times 133.3\,\text{Pa},$$
$$p_0 - p_e = 78\,\text{mmHg} = 78\times 133.3\,\text{Pa},$$
$$p_m = 32\,\text{mmHg} = 32\times 133.3\,\text{Pa},\quad u_{rm} = 0.03 r_0,$$
$$v = 1\,\text{m/s},\quad f = 3.65\,\text{cycles/s}.$$

In the static case we have

$$p - p_e = E\frac{h_0}{r_0}\frac{\tilde{\gamma}(1+a\tilde{\gamma})}{(1+\tilde{\gamma})^2}. \tag{8.3.24}$$

Figure 8.3.1 represents an experimental diagram ($p - p_e$, $\frac{r}{r_0} = 1+\tilde{\gamma}$) obtained by McDonald for the dog. From this graph and (8.3.24) we estimate $E = 5.37\times 10^5$ Pa and $a = 1.95$ for the thoracic aorta, by using $\frac{h_0}{r_0} = 0.12$. The Young's modulus thus estimated is in good agreement with (8.3.22).

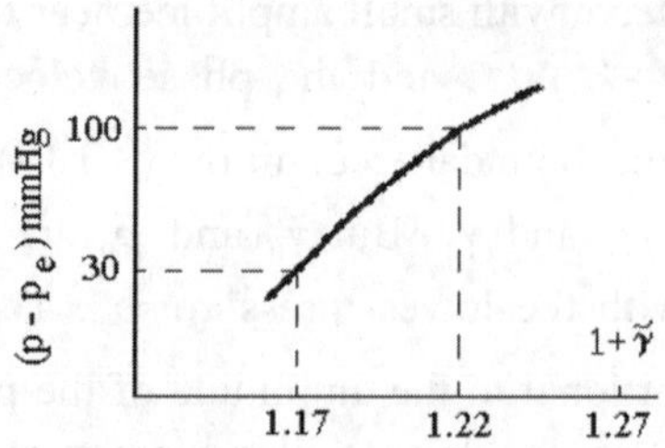

Figure 8.3.1 Experimental diagram ($p-p_e$, $\frac{r}{r_0}=1+\tilde{\gamma}$) for dog (McDonald).

For the femoral artery the same value for the nonlinear elastic coefficient a is considered. The parameters c_0 and p_0 are estimated by substituting the values of $\frac{h_0}{r_0}$, ρ and E given by (8.3.22) and (8.3.23) into (8.2.18)

$$c_0 = 5.6\,\text{m/s}\,,\ p_0 = 247\ \text{mmHg} = 247\times 133.3\ \text{Pa},\quad \text{(thoracic aorta)}\,, \tag{8.3.25}$$

$$c_0 = 8.98\,\text{m/s}\,,\ p_0 = 635\ \text{mmHg} = 635\times 133.3\ \text{Pa},\quad \text{(femoral artery)}. \tag{8.3.26}$$

The values of $\tilde{p}_0$ which correspond to the lowest pressures $p_0 - p_e = 81\,\text{mmHg}$ in the thoracic aorta, and $p_0 - p_e = 78\,\text{mmHg}$ in the femoral artery are determined from (8.2.17) and (8.3.25), (8.3.26).

$$\tilde{p}_0 = 0.328\,,\quad \text{(thoracic aorta)}, \tag{8.3.27}$$

$$\tilde{p}_0 = 0.123\,.\quad \text{(femoral artery)}. \tag{8.3.28}$$

The value of $\tilde{\gamma}_0$ is calculated from (8.2.22) and (8.3.27), (8.3.28)

$$\tilde{\gamma}_0 = 0.169,\quad \text{(thoracic aorta)}, \tag{8.3.29}$$

$$\tilde{\gamma}_0 = 0.062,\quad \text{(femoral artery)}. \tag{8.3.30}$$

The values of g, K, L and M are estimated from (8.2.28), (8.2.49) and (8.2.50)

$$g = 1.044\,,\ K = 1.024\,,\ L = 0.980\,,\ M = 1.829\,,\ \text{(thoracic aorta)}, \tag{8.3.31}$$

$$g = 1.023\,,\ K = 1.127\,,\ L = 1.102\,,\ M = 2.171\,,\ \text{(femoral artery)}. \tag{8.3.32}$$

From the soliton solutions (8.3.4), (8.3.6) and (8.3.7) we derive

$$v_m = \frac{6}{K} c_0 \tilde{k}^2\,,\ p_m = \frac{6}{L} p_0 \tilde{k}^2\,,\ (u_r)_m = \frac{6}{M} r_0 \tilde{k}^2\,, \tag{8.3.33}$$

Substituting into (8.3.33) the experimental values of v_m, p_m and $(u_r)_m$ given by (8.3.22) and (8.3.23), we estimate three values of $\tilde{k}$ for the soliton in the thoracic aorta

$$\tilde{k} = 0.129,\ \tilde{k} = 0.118,\ \tilde{k} = 0.123. \tag{8.3.34}$$

In a similar way, three values of $\tilde{k}$ are obtained for the femoral artery

$$\tilde{k} = 0.091,\ \tilde{k} = 0.096,\ \tilde{k} = 0.104. \tag{8.3.35}$$

Therefore, we can take $\tilde{k} \cong 0.12$ for the thoracic aorta, and $\tilde{k} \cong 0.09$ for the femoral artery. From Figure 8.2.1 we can estimate the amplitudes of the dicrotic flow and pressure waves in the femoral artery, denoted by v_m^d and p_m^d

$$v_m^d = 0.05 \text{ m/s},\ p_m^d = 5 \text{ mmHg}. \tag{8.3.36}$$

Then we estimate the value of $\tilde{k}$ for the dicrotic wave in the femoral artery for the first two equations (8.3.33)

$$\tilde{k}_d = 0.032,\ \tilde{k}_d = 0.038. \tag{8.3.37}$$

Therefore we can regard the wave in the femoral artery as a two-soliton wave which is characterized by $\tilde{k}_1 \cong 0.09$ and $\tilde{k}_2 \cong 0.035$.

Substituting the values of g, c_0 and $\tilde{k}$ estimated above, into $(8.3.5)_2$ we have the wave velocities

$$V \cong 6.02 \text{ m/s, (thoracic aorta),}$$

$$V \cong 9.34 \text{ m/s, (femoral artery).} \tag{8.3.38}$$

These values are reasonable compared with the experimental data (8.3.22) and (8.3.23). Since all values of $\tilde{k}$ estimated until now satisfy $\tilde{k}^2 << 1$, we can see from (8.3.5) and (8.2.28) that the solitons velocities in arteries nearly equal the sound wave velocities in them. From the profiles of the flow and pressure waves in Figure 8.2.1, we can estimate the widths of theses pulses. We have

$$\frac{2L_0}{\tilde{k}} \cong \frac{1}{15}\frac{V}{f} = 0.11 \text{m}, \text{ (thoracic aorta),}$$

$$\frac{2L_0}{\tilde{k}} \cong \frac{1}{30}\frac{V}{f} = 0.0856 \text{m}, \text{ (femoral artery),} \tag{8.3.39}$$

and

$$\frac{2L_0}{\tilde{k}V} \cong 0.02 \text{s}, \text{ (thoracic aorta),}$$

$$\frac{2L_0}{\tilde{k}V} \cong 0.01 \text{s}, \text{ (femoral artery).} \tag{8.3.40}$$

The characteristic lengths L_0 and time $T_0 = \frac{L_0}{c_0}$ are estimated by inserting the $\tilde{k}$ values into (8.3.39) and by using (8.3.25) and (8.3.26)

$$L_0 = 0.66\times10^{-2}\,\mathrm{m}\,,\; T_0 = 1.2\times10^{-3}\,\mathrm{s}\,,\text{(thoracic aorta)},$$

$$L_0 = 0.38\times10^{-2}\,\mathrm{m}\,,\; T_0 = 0.4\times10^{-3}\,\mathrm{s}\,.\text{ (femoral artery)}. \tag{8.3.41}$$

We can estimate the values of effective inertial thickness of the wall by using (8.2.16) and (8.3.22) and (8.3.41)

$$H_0 = 1.73\cdot10^{-2}\,\mathrm{m}\,,\text{ (thoracic aorta)},$$

$$H_0 = 1.93\times10^{-2}\,\mathrm{m}\,,\text{ (femoral artery)}. \tag{8.3.42}$$

The additional inertial thickness due to the mass of the tissue $h_0' = H_0 - h_0$ is estimated as

$$h_0' = 1.7\times10^{-2}\,\mathrm{m}\,,\text{ (thoracic aorta)},$$

$$h_0' = 1.9\times10^{-2}\,\mathrm{m}\,,\text{ (femoral artery)}. \tag{8.3.43}$$

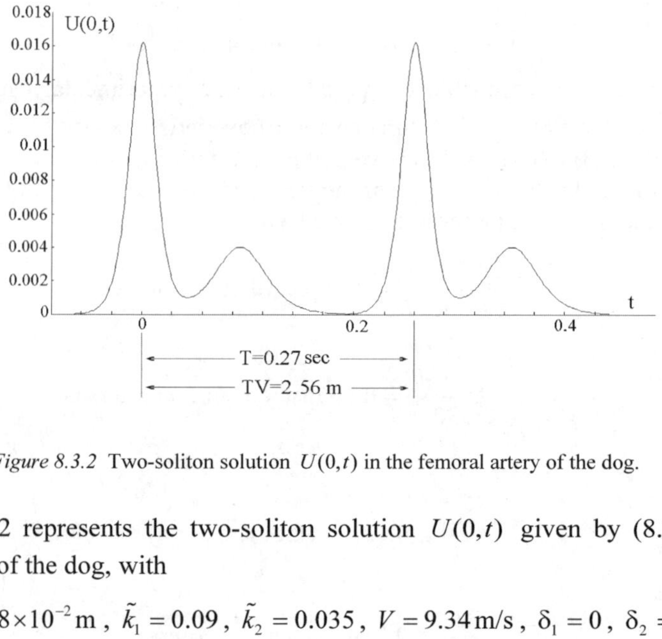

Figure 8.3.2 Two-soliton solution $U(0,t)$ in the femoral artery of the dog.

Figure 8.3.2 represents the two-soliton solution $U(0,t)$ given by (8.3.9) for the femoral artery of the dog, with

$$L_0 = 0.38\times10^{-2}\,\mathrm{m}\,,\; \tilde{k}_1 = 0.09\,,\; \tilde{k}_2 = 0.035\,,\; V = 9.34\,\mathrm{m/s}\,,\; \delta_1 = 0\,,\; \delta_2 = 2.9\,.$$

The frequency of generating the solitons is 3.65 cycles/s, the period is $T = 1/f = 0.27\,\mathrm{s}$, and the wavelength is $TV = 2.56\,\mathrm{m}$. For the thoracic aorta we have $T = 1/f = 0.27\,\mathrm{s}$, $V = 6.02\,\mathrm{m/s}$, $TV = 1.64\,\mathrm{m}$.

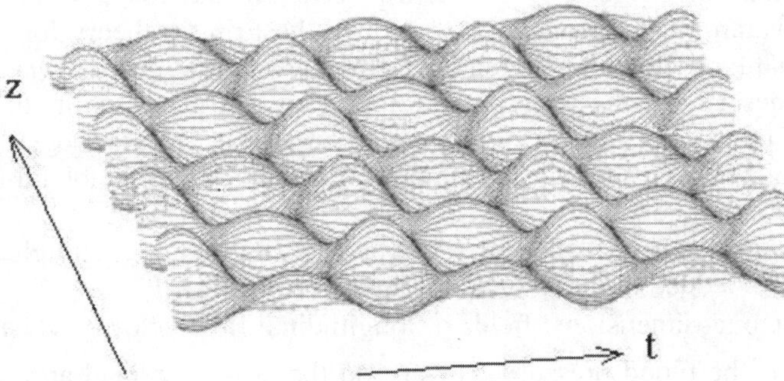

Figure 8.3.3 Two-soliton solution of flow velocity in femoral artery of the dog.

Figure 8.3.3 represents the time variation of the two-soliton solution of the flow velocity in the femoral artery of the dog, in an arbitrary location, for $\tilde{k}_1 = 0.09$, $\tilde{k}_2 = 0.035$, $V = 9.34\,\text{m/s}$, $K = 1.127$, $c_0 = 8.98\,\text{m/s}$, $\delta_1 = -1.7$, $\delta_2 = 1.3$. The frequency is 3.65 cycles/s.

Figure 8.3.4 represents the time variation of the two-soliton solution of the blood pressure in the femoral artery of the dog, in an arbitrary location, for $L_0 = 0.38 \times 10^{-2}\,\text{m}$, $\tilde{k}_1 = 0.096$, $\tilde{k}_2 = 0.041$, $V = 9.34\text{m/s}$, $L = 1.102$, $p_0 = 635\,\text{mmHg} = 635 \times 133.3\,\text{Pa}$, $\delta_1 = 0$.

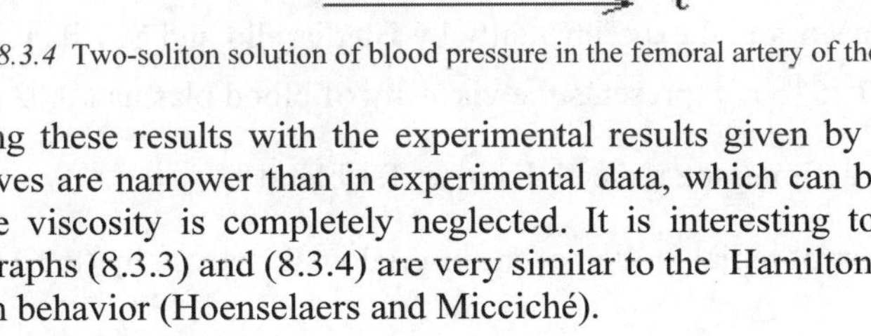

Figure 8.3.4 Two-soliton solution of blood pressure in the femoral artery of the dog.

By comparing these results with the experimental results given by McDonald we observe that waves are narrower than in experimental data, which can be explained by the fact that the viscosity is completely neglected. It is interesting to note that the features of the graphs (8.3.3) and (8.3.4) are very similar to the Hamiltonian of the sine-Gordon equation behavior (Hoenselaers and Micciché).

8.4 A micropolar model of blood flow in arteries

The blood is a rheologically complex fluid being a suspension of particles (red and white cells, platelets) undergoing unsteady flow through vessels. Eringen introduced a mathematical model for such fluids, called micropolar fluids, in 1966. This theory exhibits microrotational effects such as those experienced by blood in the larger vessels of the circulation for the dog (McDonald). Ariman and his coworkers, Turk and Sylvester, Easwaran and Majumdar, have applied the Eringen theory for describing the time-dependent blood flow. Their model formulates a new kinematic variable called microrotation describing the individual rotation of particles within the continuum, independent of the velocity field. In this theory an arbitrary pressure gradient as a sum of sine functions was considered. But the pulsatile character of the blood flow suggests the using the soliton theory.

Let us consider the thin-walled elastic tube is infinitely long, straight, circular and homogeneous, embedded in the tissue and filled with the blood

Consider the one-dimensional fields of longitudinal flow velocity $v(x,t)$, the micro-gyration $w(x,t)$, the blood pressure $p(x,t)$ and the cross-sectional area $A(x,t)$ under the assumption of uniform distributions of v and p over the cross-section of the tube. Here x is the axial distance along the vessel and t, the time. The radial component of the flow velocity is neglected in comparison with the axial component. The dimensional governing equations of the micropolar fluid dynamics are given by

$$\frac{\partial v}{\partial t}+v\frac{\partial v}{\partial x}+\frac{1}{\rho}\frac{\partial p}{\partial x}-\frac{1}{\rho}(\alpha+\mu)\frac{\partial^2 v}{\partial x^2}-\frac{\alpha}{\rho}\frac{\partial w}{\partial x}=0\,, \tag{8.4.1}$$

$$j\frac{\partial w}{\partial t}+\frac{\alpha}{\rho}\frac{\partial v}{\partial x}-\frac{\gamma}{\rho}\frac{\partial^2 w}{\partial x^2}+2\frac{\alpha}{\rho}w=0\,, \tag{8.4.2}$$

$$\frac{\partial A}{\partial t}+\frac{\partial (vA)}{\partial x}=0\,, \tag{8.4.3}$$

where ρ is the fluid density, α, μ, γ the blood coefficients and j the microinertia coefficient ($j=\dfrac{2\gamma}{2\mu+\alpha}$). The coefficient α depends on the cellular concentration (hematocrit) and is determined from the equation

$$\alpha^2+\alpha(2\mu-\gamma q^2)-\gamma q^2\mu=0\,, \tag{8.4.4}$$

where q has been found experimentally by Bugliarello and Sevilla (see Ariman *et al.*) given as $q=10^5\,\mathrm{m}^{-1}$, μ represents the viscosity of blood plasma (0.02 poise at $30^0 C$). The ratio $\dfrac{1}{q}$ is defining the red cell diameter for blood.

The equation of radial motion of the arterial wall is given by (8.2.14)

$$\rho_0\frac{\partial^2 u}{\partial t^2}=\frac{p-p_e}{H_0}(1+\frac{u}{r_0})-Eu\frac{h_0(1+a\dfrac{u}{r_0})}{H_0 r_0^2(1+\dfrac{u}{r_0})}\,, \tag{8.4.5}$$

with ρ_0 the density of the material of the wall, $r(x,t)$ the radius of the tube, h the thickness of the wall in radial direction, p_e the pressure outside the tube which can be regarded as about the same as the atmospheric pressure, σ the extending stress in the tangential direction. The density of the materials of the wall and the tissue are assumed to be equal. In (8.4.5) u is the radial displacement of the wall, E the Young's modulus of the elasticity, a a nonlinear coefficient of elasticity.

Taking account of $A = \pi(r_0 + u)^2$, we obtain from (8.4.5) the motion equation of the cross-sectional area

$$\frac{\partial^2 A}{\partial t^2} - \frac{1}{2A}\left(\frac{\partial A}{\partial t}\right)^2 = \frac{1}{\rho_0} f(A,p), \tag{8.4.6}$$

where

$$f(A,p) = 2A\frac{p-p_e}{H_0 r_0} - \frac{2\pi E h_0[r_0 + a(\sqrt{A/\pi} - r_0)]}{H_0 r_0^2}(\sqrt{A/\pi} - r_0). \tag{8.4.7}$$

We introduce dimensionless, noted by prime, variables

$$\begin{gathered} l_0 = \left(\frac{r_0 h_0 \rho_0}{2\rho}\right)^{1/2},\ t_0 = \left(\frac{r_0 \rho_0}{E}\right)^{1/2},\ c_0 = \frac{l_0}{t_0},\ x' = \frac{x}{l_0}, \\ t' = \frac{t}{t_0},\ v'(x,t) = \frac{v(x,t)}{c_0},\ p'(x,t) = \frac{p(x,t) - p_e}{\rho c_0^2}, \\ w' = t_0 w,\ A'(x,t) = \frac{A(x,t)}{A_0},\ A_0 = \pi r_0^2,\ A = \pi r^2,\ r' = \frac{r}{r_0}. \end{gathered} \tag{8.4.8}$$

The dimensionless equations can be written by dropping the prime as

$$\frac{\partial v}{\partial t} + v\frac{\partial v}{\partial x} + \frac{\partial p}{\partial x} - m\frac{\partial^2 v}{\partial x^2} - n\frac{\partial w}{\partial x} = 0, \tag{8.4.9}$$

$$b\frac{\partial w}{\partial t} + q\frac{\partial v}{\partial x} - g\frac{\partial^2 w}{\partial x^2} + dw = 0, \tag{8.4.10}$$

$$\frac{\partial A}{\partial t} + \frac{\partial (vA)}{\partial x} = 0, \tag{8.4.11}$$

$$\frac{\partial^2 A}{\partial t^2} - \frac{1}{2A}\left(\frac{\partial A}{\partial t}\right)^2 = f(A,p), \tag{8.4.12}$$

where

$$m=\frac{\alpha+\mu}{\rho c_0^2 t_0},\quad n=\frac{\alpha}{\rho c_0 l_0},\quad b=\frac{j}{l_0^2},$$
$$q=\frac{\alpha}{\rho c_0^2 t_0},\ g=\frac{\gamma t_0}{\rho l_0^4},\ d=\frac{\alpha t_0}{\rho l_0^2},$$
$$f(A,p)=2A\frac{p-p_e}{H_0 r_0}-\frac{2Eh_0[r_0+a(\sqrt{A/\pi}-r_0)]}{H_0 r_0^4}(\sqrt{A/\pi}-r_0), \tag{8.4.13}$$
$$b_1=\frac{2\rho c_0^2}{hE},\ b_2=\frac{2(2a-1)}{r_0},\ b_3=\frac{2a}{r_0},\ b_4=\frac{2(1-a)}{r_0}.$$

Let us assume an unsteady arbitrary pressure gradient of the form (Freeman)

$$p(x,t)=\sum_1^2 A_i \operatorname{sech}^2(\hat{b}_i-kx+\omega t), \tag{8.4.14}$$

with A_i , $\hat{b}_i$, k unknown constants and ω the circular frequency.

The governing flow equations (8.4.12)–(8.4.15) are solved in the condition of satisfying (8.4.16) and the initial conditions

$$v(0,0)=v_0,\ \frac{\partial v}{\partial t}(0,0)=v_1, w(0,0)=w_0,$$
$$\frac{\partial w}{\partial t}(0,0)=w_1, A(0,0)=a_0,\ \frac{\partial A}{\partial t}(0,0)=a_1. \tag{8.4.15}$$

The scenario for variation of coefficient α gives the effect of the volume concentration of the red cell upon the flow behavior of blood.

Equations (8.4.13)–(8.4.16) can be simplified into a form suitable for obtaining the periodic wave solutions. It will be of interest to obtain the wave of permanent form solutions of this system using the change of the variable

$$y=k(x-ct),\ c=\frac{\omega}{k}, \tag{8.4.16}$$

where c is the unknown wave velocity. In order to simplify equations we consider $a=1$. The system of equations are written as

$$-cv'+vv'+p'-mkv''-nw'=0,\ -bckw'+qkv'-gk^2w''+dw=0,$$
$$-cA'+(vA)'=0,\quad k^2c^2(A''-\frac{1}{2A}A'^2)=f(A,p). \tag{8.4.17}$$

The profiles of the flow velocity, the blood pressure and the cross-sectional area are numerically calculated using the data given by (8.3.22) for the thoracic aorta of the dog.

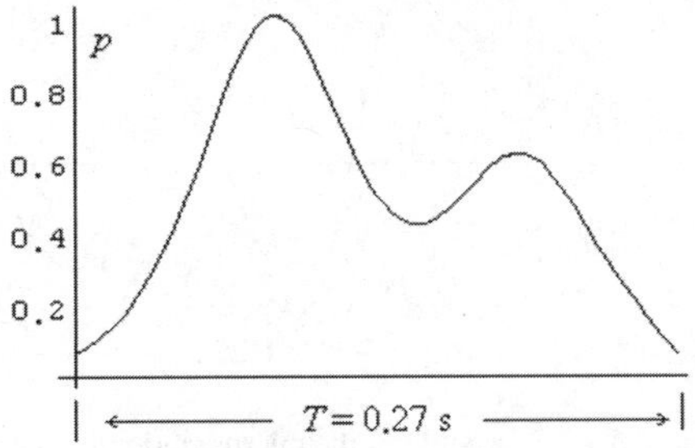

Figure 8.4.1 Variation of the blood pressure in one period.

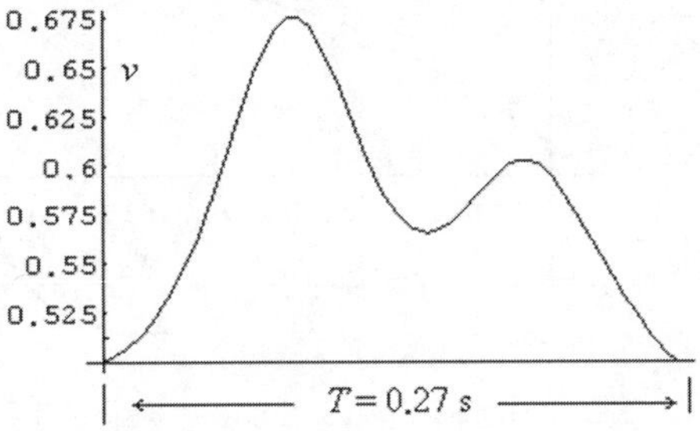

Figure 8.4.2 Variation of the flow velocity in one period.

For the blood coefficients we choose $\mu = 0.02\,\text{poise}$, $\gamma = 0.6$ mkg/s and $j = 20\,\text{m}^2$. From (8.4.4) we obtain $\alpha = 0.02\,\text{poise}$. By equating the width of the blood pressure two-soliton wave to the width (the wavelength is 162×10^{-2} m) of the pulse wave observed experimentally for the thoracic aorta of the dog, we obtain $T = 0.27\,\text{s}$ and $c = 6$ m/s. The dimensionless initial conditions are given by

$$v(0) = 0.5,\ v'(0) = 0,$$
$$w(0) = 0.2,\ w'(0) = 0,$$
$$A(0) = 1,\ A'(0) = 0.$$

Constants from (8.4.18) are found as $A_1 = 1$, $A_2 = 0.6$, $\hat{b}_1 = 2.3$, $\hat{b}_2 = 4.5$.

Figures 8.5.1–8.5.3 are plots of the blood pressure, the longitudinal flow velocity and the microgyration for one period. It is seen that the systolic and diastolic peaks exist in all graphs. The systolic and diastolic pressure peaks are related to the flow velocity peaks. The phase portraits for flow velocity and the microgyration are plotted in Figures 8.4.4–8.4.5 after four periods. For an increasing number of periods the graphs remain unchangeable. The motion becomes stable, the role of the initial conditions being very small. In Figure 8.4.6 the cross-sectional area is plotted for one period.

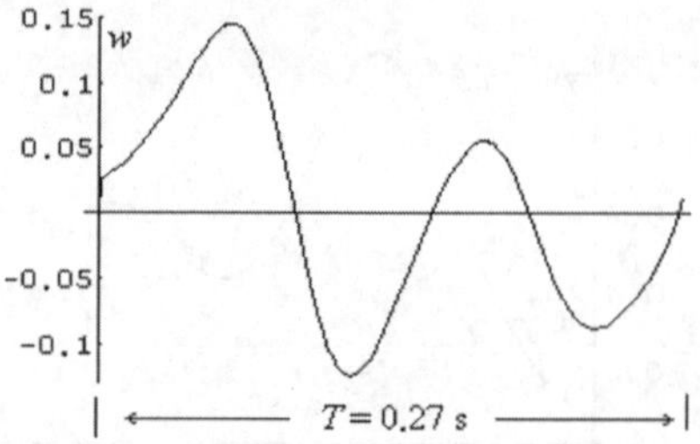

Figure 8.4.3 Variation of the microgyration in one period.

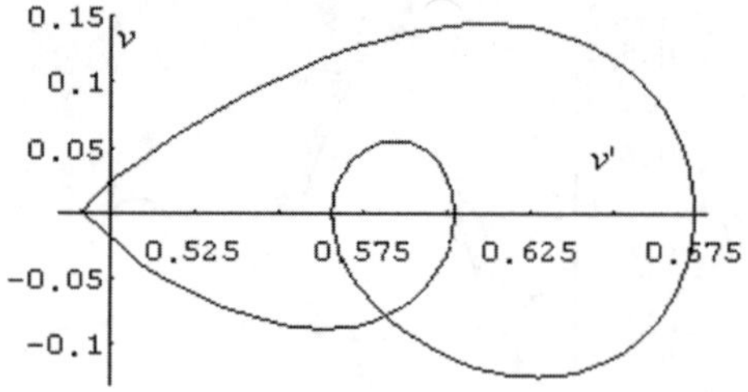

Figure 8.4.4 Phase portrait for the flow velocity in one period.

To illustrate the effect of the hematocrit on the flow velocity and on the microgyration we consider four values for the coefficient α. Figures 8.4.7–8.4.8 show a decreasing of the amplitude of the flow velocity and of the microgyration for increasing hematocrit. The results are reasonable due to the fact that the increase of viscosity overshadows the inertial effects leading to a reduction of the amplitudes and are in agreement with the experiments.

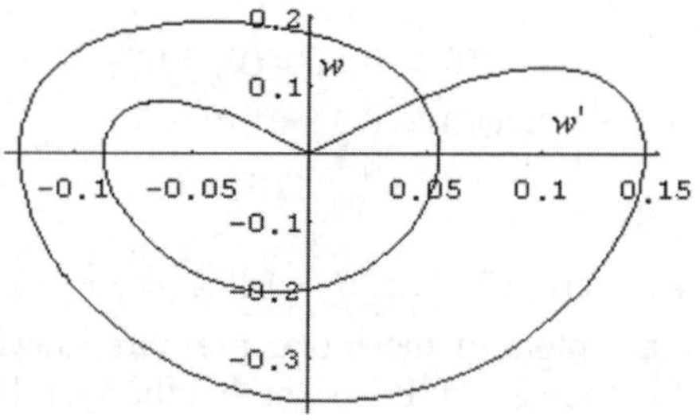

Figure 8.4.5 Phase portrait for microgyration in one period.

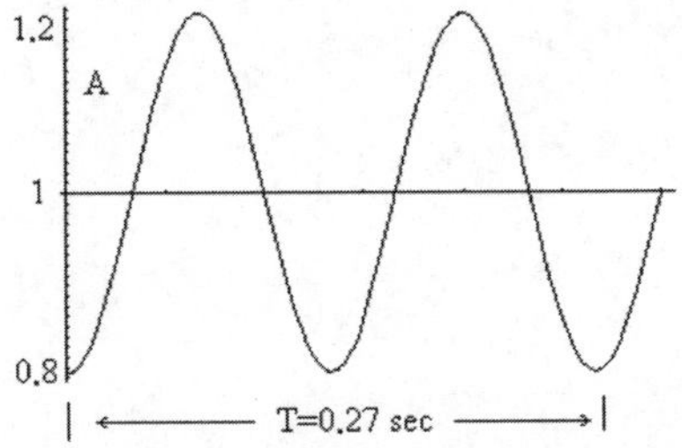

Figure 8.4.6 Variation of the area in one period.

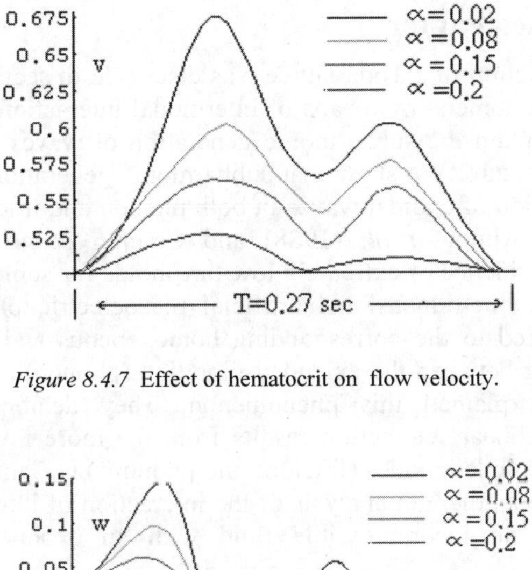

Figure 8.4.7 Effect of hematocrit on flow velocity.

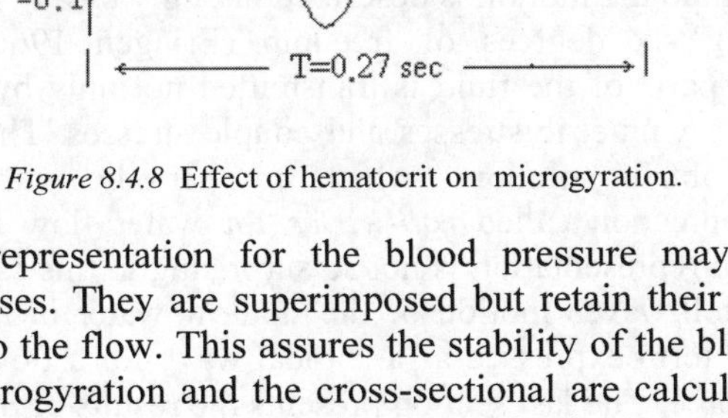

Figure 8.4.8 Effect of hematocrit on microgyration.

The two-soliton representation for the blood pressure may be viewed as an interaction of two pulses. They are superimposed but retain their identity and do not destroy each other into the flow. This assures the stability of the blood circulation. The flow velocity, the microgyration and the cross-sectional are calculated as functions of the two-soliton blood pressure pulse. The phase portraits demonstrate the stability of the blood motion (Tabor).

Chapter 9

INTERMODAL INTERACTION OF WAVES

9.1 Scope of the chapter

The modal interaction in a Toda lattice was discussed in section 6.5. In this chapter we discuss other phenomena by means of intermodal interaction of waves. In the first four sections we explain the subharmonic generation of waves in piezoelectric plates with Cantor-like structure. We show that subharmonic generation of waves is due to a nonlinear superposition of cnoidal waves in both phonon and fracton vibration regimes. Alippi (1982) and Alippi *et al.* (1988) and Crăciun *et al.* (1992) have proved experimentally the evidence of extremely low thresholds for subharmonic generation of ultrasonic waves in one-dimensional artificial piezoelectric plates with Cantor-like structure, as compared to the corresponding homogeneous and periodical plates. An anharmonic coupling between the extended-vibration (phonon) and the localized-mode (fracton) regimes explained this phenomenon. They demonstrate that the large enhancement of nonlinear interaction results from the more favorable frequency and spatial matching of coupled modes (fractons and phonons) in Cantor-like structure.

Section 9.5 presents the Yih analysis of the interaction of internal solitary waves of different modes in an incompressible fluid with an exponential stratification in densities.

The turbulent flow of a micropolar fluid downwards on an inclined open channel is studied in sections 9.6 and 9.7. The wave profile moves downstream as a linear superposition of soliton waves at a constant speed and without distortion.

In a micropolar fluid the motion is described not only by a deformation but also by a microrotation giving six degrees of freedom (Eringen 1966, Brulin 1982). The interaction between parts of the fluid is transmitted not only by a force but also by a torque, resulting in asymmetric stresses and couple stresses. The micropolar theory is employed here to obtain solutions, which are periodic with respect to distance, describing the phenomenon called *roll-waves* for water flow along a wide inclined channel. The soliton representation is not so surprising in this case because Dressler in 1949 studying the roll-waves motion of the shallow water in inclined open channels found an equivalent form expressed as a cnoidal wave for a flow subject to the Chezy turbulent resisting force. The last section presents the results of Shinbrot concerning the effect of surface tension on the solitary waves.

The chapter refers to the works by Munteanu and Donescu (2002), Chiroiu *et al.* (2001b), Yih (1994) and Shinbrot (1981).

9.2 A plate with Cantor-like structure

We consider a composite plate formed by alternating elements of nonlinear isotropic piezoelectric ceramics (PZ) and a nonlinear isotropic epoxy resin (ER), following a triadic Cantor sequence (Figure 9.2.1). Crăciun, Alippi and coworkers constructed an artificial one-dimensional Cantor structure. The Cantor set is an elementary example of a fractal. The Cantor set is generated by iteration of a single operation on a line of unit length. The operation consists of removing the middle third from each line segment of the previous set. As the number of iterations increases, the number of separate pieces tends to infinity, but the length of each one approaches zero. The property of invariance under a change of scale is called self-similarity and is common to many fractals. In contrast to a line with its infinite number of points and finite length, the Cantor set has an infinite number of points but zero length. The dimension of the Cantor set is less than 1. Denoting by N the number of segments of length ε, the dimension of the Cantor set is given by Baker and Gollub

$$d = \lim_{\varepsilon\to 0}\frac{\log N}{\log(1/\varepsilon)} = \lim_{n\to\infty}\frac{\log 2^n}{\log 3^n} = \frac{\log 2}{\log 3} < 1 .$$

We consider the same sample using a triadic Cantor sequence up to the fourth generation (31 elements). A rectangular coordinate system $Ox_1x_2x_3$ is employed. The origin of the coordinate system is located at the left end, in the middle plane of the sample, with the axis Ox_1 in-plane and normal to the layers and Ox_3 out-plane, normal to the plate. The length of the plate is l, the width of the smallest layer is $l/81$ and the thickness of the plate is h. The width of the plate is d. Let the regions occupied by the plate be $V = V^p \cup V^e$ where V^p and V^e are the regions occupied by PZ and ER layers. The boundary surface of V be S partitioned in the following way

$$S = S_1^p \cup S_1^e \cup S_2 , \quad S_1^p \cap S_1^e \cap S_2 = 0 ,$$

where

$$S_1^p = \{x_3 = \pm h/2, \quad 0 < x_1 < l\} , \text{ is the boundary surface of } V^p ,$$

$$S_1^e = \{x_3 = \pm h/2, \quad 0 < x_1 < l\} , \text{ is the boundary surface of } V^e ,$$

$$S_2 = \{x_1 = 0, \quad x_1 = l, \quad -h/2 \le x_3 \le h/2\} .$$

Let the unit outward normal of S be n_i the interfaces between constituents be I^{pe}. An index followed by a comma represents partial differentiation with respect to space variables and a superposed dot indicates differentiation with respect to time. Throughout the present paper repeated indices denote summation over the range (1,2,3).

In order to investigate theoretically the existence of multiple fracton and multiple phonon mode regimes in the displacement field for a piezoelectric plate with Cantor-like structure it is customary to consider: first, the nonlinear geometrical relations between the components of deformation and those of the displacement vector; second, to retain in the constitutive equations besides the linear terms also the nonlinear terms of lowest order. So, we have considered the piezoelectric material to be nonlinear and isotropic, characterized by two second-order elastic constants, three third-order elastic constants, two (linear and nonlinear) dielectric constants and two (linear and nonlinear) coefficients of piezoelectricity. In our first attempt to investigate the existence of multiple fracton and multiple phonon mode regimes, we considered the case of anisotropic piezoelectric material with monoclinic symmetry and we neglected the third-order constants. In spite of the bigger number of elastic, dielectric and piezoelectric constants, the results did not show clearly the existence of localized and extended modes regimes. In conclusion, a quantitative knowledge of the second-order material constants is essential for the analysis of the fracton and phonon mode regimes for a piezoelectric plate (Şoós). The dashed regions are occupied by piezoelectric ceramic of total volume V^p and boundary external surface S_1^p. The white regions are occupied by epoxy resin of total volume V^e and boundary external surface S_1^e.

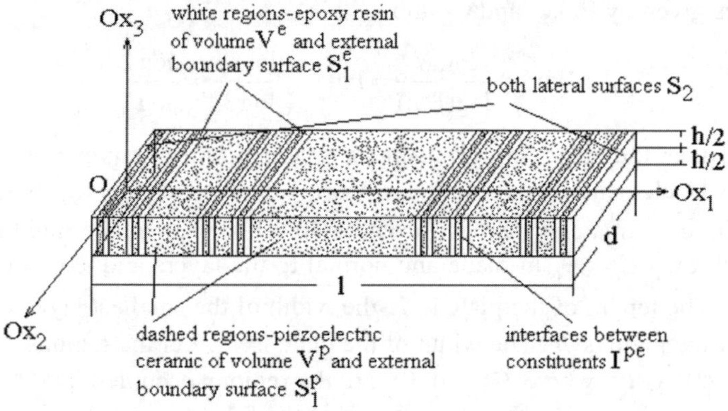

Figure 9.2.1 The plate with Cantor-like structure.

The lateral surfaces are S_2 and the interfaces between constituents I^{pe}. In the following, the basic equations are written for piezoelectric and non-piezoelectric materials:

a) *Piezoelectric material* (PZ).

1. The quasistatic motion equations

$$\rho^p \ddot{u}_i = (t_{ij} + u_{k,i} t_{kj})_{,j} \,, \text{ in } V^p \,, \tag{9.2.1}$$

$$D_{i,i} = 0 \,, \; E_i + \varphi_{,i} = 0 \,, \text{ in } V^p \,, \tag{9.2.2}$$

where ρ_p is the density, u_i is the displacement vector, t_{ij} is the stress tensor, D_i is the electric induction vector, E_i is the electric field and φ is the electric potential.

2. The constitutive equations

$$\begin{aligned} t_{ij} = {} & \lambda^p \varepsilon_{kk}\delta_{ij} + 2\mu^p \varepsilon_{ij} + A^p \varepsilon_{il}\varepsilon_{jl} + 3B^p \varepsilon_{kk}\varepsilon_{ij} + C^p \varepsilon_{kk}^2 \delta_{ij} - \\ & - e_k^p E_k \delta_{ij} - \overline{e}_k^p E_k \varepsilon_{ll}\delta_{ij} - \overline{\overline{e}}_k^p E_k \varepsilon_{ll}\delta_{ij} - \overline{\overline{e}}_k^p E_k \varepsilon_{ij}, \end{aligned} \quad \text{in } V^p \,, \tag{9.2.3}$$

$$D_i = \overline{\varepsilon}^p E_i - \frac{1}{2}\overline{\varepsilon}_i^p E^2 - e_i^p \varepsilon_{kk} - \frac{1}{2}\overline{e}_i^p \varepsilon_{kk}^2 - \frac{1}{2}\overline{\overline{e}}_i^p \varepsilon_{kl}^2 \,, \text{ in } V^p \,, \tag{9.2.4}$$

$$\varepsilon_{ij} = \frac{1}{2}(u_{i,j} + u_{j,i} + u_{l,i}u_{l,j}) \,, \text{ in } V \,, \tag{9.2.5}$$

where ε_{ij} is the strain tensor, λ^p, μ^p are the Lamé constants, A^p, B^p, C^p are the Landau constants, $\overline{\varepsilon}^p, \overline{\varepsilon}_i^p$ ($\overline{\varepsilon}_3^p = \overline{\varepsilon}_2^p = \overline{\varepsilon}_1^p$) are the linear and nonlinear dielectric constants, e_i^p ($e_1^p = e_2^p = e_3^p$), $\overline{e}_i^p$ ($\overline{e}_1^p = \overline{e}_2^p = \overline{e}_3^p$) and $\overline{\overline{e}}_i^p$ ($\overline{\overline{e}}_1^p = \overline{\overline{e}}_2^p = \overline{\overline{e}}_3^p$) are the linear and nonlinear coefficients of piezoelectricity and $E^2 = E_1^2 + E_2^2 + E_3^2$.

3. The boundary conditions

$$t_{ij}n_j = \overline{T}_{ij}n_j = \overline{T}_i \,, \text{ on } S_1^p \,, \tag{9.2.6}$$

$$D_i n_i = \overline{d}, \quad \varphi = \overline{\varphi} \,, \text{ on } S_1^p \,, \tag{9.2.7}$$

where $\overline{T}_i$, $\overline{d}$, $\overline{\varphi}$ are quantities prescribed on the boundary and $\overline{T}_{ij}$ is the Maxwell stress tensor. We consider that a periodical electric field $\overline{E}_i = \overline{E}_i^0 \exp(\mathrm{i}\omega t)$ is applied to both surfaces of the plate to excite the Lamb waves, over a wide frequency range ($10\,\text{MHz} < \omega/2\pi < 5\,\text{MHz}$).

The action of this field is described by Maxwell stress tensor $\overline{T}_{ij}$ (Kapelewski and Michalec)

$$\overline{T}_{ij} = \frac{1}{4\pi}(\overline{E}_i\overline{E}_j - \frac{1}{2}\overline{E}^2\delta_{ij}) \,, \text{ on } S_1^p \,. \tag{9.2.8}$$

The boundary conditions (9.2.6)–(9.2.7) on S_1^p are rewritten as

$$t_{13} = \frac{1}{4\pi}\overline{E}_1\overline{E}_3, \quad t_{33} = \frac{1}{8\pi}(\overline{E}_3^2 - \overline{E}_1^2) \,, \text{ on } S_1^p \,, \tag{9.2.9}$$

$$D_3 = \overline{E}_3, \; E_1 = \overline{E}_1, \text{ on } S_1^p \,. \tag{9.2.10}$$

b) *Non-piezoelectric material* (ER).

4. The motion equations

$$\rho^e \ddot{u}_i = t_{ij,j} \,, \text{ in } V^e \,. \tag{9.2.11}$$

5. The constitutive equations

$$t_{ij} = \lambda^e \varepsilon_{kk}\delta_{ij} + 2\mu^e \varepsilon_{ij} + A^e \varepsilon_{il}\varepsilon_{jl} + 3B^e \varepsilon_{kk}\varepsilon_{ij} + C^e \varepsilon_{kk}^2 \delta_{ij} \,, \text{ in } V^e \,. \tag{9.2.12}$$

6. The boundary conditions

$$t_{ij}n_j = 0 \,, \text{ on } S_1^e \,, \tag{9.2.13}$$

or

$$t_{13} = t_{31} = 0, \quad \text{on} \quad S_1^e . \tag{9.2.14}$$

7. The boundary conditions on S_2

$$-u_1 = -u_3 = 0, \text{ on } \quad S_2 . \tag{9.2.15}$$

8. The conditions on interfaces between constituents I^{pe}.

At the interfaces between constituents the displacement and the traction vectors are continuous.

$$[u_1] = [u_3] = 0, \; [t_{11}] = [t_{13}] = 0, \text{ on } \; I^{pe}, \tag{9.2.16}$$

where the bracket indicates a jump across the interface and $k = 1,3$.

The 3D problem can be reduced to a 2D problem, if we consider all quantities independent with respect to x_2, and $u_2 = 0$, $E_2 = 0$.

We have

$$u_1 = u_1(x_1, x_3, t), \; u_3 = u_3(x_1, x_3, t), \text{ in } \; V, \tag{9.2.17}$$

$$E_1 = -\varphi_{,1}, \; E_3 = -\varphi_{,3}, \; \varphi = \varphi(x_1, x_3, t), \text{ in } V^p . \tag{9.2.18}$$

We express the elastic potentials ϕ and ψ

$$u_1 = \phi_{,1} - \psi_{,3}, \; u_3 = \phi_{,3} + \psi_{,1}, \text{ in } V, \tag{9.2.19}$$

in the form

$$\begin{aligned} \phi &= [A\cos(\alpha x_3) + B\sin(\alpha x_3)]\tilde{\phi}(x_1, t), \\ \psi &= [C\cos(\beta x_3) + D\sin(\beta x_3)]\tilde{\psi}(x_1, t), \quad \text{in } V^p, \\ \varphi &= [E\cos(\gamma x_3) + F\sin(\gamma x_3)]\tilde{\varphi}(x_1, t), \end{aligned} \tag{9.2.20}$$

and

$$\begin{aligned} \phi &= [\hat{A}\cos(\hat{\alpha} x_3) + \hat{B}\sin(\hat{\alpha} x_3)]\tilde{\phi}(x_1, t), \\ \psi &= [\hat{C}\cos(\hat{\beta} x_3) + \hat{D}\sin(\hat{\beta} x_3)]\tilde{\psi}(x_1, t), \text{ in } V^e, \end{aligned} \tag{9.2.21}$$

where $\tilde{\phi}, \tilde{\psi}$ and $\tilde{\varphi}$ are unknown functions.

Kapelewski and Michalec obtained the analytical proof of the existence of solitary surface waves for a free nonlinear semi-infinite isotropic piezoelectric medium. Following this work we have observed that the governing equations (9.2.1)–(9.2.2) and (9.2.11) can support for $\phi, \psi, \varphi, \hat{\phi}, \hat{\psi}$ and $\hat{\varphi}$ certain particular functions $\tilde{\phi}, \tilde{\psi}$ and $\tilde{\varphi}$ having the form

$$\tilde{\phi} = \tilde{\phi}_0 \mathrm{sech}(\varsigma\eta)\,,\ \tilde{\varphi} = \frac{\tilde{\varphi}_0}{\tilde{\psi}_0^2}\tilde{\psi}^2\,,\ \tilde{\psi} = (\tilde{\psi}_0 + 1)\mathrm{sech}(\varsigma\eta)\exp(\frac{\mathrm{sh}(\varsigma\eta)}{\varsigma\tilde{\phi}_0}) - 1\,, \quad (9.2.22)$$

where $\eta = x_1 - vt$, v the velocity of wave.

This result plays an essential role in our strategy to determine the functions $\tilde{\phi}, \tilde{\psi}$ and $\tilde{\varphi}$ by using the cnoidal method. Using the cnoidal representations, the displacement field in V^p and V^e is given by

$$\begin{aligned} u_1 &= [A\cos(\alpha x_3) + B\sin(\alpha x_3)]\tilde{\phi}_0[\log\Theta_n]''' + \\ &+ [\beta C\sin(\beta x_3) - \beta D\cos(\beta x_3)]\tilde{\psi}_0[\log\Theta_n]'', \end{aligned} \quad (9.2.23)$$

$$\begin{aligned} u_3 &= [-\alpha A\sin(\alpha x_3) + B\alpha\cos(\alpha x_3)]\tilde{\phi}_0[\log\Theta_n]'' + \\ &+ [C\cos(\beta x_3) + D\sin(\beta x_3)]\tilde{\psi}_0[\log\Theta_n]''', \end{aligned} \quad (9.2.24)$$

in V^p, and

$$\begin{aligned} u_1 &= [\hat{A}\cos(\hat{\alpha} x_3) + \hat{B}\sin(\hat{\alpha} x_3)]\tilde{\phi}_0[\log\Theta_n]''' + \\ &+ [\hat{\beta}\hat{C}\sin(\hat{\beta} x_3) - \hat{\beta}\hat{D}\cos(\hat{\beta} x_3)]\tilde{\psi}_0[\log\Theta_n]'', \end{aligned} \quad (9.2.25)$$

$$\begin{aligned} u_3 &= [-\hat{\alpha}\hat{A}\sin(\hat{\alpha} x_3) + \hat{B}\hat{\alpha}\cos(\hat{\alpha} x_3)]\tilde{\phi}_0[\log\Theta_n]'' + \\ &+ [\hat{C}\cos(\hat{\beta} x_3) + \hat{D}\sin(\hat{\beta} x_3)]\tilde{\psi}_0[\log\Theta_n]''', \end{aligned} \quad (9.2.26)$$

in V^e. The electric field in V^p is given by

$$E_1 = -[E\cos(\gamma x_3) + F\sin(\gamma x_3)\tilde{\varphi}_0[\log\Theta_n]'''\,, \quad (9.2.27)$$

$$E_3 = [\gamma E\sin(\gamma x_3) - F\gamma\cos(\gamma x_3)\tilde{\varphi}_0[\log\Theta_n]''\,. \quad (9.2.28)$$

In these expressions, the prime means the differentiation with respect to $x \equiv x_1$. We have used here the following representations for the functions $\tilde{\phi}, \tilde{\psi}$ and $\tilde{\varphi}$

$$\begin{aligned} \tilde{\phi}(x,t) &= \tilde{\phi}_0 \frac{\partial^2}{\partial^2 x}\log\Theta_n(\eta_1,\eta_2,...,\eta_n)\,, \\ \tilde{\varphi}(x,t) &= \tilde{\varphi}_0 \frac{\partial^2}{\partial^2 x}\log\Theta_n(\eta_1,\eta_2,...,\eta_n)\,, \\ \tilde{\psi}(x,t) &= \tilde{\psi}_0 \frac{\partial^2}{\partial^2 x}\log\Theta_n(\eta_1,\eta_2,...,\eta_n)\,, \end{aligned} \quad (9.2.29)$$

where $x \equiv x_1$ and $\tilde{\phi}_0, \tilde{\psi}_0$ and $\tilde{\varphi}_0$ constants. In (9.2.29), $\Theta_n(\eta_1,\eta_2,...,\eta_n)$ is theta-function (1.4.14)

$$\Theta_n(\eta_1,\eta_2,...,\eta_n)=\sum_{\substack{M_i=-\infty\\ 1\le i\le n}}^{\infty}\exp\Big(\sum_{J=1}^{n}\mathrm{i}M_j\eta_j+\frac{1}{2}\sum_{i,j=1}^{N}M_iB_{ij}M_j\Big),$$

and $\eta_j=k_jx-\omega_jt+\beta_j$, $1\le j\le n$, where k_j are the wave numbers, ω_j the frequencies and β_j the phases. The relations (9.2.29) allow the representations

$$\tilde\phi(x,t)=\tilde\phi_{cn}(\eta)+\tilde\phi_{\mathrm{int}}(\eta),\ \tilde\varphi(x,t)=\tilde\varphi_{cn}(\eta)+\tilde\varphi_{\mathrm{int}}(\eta),\ \tilde\psi(x,t)=\tilde\psi_{cn}(\eta)+\tilde\psi_{\mathrm{int}}(\eta),\ (9.2.30)$$

where the first terms result from $\frac{\partial^2}{\partial^2 x}\log G\ (\eta)$, and the second terms from $\frac{\partial^2}{\partial^2 x}\log(1+\frac{F\ (\eta,C)}{G\ (\eta)})$, in accordance to the theorem 1.4.1.

The solutions (9.2.23)–(9.2.28) must satisfy the set of equations (9.2.1)–(9.2.10) in V^p, and the set of equations (9.2.11)–(9.2.16) in V^e.

9.3 The eigenvalue problem

In classical elasticity, there are two types of variational principles for the free vibration of an elastic body. One is associated with the potential energy, the other with the complementary energy. A variational formulation for the free vibration of a piezoelectric body is given by Yang, which is related to the internal energy.

The eigenvalue problem for the resonance of the plate is

$$\rho^p\ddot u_i=(t_{ij}+u_{k,i}t_{kj})_{,j},\ \text{in } V^p,$$

$$D_{i,i}=0,\ E_i+\varphi_{,i}=0,\ \text{in } V^p,$$

$$\begin{aligned}t_{ij}=\lambda^p\varepsilon_{kk}\delta_{ij}+2\mu^p\varepsilon_{ij}+A^p\varepsilon_{il}\varepsilon_{jl}+3B^p\varepsilon_{kk}\varepsilon_{ij}+C^p\varepsilon_{kk}^2\delta_{ij}-\\ -e_k^pE_k\delta_{ij}-\overline{e}_k^pE_k\varepsilon_{ll}\delta_{ij}-\overline{\overline{e}}_k^pE_k\varepsilon_{ll}\delta_{ij}-\overline{e}_k^pE_k\varepsilon_{ij},\end{aligned}\quad \text{in } V^p,$$

$$D_i=\overline{\varepsilon}^pE_i-\frac{1}{2}\overline{\varepsilon}_i^pE^2-e_i^p\varepsilon_{kk}-\frac{1}{2}\overline{e}_i^p\varepsilon_{kk}^2-\frac{1}{2}\overline{\overline{e}}_i^p\varepsilon_{kl}^2,\quad \text{in } V^p,$$

$$t_{ij}=\lambda^e\varepsilon_{kk}\delta_{ij}+2\mu^e\varepsilon_{ij}+A^e\varepsilon_{il}\varepsilon_{jl}+3B^e\varepsilon_{kk}\varepsilon_{ij}+C^e\varepsilon_{kk}^2\delta_{ij},\ \text{in } V^e,$$

$$(t_{ij}+u_{k,i}t_{kj})n_j=0,\ \text{on } S,\ D_in_i=0,\ \varphi=0,\ \text{on } S_1^p,$$

$$\rho^e\ddot u_i=(t_{ij}+u_{k,i}t_{kj})_{,j},\ \text{in } V^e,\ -u_1=-u_3=0,\ \text{on } S_2,\qquad(9.3.1)$$

$$[u_1]=[u_3]=0,\ [t_{11}+u_{k,1}t_{k1}]=[t_{13}+u_{k,3}t_{k3}]=0,\ \text{on } I^{pe},$$

$$\varepsilon_{ij} = \frac{1}{2}(u_{i,j} + u_{j,i} + u_{l,i}u_{l,j}), \text{ in } V, \ i,j = 1,3.$$

For the eigenvalue problem (9.3.1) for the resonance of the plate we give a formulation of a variational principle by introducing the Yang functional

$$\Pi(u_i, \varepsilon_{ij}, t_{ij}, \varphi, D_i) = \frac{\Lambda(u_i, \varepsilon_{ij}, t_{ij}, \varphi, D_i)}{\Gamma(u_i)}, \tag{9.3.2}$$

where

$$\Lambda(u_i, \varepsilon_{ij}, t_{ij}, \varphi, D_i) = \int_V [t_{ij}u_{i,j} - D_{i,i}\varphi + U(\varepsilon_{ij}, D_i) - t_{ij}\varepsilon_{ij}]\mathrm{d}V - \\ - \int_{S_2} t_{ji}n_j u_i \mathrm{d}S + \int_{S_1^p} D_i n_i \varphi \mathrm{d}S, \tag{9.3.3}$$

$$\Gamma(u_i) = \int_V \frac{1}{2}\rho u_i u_i \mathrm{d}V, \tag{9.3.4}$$

and U the internal energy

$$U = U(\varepsilon_{ij}, D_i) = \frac{1}{2}[t_{ij}\varepsilon_{ij} + E_i D_i]. \tag{9.3.5}$$

For free vibrations, the three displacement functions are expressed as (9.2.23)–(9.2.26) and the components of the electric field by (9.2.27)–(9.2.28).

The stationary condition on Π gives the eigenvalue problem (9.3.1) with the stationary value of ω^2. The values of ω^2 are sought corresponding to which nontrivial solutions of (9.3.1) exist. The angular frequency ω is related to he circular frequency f by $\omega = 2\pi f$.

9.4 Subharmonic waves generation

Given a set of measured frequencies $f_n = \omega_n / 2\pi$, $n = 1,2,...,N$, of the plate we aim at determining the unknown system parameters

$$P = \{ k_j, \omega_j, \beta_j, B_{ij}, A, B, C, D, E, F, \hat{A}, \hat{B}, \hat{C}, \hat{D}, \alpha, \beta, \gamma, \hat{\alpha}, \hat{\beta}, \ i,j = 1,2,3,4,5 \}, \tag{9.4.1}$$

necessary to analytically construct the solutions of the set of governing equations.

The inverse problem we want to consider is, given the solutions representations (9.2.23)–(9.2.28), to find the 55 parameters P given by (9.4.1) by inversion of the measured natural frequencies of the plate. Because we do not have an experimental set of eigenfrequencies we will use for the inverse problem the set of eigenfrequencies computed from the direct problem (9.3.2)–(9.3.5).

To determine the parameters P from the *measured* natural frequencies data by the nonlinear least-squares optimization technique, an objective function $\Im$ must be chosen that measures the agreement between theoretical and experimental data

$$\Im(P) = \sum_{i=1}^{N} \{[f_i^e - f_i\ (P)]^2\}\ \to \min\,, \tag{9.4.2}$$

where f_i^e are the measured i-th eigenfrequency, f_i the corresponding model prediction, and N the number of measurements of frequencies $N > K$, with K the number of unknown parameters which are given by P (for our case this number is 55).

The primary goal of the optimization procedure is the minimization of the objective function $\Im(P)$, where P are the design variables which make up the solutions of the problem. The signs "$\to \min$", "$\to \max$" mean the minimization or maximization of the objective function with some required precision: suppose six decimal places for the variable values are desirable. To measure the accuracy of the identification of P we introduce an error indicator δ to estimate the verification of the governing equations, that is

$$\delta(P_{opt}) = \sum_{k=1}^{6} \delta_k\,, \tag{9.4.3}$$

$$\delta_1 = \int_{V^p} \{(\rho^p \ddot{u}_i - t_{ij,j})^2 + D_{i,i}^2 + (E_i + \varphi_{,i})^2\} \mathrm{d}V^p\,, \tag{9.4.4}$$

$$\delta_2 = \int_{V^p} (\rho^e \ddot{u}_i - t_{ij,j})^2 \mathrm{d}V^e\,, \tag{9.4.5}$$

$$\delta_3 = \int_{S_1^p} \{(t_{13} - \frac{1}{4\pi}\overline{E}_1\overline{E}_3)^2 + [t_{33} - \frac{1}{8\pi}(\overline{E}_3^2 - \overline{E}_1^2)]^2 + (D_3 - \overline{E}_3)^2 + (E_1 - \overline{E}_1)^2\} \mathrm{d}S_1^p\,, \tag{9.4.6}$$

$$\delta_4 = \int_{S_1^e} [t_{13}^2 + t_{33}^2] \mathrm{d}S_1^e\ , \tag{9.4.7}$$

$$\delta_5 = \int_{S_1^e} [u_1^2 + u_3^2] \mathrm{d}S_2\ , \tag{9.4.8}$$

$$\delta_6 = \int_{I^{pe}} [u_1^2 + u_3^2 + t_{11}^2 + t_{13}^2] \mathrm{d}I^{pe}\,, \tag{9.4.9}$$

where P_{opt} is the solution of the inverse algorithm (9.4.2), t_{ij} are given by (9.2.3) in V^p and (9.2.12) in V^e, D_i are given by (9.2.4) and ε_{ij} by (9.2.5).

We define fitness as follows

,

$$F = \frac{\Im_0}{\Im}\,, \Im_0 = \sum_{i=1}^{m} (f_i^e)^2\,. \tag{9.4.10}$$

As the convergence criterion of iterative computations we use the expression Z to be maximum

$$Z = \frac{1}{2}\log_{10}\frac{\Im_0}{\Im} \quad \rightarrow \max . \qquad (9.4.11)$$

The quality of the model depends on the maximum value of the function Z and the associated value of $\delta(P_{opt})$. The integrals (9.4.4)–(9.4.8) can be analytically computed.

We use the genetic algorithm described in Chapter 7, in this case the binary vector having 55 genes to represent the real values of parameters P given by (9.4.1).

The numerical computation of theta-function is drastic by noting that the number of complex exponentials is $(2M+1)^n - 1$ (Osborne). For $M=5$, $n=5$ this number is 161050 and for $M=10$, $n=100$ this number is $\approx 10^{132}$. We consider here the case of five degrees of freedom solutions $n=5$ and $M=5$ $(-5 \le M_n \le 5)$. The calculus was carried out for $l=$ 67.5mm and $h=$ 0.3mm. The material constants are shown in Table 9.4.1. The calculus was carried out for $l=$ 67.5mm and $h=$ 0.3mm. The material constants are shown in Table 9.4.1 ($\tilde{\phi}_0 = \tilde{\varphi}_0 = \tilde{\psi}_0 = 1$).

Table 9.4.1 The material constants for piezoelectric ceramics and epoxy resin (Rogacheva)

	piezoelectric ceramics	epoxy resin
λ	71.6 GPa	42.31 GPa
μ	35.8 GPa	3.76 GPa
A	-2000 GPa	2.8 GPa
B	-1134 GPa	9.7 GPa
C	-900 GPa	-5.7 GPa
$\overline{\varepsilon}$	4.065 nF/m	-
$\overline{\varepsilon}_1$	2.079 nF/m	-
e_1	-0.218 nm/V	-
$\overline{e}_1 = \overline{\overline{e}}$	-0.435 nm/V	-
ρ	7650 Kg/m ^{3}	1170Kg/m ^{3}

Table 9.4.2 Estimation results: computed eigenfrequencies

$\omega_n/2\pi$	100.2 ± 0.05	167 ± 0.01	217.1 ± 0.03	250.5 ± 0.1	334 ± 0.01	367.4 ± 0.01	417.5 ± 0.1	501 ± 0.02	584.5 ± 0.03
	617.9 ± 0.01	668 ± 0.03	835 ± 0.06	935.2 ± 0.06	1085.5 ± 0.1	1169 ± 0.07	1269.2 ± 0.02	1503 ± 0.05	1670 ± 0.4
	1770.2 ± 0.2	1987.3 ± 0.12	2120.9 ± 0.02	2250 ± 0.1	2471.6 ± 0.3	2655.3 ± 0.01	2672 ± 0.01	2972.6 ± 0.2	3340 ± 0.4
	3540.4 ± 0.04	3577.4 ± 0.02	3690.7 ± 0.01	3774.2 ± 0.15	3974.6 ± 0.07	3991.3 ± 0.24	4241.8 ± 0.07	4250 ± 0.03	4291.9 ± 0.06
	4322 ± 0.04	4525.7 ± 0.2	4655 ± 0.1	4698.6 ± 0.02	4766 ± 0.2	4798.4 ± 0.03	4826.3 ± 0.01	4856 ± 0.04	4881.7 ± 0.04
	4899.4 ± 0.01	4901 ± 0.04	4943.2 ± 0.1	5003.5 ± 0.1	5019.4 ± 0.15	5122.3 ± 0.07	5146.6 ± 0.16	5233 ± 0.1	5256.9 ± 0.3
	5298.6 ± 0.1	5308 ± 0.06	5310.6 ± 0.02	5319.5 ± 0.5	5344 ± 0.15	5367.7 ± 0.51	5401.9 ± 0.55	5423 ± 0.01	5436.7 ± 0.01

The eigenfrequencies $\omega_n / 2\pi$ are determined from the stationary condition on Π (9.3.2). Table 9.4.2 shows the computed frequencies and the errors obtained by the eigenvalue problem.

Resonant vibration modes are excited by applying an external electric field $\bar{E}_1 = \bar{E}_3 = \bar{E}^0 \exp(i\omega_0 t)$ on both sides of the plate with $\omega = \omega_n$. The undetermined system parameters P are computed by using a genetic algorithm (Chiroiu C. *et al.* 2000). The number of the *measured* eigenfrequencies is $N = 63$ Table 9.4.2).

The genetic parameters are: number of populations = 45, ratio of reproduction = 1, number of multi-point crossovers = 1, probability of mutation = 0.25 and maximum number of generations = 550. The genetic algorithm exhibits very good convergence and accuracy. For example, for $\varepsilon = 0$ the maximum values of Z after 344 iteration is $Z_{\max} = 0.51\times 10^4$ and $\delta = 0.52\times 10^{-6}$ in the case of normal modes $\omega / 2\pi$ =334 kHz, and $Z_{\max} = 0.33\times 10^4$ and $\delta = 0.34\times 10^{-6}$ after 275 iteration in the case of $\omega / 2\pi$ =501 kHz. The minimum value for $Z_{\max}$ was found to be $\omega / 4\pi$ and the maximum value for $\max \delta = 0.11\times 10^{-5}$ for $\varepsilon = 0$. Number of iterations varies from 155 to 500.

In all computations, the *measured* eigenfrequencies are computable from the direct problem. A *measurement noise* has been artificially introduced by multiplication of the data values by $1 + r$, r being random numbers uniformly distributed in $[-\varepsilon, \varepsilon]$, with $\varepsilon = 10^{-1}, 10^{-2}, 10^{-3}$. Results for the case of normal modes $\omega / 2\pi$ =334 kHz are displayed in Table 9.4.3.

Table 9.4.3 Maximum value of Z and number of iterations for the normal case $\omega / 2\pi$ =334 kHz.

	$\varepsilon = 0$	$\varepsilon = 10^{-1}$	$\varepsilon = 10^{-2}$	$\varepsilon = 10^{-3}$
Z_{max}	0.33×10^4	0.256×10^2	0.193×10^3	0.673×10^4
δ	0.52×10^{-6}	0.119×10^{-2}	0.452×10^{-4}	0.92×10^{-6}
number of iterations	344	386	277	289

Table 9.4.3 shows that $Z_{\max}$ varies linearly with ε and δ quadratically with ε. Other cases have shown also a linear variation of $Z_{\max}$ and a quadratic variation of δ with respect to measurement noise.

In Figure 9.4.1 the admittance curve ($k / \rho\omega$ vs. $\omega / 2\pi$) in the linear regime ($\bar{E}^0 \cong 0.1\text{V}$) marks by peaks the frequencies $\omega = \omega_n$ of the modes. The agreement between the eigenfrequencies given by this curve and by the eigenvalue problem results (Table 9.4.2) is noted to be excellent. On comparison of the results given by the admittance curve (Figure 9.4.1) with the results obtained from the similar experimental curve derived by Alippi and Crăciun, the deviation between them is found to be 5–15% for low natural frequencies, and less than 4% for high natural frequencies. If $\bar{E}^0$ is increased above a threshold value $\bar{E}^0_{th}$ = 5.27 V the $\omega / 2$ subharmonic generation is observed. Note that Alippi and Crăciun obtain in the Cantor-like sample typical values of the lowest threshold voltages of 3–5 V. The amplitude of waves is calculated at the surface of the plate as a function of $\bar{E}^0$.

Figures 9.4.2–9.4.4 show the displacements of the normal modes $\omega/2\pi$ =334 kHz, 501 kHz, 835 kHz and respectively of the subharmonic modes $\omega/4\pi$ =167 kHz, 250.5 kHz, 417.5 kHz. Two kinds of vibration regimes are found: a localized mode (fracton) regime represented in Figure 9.4.5 for $\omega/2\pi$ =1169 kHz, 2672 kHz and 3340 kHz and an extended vibration (phonon) regime represented in Figure 9.4.6 for $\omega/2\pi$ = 4175 kHz and 4250 kHz. A sketch of the plate geometry is given on the abscissa (dashed, piezoelectric ceramic and white, epoxy resin).

The fracton vibrations are mostly localized on a few elements, while the phonon vibrations essentially extend to the whole plate. In the case of a periodical plate the dispersion prevents good frequency matching between the fundamental and appropriate subharmonic modes. For the homogeneous plate the mismatch $\omega_n - \omega/2$ is due to the symmetry of fundamental modes with respect to x. Only symmetric odd n can induce a subharmonic, but $\omega/2$ never coincides with a plate vibration mode.

For a Cantor plate, we have obtained qualitatively the same result as Crăciun *et al.*: given a normal mode ω_n, for excitation at $\omega = \omega_n$, the value of the expected threshold E_{th}, i. e. the ability of generating the $\omega/2$ subharmonic, is determined by the existence of a normal mode with: (i) small frequency mismatch $\omega_n - \omega/2$, and, (ii) large spatial overlap between the fundamental and subharmonic displacement field.

Concerning the values of amplitudes, the results obtained by us verify experimental predictions of resonant amplitudes with accuracy better than six percent for similar fundamental modes.

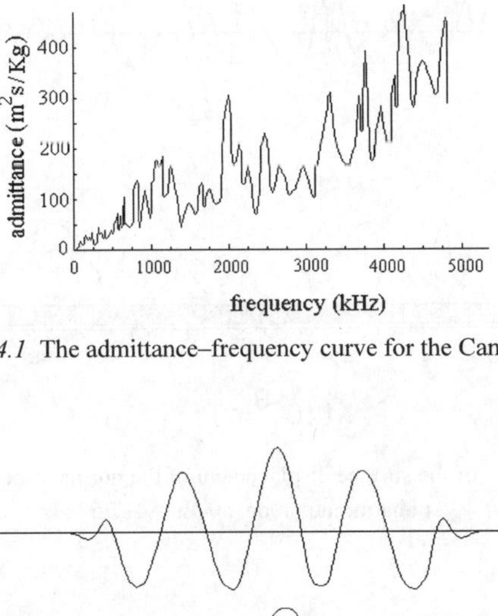

Figure 9.4.1 The admittance–frequency curve for the Cantor plate.

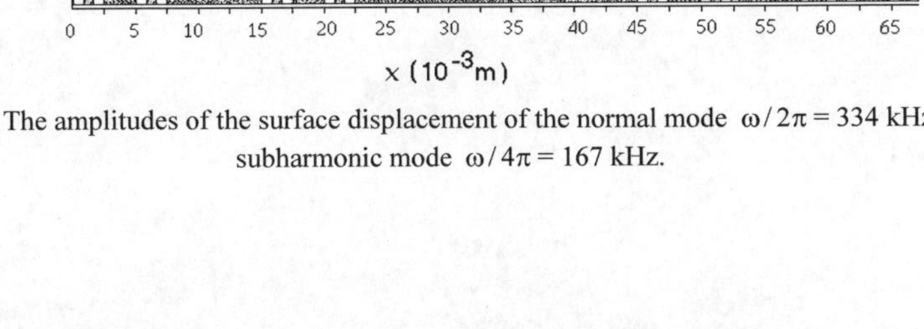

Figure 9.4.2 The amplitudes of the surface displacement of the normal mode $\omega/2\pi = 334$ kHz and of the subharmonic mode $\omega/4\pi = 167$ kHz.

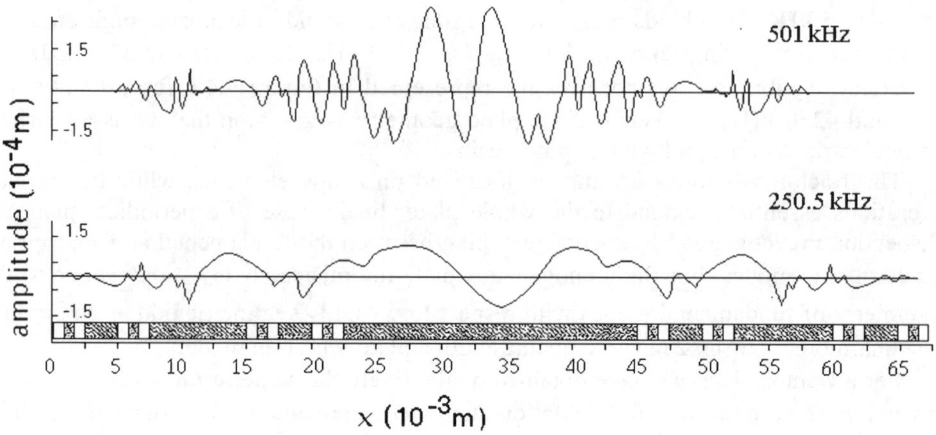

Figure 9.4.3 The amplitudes of the surface displacement of the normal mode $\omega / 2\pi = 501$ kHz and of the subharmonic mode $\omega / 4\pi = 250.5$ kHz.

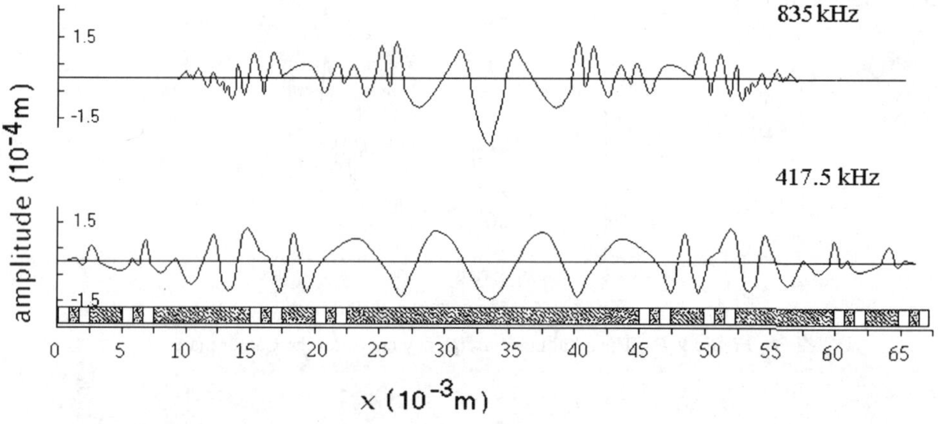

Figure 9.4.4 The amplitudes of the surface displacement of the normal mode $\omega / 2\pi = 835$ kHz and of the subharmonic mode $\omega / 4\pi = 417.5$ kHz.

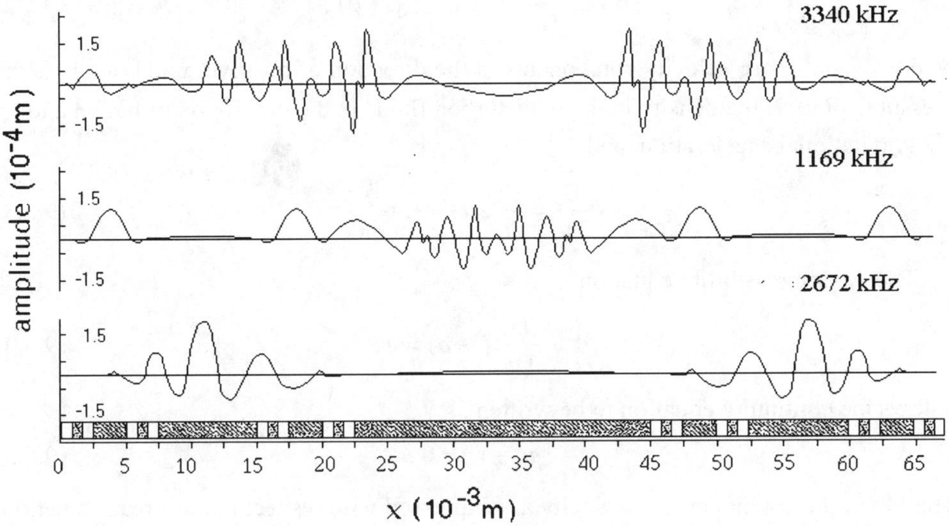

Figure 9.4.5 The normal amplitudes for three localized vibration modes ($\omega/2\pi$ =1169 kHz, $\omega/2\pi$ =2672 kHz and $\omega/2\pi$ =3340 kHz).

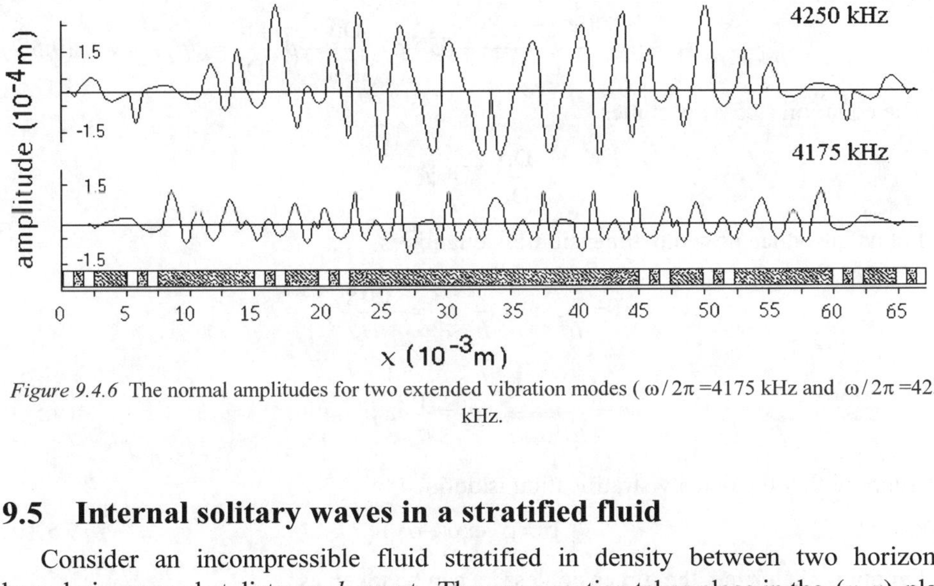

Figure 9.4.6 The normal amplitudes for two extended vibration modes ($\omega/2\pi$ =4175 kHz and $\omega/2\pi$ =4250 kHz.

9.5 Internal solitary waves in a stratified fluid

Consider an incompressible fluid stratified in density between two horizontal boundaries spaced at distance h apart. The wave motion takes place in the (x, y) plane of a Cartesian system of coordinates with the origin in the lower boundary and y vertically upward. We review the Yih results in the case of an exponential stratification. The Euler motion equations are given by

$$(\bar{\rho}+\rho)\frac{\mathrm{D}u}{\mathrm{D}t} = -p_x , \qquad (9.5.1)$$

$$(\overline{\rho}+\rho)\frac{\mathrm{D}v}{\mathrm{D}t}=-p_y-g(\overline{\rho}+\rho)\,, \tag{9.5.2}$$

where u and v are velocity components in the direction of positive x and y, p is the pressure, $\overline{\rho}(y)$ is the density in the undisturbed fluid, ρ the density perturbation and g the gravitational acceleration, and

$$\frac{\mathrm{D}}{\mathrm{D}t}=\frac{\partial}{\partial t}+u\frac{\partial}{\partial x}+v\frac{\partial}{\partial y}\,. \tag{9.5.3}$$

The incompressibility equation

$$\frac{\mathrm{D}}{\mathrm{D}t}(\overline{\rho}+\rho)=0\,, \tag{9.5.4}$$

allows the continuity equation to be written as

$$u_x+v_y=0\,. \tag{9.5.5}$$

Therefore, the components of velocity are expressed with respect to the stream function

$$u=\psi_y\,,\ v=-\psi_x\,. \tag{9.5.6}$$

From (9.5.1) and (9.5.2) we have

$$(\overline{\rho}+\rho)\frac{\mathrm{D}(\psi_{xx}+\psi_{yy})}{\mathrm{D}t}+(\overline{\rho}+\rho)_y\frac{\mathrm{D}u}{\mathrm{D}t}-p_x\frac{\mathrm{D}v}{\mathrm{D}t}=gp_x\,. \tag{9.5.7}$$

The equation (9.5.4) becomes

$$\frac{\mathrm{D}\rho}{\mathrm{D}t}=\overline{\rho}_y\psi_x\,. \tag{9.5.8}$$

Let us introduce now the dimensionless quantities

$$x'=\frac{x}{h}\,,\ y'=\frac{y}{h}\,,\ t'=\frac{t\sqrt{\alpha g}}{\sqrt{h}}\,,$$

$$\psi'=\frac{\psi}{h\sqrt{\alpha g h}}\,,\ \rho'=\frac{\rho}{\rho_0}\,,\ \overline{\rho}'=\frac{\overline{\rho}}{\rho_0}\,, \tag{9.5.9}$$

and suppose that the density stratification is defined by

$$\overline{\rho}=\rho_0\exp(-\alpha y)\,. \tag{9.5.10}$$

Dropping the prime, the dimensionless ρ can be taken as

$$\rho=\alpha\exp(-\alpha y)\theta\,. \tag{9.5.11}$$

The equations (9.5.7) and (9.5.8) are written in terms of dimensionless terms

$$(1+\alpha\theta)\frac{\mathrm{D}(\psi_{xx}+\psi_{yy})}{\mathrm{D}t}+(\alpha+\alpha\theta_y)_y\frac{\mathrm{D}u}{\mathrm{D}t}-\alpha\theta_x\frac{\mathrm{D}v}{\mathrm{D}t}=\theta_x\,, \tag{9.5.12}$$

$$\theta_t + u\theta_x + v\theta_y = -\psi_x . \tag{9.5.13}$$

The solutions for the m-th and n-th modes of the equations (9.5.12) and (9.5.13) are, to the lowest order

$$\psi_m = (-1)^m a_m \text{sech}^2[\gamma_m(x - \frac{1}{m\pi}t)]\{\sin m\pi y + \frac{\alpha}{2} y \sin m\pi y\}, \quad \theta_m = m\pi\psi_m , \tag{9.5.14}$$

$$\psi_n = (-1)^n a_n \text{sech}^2[\gamma_n(x - \frac{1}{n\pi}t)]\{\sin n\pi y + \frac{\alpha}{2} y \sin n\pi y\}, \quad \theta_n = n\pi\psi_n , \tag{9.5.15}$$

$$\gamma_m^2 = \frac{m\pi\alpha a_m}{9}, \text{ for odd } m, \quad \gamma_m^2 = \frac{m\pi\alpha^2 a_m}{36}, \text{ for even } m, \tag{9.5.16}$$

and similarly for γ_n^2.

Consider the simple case when $\alpha = 0$. In this case, let us assume

$$\psi = \psi_m + \psi_n + \phi, \quad \theta = \theta_m + \theta_n + \hat{\theta} . \tag{9.5.17}$$

Neglecting the higher order terms, the linearised forms of (9.5.12) and (9.5.13) are

$$\phi_{yyt} - \hat{\theta}_x = (n^2 - m^2)\pi^2 W ,$$

$$\phi_x + \hat{\theta}_t = (n - m)\pi W , \tag{9.5.18}$$

$$W = \psi_{ny}\psi_{mx} - \psi_{my}\psi_{nx} = P\sin(n-m)\pi y + Q\sin(n+m)\pi y ,$$

$$P = (-1)^{m+n} a_m a_n \pi\{n\gamma_m \text{sech}^2[\gamma_n(x - \frac{1}{n\pi}t)] \text{ sech}^2[\gamma_m(x - \frac{1}{m\pi}t)]\times$$

$$\times \tanh[\gamma_m(x - \frac{1}{m\pi}t)] m\gamma_n \text{sech}^2[\gamma_m(x - \frac{1}{m\pi}t)] \text{sech}^2[\gamma_n(x - \frac{1}{n\pi}t)]\times \tag{9.5.19}$$

$$\times \tanh[\gamma_n(x - \frac{1}{n\pi}t)] \},$$

$$Q = (-1)^{m+n} a_m a_n \pi\{-n\gamma_m \text{sech}^2[\gamma_n(x - \frac{1}{n\pi}t)] \text{ sech}^2[\gamma_m(x - \frac{1}{m\pi}t)]\times$$

$$\times \tanh[\gamma_m(x - \frac{1}{m\pi}t)] m\gamma_n \text{sech}^2[\gamma_m(x - \frac{1}{m\pi}t)] \text{ sech}^2[\gamma_n(x - \frac{1}{n\pi}t)]\times$$

$$\times \tanh[\gamma_n(x - \frac{1}{n\pi}t)]\}.$$

From (9.5.18) and (9.5.19) we see that we can take

$$\phi = f_{n-m} \sin(n-m)\pi y + f_{n+m} \sin(n+m)\pi y ,$$

$$\hat{\theta} = g_{n-m} \sin(n-m)\pi y + g_{n+m} \sin(n+m)\pi y . \tag{9.5.20}$$

We treat only the $(n-m)$ mode, since the $(n+m)$ mode is similar. The terms containing $\sin(n-m)\pi y$ in (9.5.18) are

$$(n-m)^2\pi^2 f_{(n-m)t} + g_{(n-m)x} = -(n^2-m^2)\pi^2 P\,,$$

$$f_{(n-m)x} + g_{(n-m)t} = (n-m)\pi P\,. \tag{9.5.21}$$

By eliminating g_{n-m} from (9.5.21), we have

$$(n-m)^2\pi^2 f_{(n-m)tt} - f_{(n-m)xx} = -(n^2-m^2)\pi^2 P_t - (n-m)\pi P_x\,, \tag{9.5.22a}$$

and similarly, by eliminating f_{n-m} from these equations

$$(n-m)^2\pi^2 g_{(n-m)tt} - g_{(n-m)xx} = (n-m)^3\pi^3 P_t + (n^2-m^2)\pi^2 P_x\,. \tag{9.5.22b}$$

The hyperbolic equations (9.5.22) can be solved by integration along the characteristics. We solve only (9.5.22a), the other equation being solved in a similar way. With new variables

$$\xi = x - \frac{1}{(n-m)\pi}t\,,\quad \eta = x + \frac{1}{(n-m)\pi}t\,, \tag{9.5.23}$$

the equation (9.5.22a) yields

$$-4f_{(n-m)\xi\eta} = \pi[2mP_\xi - 2nP_\eta]\,. \tag{9.5.24}$$

By integrating from $t=-\infty$, we obtain

$$2f_{n-m} = \pi\int_{-\infty}^{\eta} mP\mathrm{d}\eta - \pi\int_{\infty}^{\xi} nP\mathrm{d}\xi\,, \tag{9.5.25}$$

or

$$4f_{n-m} = (-1)^{m+n}a_m a_n\pi^2\left(n(n-m)\gamma_m I_1 - \frac{n^2(n-m)}{2n-m}\gamma_m I_2 - m\frac{n(n-m)}{2n-m}H\right), \tag{9.5.26}$$

where

$$I_1 = \int_{-\infty}^{\eta}\mathrm{sech}^2[\gamma_n(x-\frac{1}{n\pi}t)]\,\mathrm{sech}^2[\gamma_m(x-\frac{1}{m\pi}t)]\tanh[\gamma_m(x-\frac{1}{m\pi}t)]\mathrm{d}\eta\,,$$

$$I_2 = \int_{\infty}^{\xi}\mathrm{sech}^2[\gamma_n(x-\frac{1}{n\pi}t)]\,\mathrm{sech}^2[\gamma_m(x-\frac{1}{m\pi}t)]\tanh[\gamma_m(x-\frac{1}{m\pi}t)]\mathrm{d}\xi\,, \tag{9.5.27}$$

$$H = \mathrm{sech}^2[\gamma_m(x-\frac{1}{m\pi}t)]\,\mathrm{sech}^2[\gamma_n(x-\frac{1}{n\pi}t)]\,.$$

The computations have shown that integrals I_1 and I_2 are convergent. In particular, for $n=2m$, we obtain

$$I_1 = 8m^2 \operatorname{sech}^2\{\gamma_m(x - \frac{1}{m\pi}t)\}\tanh[\gamma_m(x - \frac{1}{m\pi}t)]\tanh[\gamma_n(x - \frac{1}{n\pi}t)]\}. \quad (9.5.28)$$

The interaction of waves can be discussed looking at (9.5.26). The first part in (9.5.26) may be eventually independent of η and represents a wave propagating in the positive direction of x. The second term may be independent of ξ and represents a wave propagating in the opposite direction. The last term in (9.5.26) is zero as t increases, since it is the product of two squares of the sech function with different arguments. These waves propagate with velocities $\pm\frac{1}{(n-m)\pi}$.

It is convenient to discuss the interaction of waves for the particular case of $n = 2m$. These waves may propagate in the same direction or in opposite directions. The result of interactions of waves are two pairs of solitary waves, each pair consisting of two opposite directions propagating solitary waves of the same mode. The modes of the two pairs are different from each other, and are different from the *m*-th and *n*-th modes of the waves. The original waves propagate after interaction without changing their identities, but only the m-wave suffers a shift of phase.

In the same manner, the analysis may be applied for the general case of $\alpha \neq 0$. The interaction of more than two waves can be dealt with pair by pair (Hirota and Satsuma).

9.6 The motion of a micropolar fluid in inclined open channels

Consider a two-dimensional flow of a micropolar, isotropic, incompressible, viscous fluid in a wide channel over a rigid bottom (Chiroiu *et al.* 2003b). We have chosen a wide channel to be sure that the motion will be two-dimensional only. The *x*-axis is horizontal and the bottom is given by *h*(*x*). The channel bed is linear and is inclined at an angle $\theta > 0$ below the horizontal, that is $y = -mx$, with $m = \tan\theta > 0$ (Figure 9.6.1). The vertical distance of the surface above the *x*-axis is denoted by $\eta(x)$. The fluid domain Ω is a two-dimensional strip Ω : $-\infty < x < \infty$, $0 < y < \eta(x)$, bounded by a free surface Γ : $y = \eta$, and a lower rigid bottom S : $h(x) = -mx$.

In the shallow flow the vertical dimensions are small compared to the horizontal dimensions. The motion equations of a micropolar, viscous fluid are given by Eringen (1966, 1970):

$$\begin{aligned}\rho\dot{v} + \rho v \,\mathrm{grad}\, v = X - \mathrm{grad}\,\pi - (\mu+\alpha)\mathrm{curl}\,\mathrm{curl}\, v + \\ + (2\mu+\lambda)\,\mathrm{grad}\,\mathrm{div}\, v + 2\alpha\,\mathrm{curl}\, w,\end{aligned} \quad (9.6.1)$$

$$\begin{aligned}\rho J\dot{w} + \rho J v\,\mathrm{grad}\, w = Y - (\gamma+\varepsilon)\,\mathrm{curl}\,\mathrm{curl}\, w + \\ + (2\gamma+\zeta)\,\mathrm{grad}\,\mathrm{div}\, w - 4\alpha w + 2\alpha\,\mathrm{curl}\, v,\end{aligned} \quad (9.6.2)$$

where $X = (X_1, X_2)$ is the exterior body force, $Y = (Y_1, Y_2)$ is the exterior body couple, π is the thermodynamic pressure, ρJ is the inertia tensor density, $v = (v_1, v_2)$ is the velocity vector $v = \frac{\partial}{\partial t}u$, $u = (u_1, u_2)$ the displacement vector, $\phi = (\varphi_1, \varphi_2)$ the microrotation vector, $w=(w_1,w_2)$ the microrotation velocity $w=\frac{\partial}{\partial t}\phi$, ρ the fluid density. The superposed dot indicates the partial differentiation with respect to time $\dot{a}=\frac{\partial}{\partial t}a$. In (9.6.1) and (9.6.2) λ and μ are the classical viscosities coefficients of the Navier–Stokes theory. The constants α,ζ,γ and ε are the micropolar coefficients of viscosity.

The elastic coefficients must fulfil the condition (Eringen 1970)

$$\begin{aligned} &\mu\ge 0,\quad 2\mu+3\lambda\ge 0,\quad \alpha\ge 0,\\ &\gamma\ge 0,\quad 2\gamma+3\zeta\ge 0,\quad \varepsilon\ge 0. \end{aligned} \tag{9.6.3}$$

Equations (9.6.2) and (9.6.3) are six equations with unknown vector fields the velocity v and microrotation w. These equations must be supplemented by the equation of continuity

$$\dot{\rho}+\operatorname{div}(\rho v)=0\,. \tag{9.6.4}$$

Here we consider an incompressible fluid with ρ constant. From (2.5) we obtain

$$\operatorname{div} v=0\,. \tag{9.6.5}$$

In this case, the thermodynamic pressure π must be replaced by an unknown pressure p to be determined through the solution of each problem.

The constitutive relations are

$$\sigma_{ij}=(-p+\lambda v_{k,k})\delta_{ij}+(\mu+\alpha)v_{j,i}+(\mu-\alpha)v_{i,j}-2\alpha\varepsilon_{kij}w_k\,, \tag{9.6.6}$$

$$\mu_{ij}=\zeta w_{k,k}\delta_{ij}+(\gamma+\varepsilon)w_{j,i}+(\gamma-\varepsilon)w_{i,j}\,, \tag{9.6.7}$$

where σ_{ij} is the stress tensor and μ_{ij} is the couple stress tensor.

The field of equations (9.6.1), (9.6.2) and (9.6.5) are subject to certain boundary and initial conditions:

– Traction conditions on Γ

$$\sigma_{kl}n_k=t_l\,,\ \mu_{kl}n_k=\mu_l\,, \tag{9.6.8}$$

where t_l are the surface traction and μ_l the surface couple acting on Γ. Here n_k, $k=1,2$ are the unit vectors of the coordinate system (x,y).

– The condition for a particle at the surface Γ to remain at the surface Γ

$$\dot{\eta}+v_1\eta_{,x}=v_2\,. \tag{9.6.9}$$

– Velocity conditions of adherence of the fluid to S

$$v_k(\overline{x},t)=v_k^0,\quad w_k(\overline{x},t)=w_k^0, \tag{9.6.10}$$

where $\overline{x}\in S$ and v_k^0, w_k^0 the given values for velocity and microrotation velocity.

– The initial conditions at $t=0$

$$v=v_0,\ w=w_0,\ \eta=\eta_0\,. \tag{9.6.11}$$

We consider that $X_1 = -\frac{r^2\rho v_1 |v_1|}{R} = -\frac{r^2\rho v_1 |v_1|}{\eta}$, $X_2 = -\rho g$, $Y_1 = Y_2 = 0$ in (9.6.1) and (9.6.2). Therefore, equations (9.6.1), (9.6.2) and (9.6.5) are then

$$\rho\dot{v}_1 + \rho v_1 v_{1,x} + \rho v_2 v_{1,y} = -p_{,x} + (\mu+\alpha)\Delta v_1 - \frac{r^2\rho v_1 |v_1|}{\eta}, \tag{9.6.12}$$

$$\rho\dot{v}_2 + \rho v_1 v_{2,x} + \rho v_2 v_{2,y} = -p_{,y} + (\mu+\alpha)\Delta v_2 - \rho g, \tag{9.6.13}$$

$$\begin{aligned} \rho J\dot{w}_1 + \rho J v_1 w_{1,x} + \rho J v_2 w_{1,y} &= (2\gamma+\zeta) w_{1,xx} + \\ &+ (\gamma+\zeta-\varepsilon) w_{2,xy} + (\gamma+\varepsilon) w_{1,yy} - 4\alpha w_1, \end{aligned} \tag{9.6.14}$$

$$\begin{aligned} \rho J\dot{w}_2 + \rho J v_1 w_{2,x} + \rho J v_2 w_{2,y} &= (2\gamma+\zeta) w_{2,yy} + \\ &+ (\gamma+\zeta-\varepsilon) w_{1,xy} + (\gamma+\varepsilon) w_{2,xx} - 4\alpha w_2, \end{aligned} \tag{9.6.15}$$

$$v_{2,x} - v_{1,y} = 0, \tag{9.6.16}$$

$$w_{2,x} - w_{1,y} = 0, \tag{9.6.17}$$

$$v_{1,x} + v_{2,y} = 0. \tag{9.6.18}$$

The comma represents the differentiation with respect to the shown variable. In (9.6.13), g is the constant gravitational acceleration. In the right side of (9.6.12) the term $-\frac{r^2 v_1 |v_1|}{R}$ represents the resisting body force always acting opposite to that of the flow, where r^2 is a constant depending upon the roughness of the channel walls and R is the hydraulic radius.

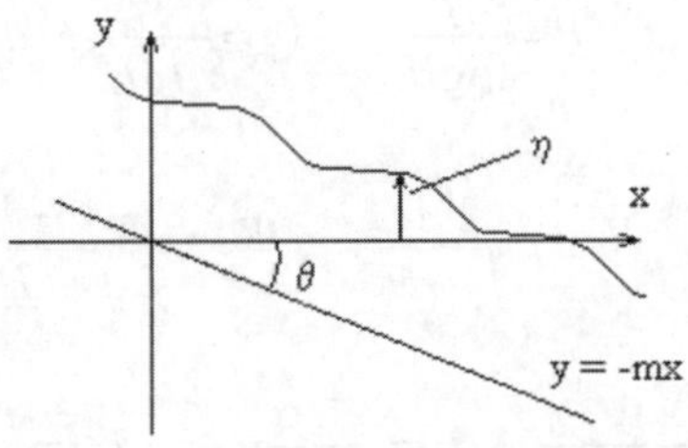

Figure 9.6.1 Geometry of the flow (Dressler).

According to this formula, the turbulent fluctuations exert on the main flow a resistive body force at every point of magnitude $\frac{r^2 v_1^2}{R}$. Since the most flows in practice are highly turbulent we take account of the resistive force due to the momentum transport of the secondary flow exerted on the average flow at each point. The resistance effects due to the dynamic viscosity of the water are neglected. The above expression of the resisting force was given by Chezy (Dressler). The hydraulic radius is defined as the ratio of the area of a cross-section of the water normal to the channel to its *wetted perimeter*, that is that part of the perimeter excluding the free surface of the water. The Chezy formula expresses the fact that the resistance will be greater in shallow regions where all of the water is closer to the rough boundary. The Chezy formula is valid only for uniform flows, and although it is used for non-uniform flows when the flow varies slowly with respect to x, y and t. In our case $R = \eta$. In (9.6.12)–(9.6.18) the unknown functions are v_i, w_i, $i=1,2$, p and η.

We attach to (9.6.12)–(9.6.18) the conditions (9.6.8)–(9.6.11)

$$\sigma_{11}n_1+\sigma_{21}n_2=t_1\,,\ \sigma_{12}n_1+\sigma_{22}n_2=t_2\,,\ \text{ on }\ \Gamma\,, \tag{9.6.19}$$

$$\mu_{11}n_1+\mu_{21}n_2=\mu_1\,,\ \mu_{12}n_1+\mu_{22}n_2=\mu_2\,,\ \text{ on }\ \Gamma\,, \tag{9.6.20}$$

$$\dot\eta+v_1\eta_{,x}=v_2\,,\ \text{ on }\ \Gamma\,, \tag{9.6.21}$$

$$v_1=v_1^0,\quad v_2=v_2^0,\quad w_1=w_1^0,\quad w_2=w_2^0\ \text{ on }\ S\,, \tag{9.6.22}$$

$$v_1=v_{10},\ v_2=v_{20},\quad w_1=w_{10},\ w_2=w_{20},\quad \eta=\eta_0\ \text{ at }\ t=0\,. \tag{9.6.23}$$

In (9.6.19)–(9.6.20) the stress and couple stress components are given by (9.6.6)–(9.6.7).

Let the constant H be a typical vertical dimension of the resulting flow, and L a typical horizontal dimension. Let $\frac{H^2}{L^2}=\delta$ be a small positive parameter, which we will use for a perturbation procedure. New dimensionless variables are defined by

$$\alpha=\frac{x}{L},\quad \beta=\frac{y}{H},\quad \tau=\frac{\sqrt{gH}}{L}t,$$

$$\mathrm{Y}=\frac{\eta}{H},\quad P=\frac{p}{\rho gH},\quad V_1=\frac{v_1}{\sqrt{gH}},\quad d=\frac{h}{H},$$

$$V_2=\frac{H}{L}\frac{v_2}{\sqrt{gH}},\quad W_1=\frac{w_1}{\sqrt{g/H}},\quad W_2=\frac{Hw_2}{L\sqrt{g/H}}, \tag{9.6.24}$$

$$\bar\mu=\frac{\mu}{\rho L\sqrt{gH}},\quad \bar\alpha=\frac{\alpha}{\rho L\sqrt{gH}},\ \bar r^2=\frac{r^2H}{L},$$

$$\bar\zeta=\frac{\zeta}{\rho LJ\sqrt{gH}},\quad \bar\gamma=\frac{\gamma}{\rho LJ\sqrt{gH}},\quad \bar\varepsilon=\frac{\varepsilon}{\rho LJ\sqrt{gH}}.$$

In terms of the dimensionless variables (9.6.12)–(9.6.23) become

$$\delta Y[V_{1,\tau}+V_1V_{1,\alpha}-(\bar\mu+\bar\alpha)V_{1,\alpha\alpha}+P_{,\alpha}]+V_2V_{1,\beta}Y-(\bar\mu+\bar\alpha)V_{1,\beta\beta}Y+\bar r^2V_1^2=0\,, \tag{9.6.25}$$

$$\delta[V_{2,\tau}+V_1V_{2,\alpha}-(\bar\mu+\bar\alpha)V_{2,\alpha\alpha}+P_{,\beta}+1]+V_2V_{2,\beta}-(\bar\mu+\bar\alpha)V_{2,\beta\beta}=0\,, \tag{9.6.26}$$

$$\delta[W_{1,\tau}+V_1W_{1,\alpha}-(2\overline{\gamma}+\overline{\zeta})W_{1,\alpha\alpha}+4\overline{\alpha}W_1]+V_2W_{1,\beta}- \\ -(\overline{\gamma}+\overline{\varepsilon})W_{1,\beta\beta}-(\overline{\gamma}+\overline{\zeta}-\overline{\varepsilon})W_{2,\alpha\beta}=0, \tag{9.6.27}$$

$$\delta[W_{2,\tau}+V_1W_{2,\alpha}-(\overline{\gamma}+\overline{\varepsilon})W_{2,\alpha\alpha}+4\overline{\alpha}W_2]+V_2W_{2,\beta}- \\ -(2\overline{\gamma}+\overline{\zeta})W_{2,\beta\beta}-(\overline{\gamma}+\overline{\zeta}-\overline{\varepsilon})W_{1,\alpha\beta}=0, \tag{9.6.28}$$

$$V_{2,\alpha}-V_{1,\beta}=0\,, \tag{9.6.29}$$

$$W_{2,\alpha}-W_{1,\beta}=0\,, \tag{9.6.30}$$

$$\delta V_{1,\alpha}+V_{2,\beta}=0\,, \tag{9.6.31}$$

$$\sigma_{11}n_1+\sigma_{21}n_2=t_1\,,\ \sigma_{12}n_1+\sigma_{22}n_2=t_2\,,\ \text{at}\quad \beta=Y\,, \tag{9.6.32}$$

$$\mu_{11}n_1+\mu_{21}n_2=\mu_1\,,\ \mu_{12}n_1+\mu_{22}n_2=\mu_2\,,\ \text{at}\quad \beta=Y\,, \tag{9.6.33}$$

$$\delta(Y_{,\tau}+V_1Y_{,\alpha})=V_2\,,\ \text{at}\quad \beta=Y\,, \tag{9.6.34}$$

$$v_1=v_1^0,\quad v_2=v_2^0,\quad w_1=w_1^0,\quad w_2=w_2^0,\ \text{at}\quad \beta=d\,, \tag{9.6.35}$$

$$\delta V_1 m/H=V_2\,,\ \text{at}\quad \beta=d\,, \tag{9.6.36}$$

$$v_1=v_{10},\quad v_2=v_{20},\quad w_1=w_{10},\quad w_2=w_{20},\quad \eta=\eta_0,\ \text{at}\ \ t=0\,. \tag{9.6.37}$$

We assume that the unknowns can be expressed as power series in terms of δ

$$V_i=\sum_{k=0}^{\infty}V_i^{(k)}(\alpha,\beta,\tau)\delta^k\ ,\ i=1,2\ ,\ W_i=\sum_{k=0}^{\infty}W_i^{(k)}(\alpha,\beta,\tau)\delta^k\ ,\ i=1,2\ ,$$

$$P=\sum_{k=0}^{\infty}P^{(k)}(\alpha,\beta,\tau)\delta^k\ ,\quad Y=\sum_{k=0}^{\infty}Y^{(k)}(\alpha,\tau)\delta^k\ . \tag{9.6.38}$$

We introduce (9.6.38) into (9.6.25)–(9.6.37) and the resulting coefficients of like powers of δ are equated (Nayfeh, Kamel). We obtain

1. From the coefficients for δ^0

$$L_1^{(0)}\equiv Y^{(0)}V_2^{(0)}V_{1,\beta}^{(0)}-(\overline{\mu}+\overline{\alpha})Y^{(0)}V_{1,\beta\beta}^{(0)}+\overline{r}^2V_1^{(0)2}=0\,,$$

$$L_2^{(0)}\equiv V_2^{(0)}V_{2,\beta}^{(0)}-(\overline{\mu}+\overline{\alpha})Y^{(0)}V_{2,\beta\beta}^{(0)}=0\,,$$

$$L_3^{(0)}\equiv V_2^{(0)}W_{1,\beta}^{(0)}-(\overline{\gamma}+\overline{\varepsilon})W_{1,\beta\beta}^{(0)}+(\overline{\gamma}+\overline{\zeta}-\overline{\varepsilon})W_{2,\alpha\beta}^{(0)}=0\,,$$

$$L_4^{(0)}\equiv V_2^{(0)}W_{2,\beta}^{(0)}-(2\overline{\gamma}+\overline{\zeta})W_{2,\beta\beta}^{(0)}+(\overline{\gamma}+\overline{\zeta}-\overline{\varepsilon})W_{1,\alpha\beta}^{(0)}=0\,,$$

$$L_5^{(0)} \equiv V_{2,\alpha}^{(0)} - V_{1,\beta}^{(0)} = 0\,, \quad L_6^{(0)} \equiv W_{2,\alpha}^{(0)} - W_{1,\beta}^{(0)} = 0\,, \tag{9.6.39}$$

$$L_7^{(0)} \equiv V_{2,\beta}^{(0)} = 0\,,\; L_8^{(0)} \equiv V_2^{(0)} - Y^{(0)} + \beta = 0\,,$$

$$L_9^{(0)} \equiv P^{(0)} - Y^{(0)} + \beta = 0\,,\; L_{10}^{(0)} \equiv V_2^{(0)} - d + \beta = 0\,,\; L_{11}^{(0)} \equiv W_1^{(0)} - d + \beta = 0\,.$$

2. From the coefficients for δ^1

$$L_1^{(1)} \equiv Y^{(0)} V_{1,\tau}^{(0)} + V_1^{(0)} V_{1,\alpha}^{(0)} Y^{(0)} - (\overline{\mu} + \overline{\alpha}) Y^{(0)} V_{1,\alpha\alpha}^{(0)} + P_{,\beta}^{(0)} Y^{(0)} +$$
$$+ V_2^{(1)} V_{1,\beta}^{(0)} Y^{(0)} + V_2^{(0)} V_{1,\beta}^{(1)} Y^{(0)} - (\overline{\mu} + \overline{\alpha}) Y^{(0)} V_{1,\beta\beta}^{(1)} +$$
$$+ V_2^{(0)} V_{1,\beta}^{(0)} Y^{(1)} - (\overline{\mu} + \overline{\alpha}) Y^{(1)} V_{1,\beta\beta}^{(0)} + 2\overline{r}^2 V_1^{(0)} V_1^{(1)} = 0,$$

$$L_2^{(1)} \equiv V_{2,\tau}^{(0)} + V_1^{(0)} V_{2,\alpha}^{(0)} - (\overline{\mu} + \overline{\alpha}) V_{2,\alpha\alpha}^{(0)} + P_{,\beta}^{(0)} + 1 +$$
$$+ V_2^{(1)} V_{2,\beta}^{(0)} + V_2^{(0)} V_{2,\beta}^{(1)} - (\overline{\mu} + \overline{\alpha}) V_{2,\beta\beta}^{(1)} = 0,$$

$$L_3^{(1)} \equiv W_{1,\tau}^{(0)} + V_1^{(0)} W_{1,\alpha}^{(0)} - (2\overline{\gamma} + \overline{\zeta}) W_{1,\alpha\alpha}^{(0)} + 4\overline{\alpha} W_1^{(0)} + V_2^{(1)} W_{1,\beta}^{(0)} +$$
$$+ V_2^{(0)} W_{1,\beta}^{(1)} - (\overline{\gamma} + \overline{\varepsilon}) W_{1,\beta\beta}^{(1)} + (\overline{\gamma} + \overline{\zeta} - \overline{\varepsilon}) W_{2,\alpha\beta}^{(1)} = 0,$$

$$L_4^{(1)} \equiv W_{2,\tau}^{(0)} + V_1^{(0)} W_{2,\alpha}^{(0)} - (\overline{\gamma} + \overline{\varepsilon}) W_{2,\alpha\alpha}^{(0)} + 4\overline{\alpha} W_2^{(0)} + V_2^{(0)} W_{2,\beta}^{(1)} +$$
$$+ V_2^{(1)} W_{2,\beta}^{(0)} - (2\overline{\gamma} + \overline{\zeta}) W_{2,\beta\beta}^{(1)} + (\overline{\gamma} + \overline{\zeta} - \overline{\varepsilon}) W_{1,\alpha\beta}^{(1)} = 0,$$

$$L_5^{(1)} \equiv V_{2,\alpha}^{(1)} - V_{1,\beta}^{(1)} = 0\,, \quad L_6^{(1)} \equiv W_{2,\alpha}^{(1)} - W_{1,\beta}^{(1)} = 0\,, \quad L_7^{(1)} \equiv V_{1,\alpha}^{(0)} + V_{2,\beta}^{(1)} = 0\,, \tag{9.6.40}$$

$$L_8^{(1)} \equiv Y_{,\tau}^{(0)} + V_1^{(0)} Y_{,\alpha}^{(0)} - V_2^{(1)} - Y^{(1)} = 0\,,\; L_9^{(1)} \equiv P^{(1)} - Y^{(1)} + \beta = 0\,,$$

$$L_{10}^{(1)} \equiv V_1^{(0)} m / H - V_2^{(1)} = 0\,,\; L_{11}^{(1)} \equiv W_1^{(1)} - d + \beta = 0\,,\; L_{12}^{(1)} \equiv W_2^{(1)} - d + \beta = 0\,.$$

3. From the coefficients for δ^2

$$L_1^{(2)} \equiv Y^{(0)} V_{1,\tau}^{(1)} + V_{1,\tau}^{(0)} Y^{(1)} + V_1^{(0)} V_{1,\alpha}^{(0)} Y^{(1)} + V_1^{(1)} V_{1,\alpha}^{(0)} Y^{(0)} +$$
$$+ V_1^{(0)} V_{1,\alpha}^{(1)} Y^{(0)} + V_2^{(1)} V_{1,\beta}^{(0)} Y^{(1)} + V_2^{(0)} V_{1,\beta}^{(1)} Y^{(1)} + V_2^{(0)} V_{1,\beta}^{(0)} Y^{(2)} -$$

$$-(\overline{\mu} + \overline{\alpha}) Y^{(1)} V_{1,\alpha\alpha}^{(0)} - (\overline{\mu} + \overline{\alpha}) Y^{(0)} V_{1,\alpha\alpha}^{(1)} +$$
$$+ P_{,\alpha}^{(1)} Y^{(0)} + P_{,\alpha}^{(0)} Y^{(1)} + V_2^{(0)} V_{1,\beta}^{(2)} Y^{(0)} - (\overline{\mu} + \overline{\alpha}) Y^{(2)} V_{1,\beta\beta}^{(0)} -$$
$$-(\overline{\mu} + \overline{\alpha}) Y^{(1)} V_{1,\beta\beta}^{(1)} - (\overline{\mu} + \overline{\alpha}) Y^{(0)} V_{1,\beta\beta}^{(2)} + \overline{r}^2 (V_1^{(1)2} + 2 V_1^{(0)} V_1^{(2)}) = 0,$$

$$L_2^{(2)} \equiv V_{2,\tau}^{(1)} + V_1^{(0)} V_{2,\alpha}^{(1)} + V_1^{(1)} V_{2,\alpha}^{(0)} + V_2^{(0)} V_{2,\beta}^{(2)} +$$
$$+ V_2^{(2)} V_{2,\beta}^{(0)} + V_2^{(1)} V_{2,\beta}^{(1)} - (\overline{\mu} + \overline{\alpha}) V_{2,\alpha\alpha}^{(1)} - (\overline{\mu} + \overline{\alpha}) Y^{(0)} V_{1,\beta\beta}^{(2)} = 0,$$

$$L_3^{(2)} \equiv W_{1,\tau}^{(1)} + V_1^{(0)} W_{1,\alpha}^{(0)} + V_1^{(0)} W_{1,\alpha}^{(1)} + V_2^{(2)} W_{1,\beta}^{(0)} +$$
$$+V_2^{(1)} W_{1,\beta}^{(1)} + V_2^{(0)} W_{1,\beta}^{(2)} - (2\bar{\lambda} + \bar{\zeta}) W_{1,\alpha\alpha}^{(1)} + 4\bar{\alpha} W_1^{(1)} - (\bar{\gamma} + \bar{\varepsilon}) W_{1,\beta\beta}^{(2)} +$$
$$+(\bar{\gamma} + \bar{\zeta} - \bar{\varepsilon}) W_{2,\beta\beta}^{(2)} = 0,$$

$$L_4^{(2)} \equiv W_{2,\tau}^{(1)} + V_1^{(1)} W_{2,\alpha}^{(0)} + V_1^{(0)} W_{2,\alpha}^{(1)} + V_2^{(0)} W_{2,\beta}^{(2)} +$$
$$+V_2^{(2)} W_{2,\beta}^{(0)} + V_2^{(1)} W_{2,\beta}^{(1)} - (2\bar{\gamma} + \bar{\zeta}) W_{2,\beta\beta}^{(2)} + 4\bar{\alpha} W_2^{(1)} - (\bar{\gamma} + \bar{\varepsilon}) W_{2,\alpha\alpha}^{(1)} +$$
$$+(\bar{\gamma} + \bar{\zeta} - \bar{\varepsilon}) W_{1,\alpha\beta}^{(2)} = 0,$$

$$L_5^{(2)} \equiv V_{2,\alpha}^{(2)} - V_{1,\beta}^{(2)} = 0, \tag{9.6.41}$$

$$L_6^{(2)} \equiv W_{2,\alpha}^{(2)} - W_{1,\beta}^{(2)} = 0,$$

$$L_7^{(2)} \equiv V_{1,\alpha}^{(1)} + V_{2,\beta}^{(2)} = 0,$$

$$L_8^{(2)} \equiv Y_{,\tau}^{(1)} + V_1^{(0)} Y_{,\alpha}^{(1)} + V_1^{(1)} Y_{,\alpha}^{(0)} - V_2^{(2)} - Y^{(2)} = 0,$$

$$L_9^{(2)} \equiv P^{(2)} - Y^{(2)} + \beta = 0, \; L_{10}^{(2)} \equiv V_1^{(1)} m / H - V_2^{(2)} = 0,$$

$$L_{11}^{(2)} \equiv W_1^{(2)} - d + \beta = 0, \; L_{12}^{(2)} \equiv W_2^{(2)} - d + \beta = 0.$$

9.7 Cnoidal solutions

Consider that

$$M_i^{(k)} = \frac{\partial^k}{\partial \alpha^k} \log f_i^{(k)}(\alpha,\beta,\tau), \; k = 1,2,...,N, \; i = 1,2,...,6, \tag{9.7.1}$$

where $M = (V_1, V_2, W_1, W_2, P, Y)$ and

$$f_i^{(1)}(\alpha,\beta,\tau) = 1 + \exp\theta_{1i},$$

$$f_i^{(2)}(\alpha,\beta,\tau) = 1 + \exp\theta_{1i} + \exp\theta_{2i} + \exp(\theta_{1i} + \theta_{2i}), \tag{9.7.2}$$

....

$$f_i^{(N)}(\alpha,\beta,\tau) = 1 + \sum_{j=1}^{N} \exp\theta_{ji} + \sum_{j \neq l = 1}^{N} \exp(\theta_{ji} + \theta_{li}) + \sum_{j \neq l \neq r = 1}^{N} \exp(\theta_{ji} + \theta_{li} + \theta_{ri}) + ...,$$

with

$$\theta_{ki} = a_{ki}\alpha + b_{ki}\beta - \omega_{ki}\tau + \varsigma_{ki}, \; k = 1,2,...,N, \; i = 1,2,...,6, \tag{9.7.3}$$

and a_{ki}, b_{ki} the dimensionless wave numbers, ω_{ki} the dimensionless frequencies and ς_{ki} the dimensionless phases. We find that asymptotically the solutions become

$$M_i^{(k)} = A_{ik}\mathrm{sech}^2(a_{ki}\alpha + b_{ki}\beta - \omega_{ki}t - 2\Delta_{ki}), \tag{9.7.4}$$

with $k = 1,2,...,N$, $i = 1,2,...,6$, as $t \to \pm\infty$.

The functions $M_i^{(k)}$ are periodic with the period $2\Delta_{ki}$. These solutions represent a linear superposition of soliton waves, which is a row of solitons, spaced $2\Delta_{ki}$ apart.

The $23 \times \mathrm{N}$ parameters in this formulation a_{ki}, b_{ki}, ω_{ki} and ς_{ki}, $k = 1,2,...,N$, $i = 1,2,...,6$ are computable by substituting (9.7.1) in (9.6.39)–(9.6.41).

The parameters are defined by

$$p = \{a_{ki}, b_{ki}, \omega_{ki}, \varsigma_{ki}\}, \ k = 1,2,...N, \ i = 1,2,...,6. \tag{9.7.5}$$

The wave numbers, frequencies and constant phases are also vectors

$$\begin{aligned} a_{ki} &= (a_{11}, a_{12}, a_{13}, ...a_{N6}), \\ b_{ki} &= (b_{11}, b_{12}, b_{13}, ...b_{N5}), \\ \omega_{ki} &= (\omega_{11}, \omega_{12}, \omega_{13}, ...\omega_{N6}), \\ \varsigma_{ki} &= (\varsigma_{11}, \varsigma_{12}, \varsigma_{13}, ...\varsigma_{N6}). \end{aligned} \tag{9.7.6}$$

The resulting system is a system of 36 equations to determine a number of $23 \times \mathrm{N}$ unknowns. Details of the genetic algorithm can be found in Goldberg. It is assumed the parameters p are discretized into discrete values with the step width $\Delta p = \{\Delta a_{ki}, \Delta b_{ki}, \Delta\omega_{ki}, \Delta\varsigma_{ki}\}$. The set of parameters for arbitrary values $p = \{a_{ki,m}, b_{ki,n}, \omega_{ki,q}, \varsigma_{ki,s}\}$ can be expressed as 6N numbers

$$N_{ikmnqs} = (m-1)N_{ik}Q_{ik}S_{ik} + (n-1)Q_{ik}S_{ik} + (q-1)S_{ik} + s,$$

where M_{ki}, N_{ki}, Q_{ki} and S_{ki} denote the total number of discretized values for each parameter p. These numbers represent an individual in a population and for the discretized parameters indicate a specific solution. An individual is expressed as a row of the integer number with $N_{gen} = 6\mathrm{N}$ genes.

A fitness value is evaluated for each individual and in the total population only individuals with a higher fitness remain at the next generation.

The alternation of generations is stopped when convergence is detected. If no convergence the iteration process continues until the specified maximum number of generations is reached.

To compute the fitness F we write (9.6.39)–(9.6.41) in the form

$$L_k^{(m)} = \pi_k^{(m)}, \ m = 0,1,2, \ k = 1,2,...,12. \tag{9.7.7}$$

and note the square sum of differences $L_k^{(m)} - \pi_k^{(m)}$ by $\Im$

$$\Im = \sum_{j=0}^{2}\sum_{k=1}^{12}(L_k^{(j)} - \pi_k^{(j)})^2. \tag{9.7.8}$$

We define fitness as follows

$$F = \frac{\Im_0}{\Im}, \tag{9.7.9}$$

with

$$\Im_0 = \sum_{j=0}^{2}\sum_{k=1}^{12}(\pi_k^{(j)})^2 . \tag{9.7.10}$$

As the convergence criterion of iterative computations we use (9.4.14)

$$Z = \frac{1}{2}\log_{10}\frac{\Im_0}{\Im} \rightarrow \max .$$

Numerical simulation is carried out for $\lambda = 1.055\times 10^{-3}$ kg/ms, and $\mu = 1.205\times 10^{-3}$ kg/ms. The micropolar coefficients of viscosity have values $\alpha = \zeta = \varepsilon = 1.035\times 10^{-3}$ mkg/s (Gauthier). We consider $m = \tan\theta$, with $m \in [0.2, 0.8]$. The value $m = 0.8$ represents an upper limit on the slopes for which the shallow fluid theory furnish a good approximation.

The number r^2 must satisfy the condition (Dressler)

$$4r^2 \leq 0.7m , \tag{9.7.11}$$

which is important for existence of waves. If the resistance is too large, the waves cannot form. This condition is obtained numerically. We take $r^2 \in [0.035, 0.14]$. The value $r^2 = 0.14$ was chosen as the greatest value for the resistance since it satisfies the condition (9.7.11) for $m = 0.8$. The intervals for the model parameters are evaluated from the condition that the total mass of fluid per wavelength is constant and the same in all approximations.

In order to illustrate the results three cases are considered ($N = 4$):

$$\begin{aligned} &- \textit{Case 1.}\quad \theta = 45^\circ\ (m = 1),\ r^2 = 0.17,\\ &- \textit{Case 2.}\quad \theta = 31^\circ\ (m = 0.6),\ r^2 = 0.1,\\ &- \textit{Case 3.}\quad \theta = 22^\circ\ (m = 0.4),\ r^2 = 0.06. \end{aligned} \tag{9.7.12}$$

In all cases we have assumed that the number of populations is 25, ratio of reproduction is 1, number of multi-point crossovers is 1, probability of mutation is 0.2 and maximum number of generations is 250.

The linear summation of the solution $Y(\alpha,\tau)$ for $\tau \rightarrow \infty$ is given in Figures 9.7.1–9.7.3 ($\tau \rightarrow \infty$ means in the numerical simulation the time interval after which the solutions have a permanent profile in time). In all cases the fluid velocity is greater in the region of the crests than in the shallower regions, but nowhere will the fluid velocity be as great as the wave speed. For example, in Case 1 the average fluid velocity is about 3.05 m/s while the wave velocity is about 4.1 m/s.

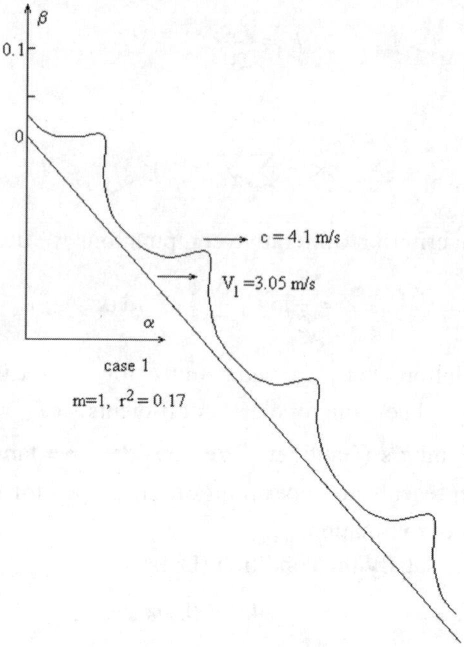

Figure 9.7.1 Profile of the wave $Y(\alpha, \tau)$ in Case 1.

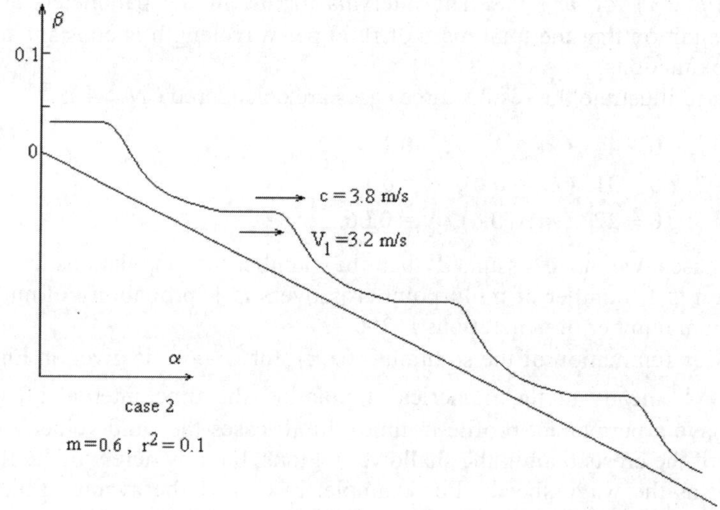

Figure 9.7.2 Profile of the wave $Y(\alpha, \tau)$ in Case 2

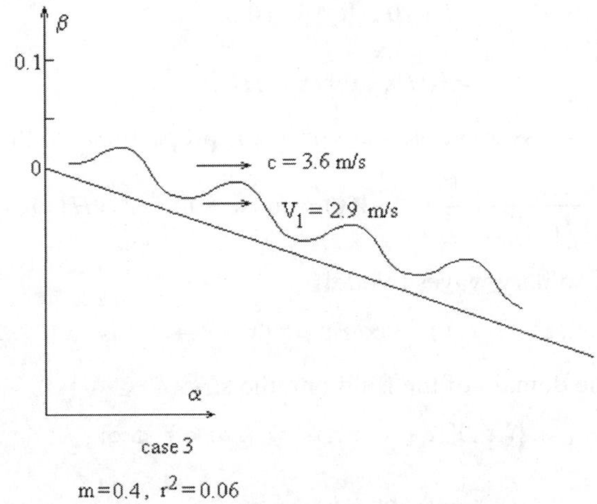

Figure 9.7.3 Profile of the wave $Y(\alpha, \tau)$ in Case 3

From numerical simulations we conclude that the remaining solutions have a similar evolution with respect to α : they increase and decrease in the same manner as Y .

The microrotation components and the vertical component of the fluid velocity are greater in the crest regions than in the shallower regions. The model parameters were obtained after 149 iterations in Case 1, 167 iterations in Case 2 and 187, in the last case.

In conclusion, the solutions describe the phenomenon called *roll-waves* for fluid flow along a wide inclined channel. This phenomenon appears in hydraulic applications, such as run-off channels and open aquaducts.

When a liquid flows turbulently downwards on an inclined open channel, the wave profile represented as sums of solitons moves downstream as a progressing wave at a constant speed and without distortion, and such that the velocities of the fluid particles are everywhere less than the wave velocity.

9.8 The effect of surface tension on the solitary waves

In this section we discuss the effect of surface tension on the solitary waves. This effect may be important when the depth is small enough and the surface tension is present. This effect was analyzed by Shinbrot in 1981, and the present discussion is principally based on his results. Let h be the depth of the fluid at infinity, ρg the weight density of the fluid, and T the surface tension. The fluid moves over a flat bottom being acted by the gravity force. The system of coordinates is moving horizontally with the phase speed of the flow, the x- axis is lying at the bottom and the y- axis is vertical. The free surface is described by the function $y = H(x)$. The velocity of the fluid is denoted by $S(x, y) = (U, V)$. The uniform flow of the fluid is described by the equation of irrotationality and incompressibility of the fluid (Camenschi)

$$U_y = V_x, \quad U_x + V_y = 0, \text{ for } 0 < y < H(x), \tag{9.8.1}$$

by the streamlines at the top and bottom

$$V = 0, \text{ for } y = 0, \tag{9.8.2}$$

$$V = UH_x, \text{ for } y = H(x), \tag{9.8.3}$$

the condition that the pressure at the free surface is proportional to the curvature

$$gH - \frac{T}{\rho}\frac{H_{xx}}{(1+H_x^2)^{3/2}} + \frac{1}{2}(U^2 + V^2) = \text{const.}, \text{ for } y = H(x), \tag{9.8.4}$$

and the condition of solitary waves (Nettel)

$$H(x) \to \text{const.}, \ U(x,y) \to \text{const.}, \ V(x,y) \to 0, \text{ as } |x| \to \infty. \tag{9.8.5}$$

Shinbrot maps the domain of the fluid into the strip

$$\{(X,Y): -\infty < X < \infty, -\infty < Y < \infty\},$$

$$X = \frac{\varepsilon x}{h}, \ Y = \frac{y}{H(x)}, \tag{9.8.6}$$

where $h = \lim_{|x|\to\infty} H(x)$, and ε is a small parameter, measuring the ratio of the amplitude of the wave to the depth at infinity. Suppose $H(x)$, $U(x,y)$ and $V(x,y)$ have the form

$$H(x) = h[1 + \varepsilon^2\eta(X)], \tag{9.8.7}$$

$$U(x,y) = U_0[1 + \varepsilon^2 u(X,Y)], \tag{9.8.8}$$

$$V(x,y) = U_0\varepsilon^3 v(X,Y). \tag{9.8.9}$$

Here, the function η is continuous as ε tends to zero, and suppose we have the normalized condition

$$\sup_X |\eta(X)| = 1, \tag{9.8.10}$$

$$U_0 = \lim_{|x|\to\infty} U(x,y). \tag{9.8.11}$$

The equations (9.8.1)–(9.8.5) become

$$u_Y = \varepsilon^2 v_X + \varepsilon^4(\eta v_X - Y\eta_X v_Y), \text{ for } 0 < Y < 1, \tag{9.8.12}$$

$$u_X + v_Y = \varepsilon^2(Y\eta_X u_Y - \eta u_X), \text{ for } 0 < Y < 1, \tag{9.8.13}$$

$$v = 0, \text{ for } Y = 0, \tag{9.8.14}$$

$$v = \eta_X + \varepsilon^2\eta_X u, \text{ for } Y = 1, \tag{9.8.15}$$

$$\eta - \frac{\varepsilon^2\tau\eta_{XX}}{(1+\varepsilon^6\eta_X^2)^{3/2}} + F(u + \frac{\varepsilon^2}{2}u^2 + \frac{\varepsilon^4}{2}v^2) = 0, \text{ for } Y = 1, \tag{9.8.16}$$

$$\eta(X),\ u(X,Y),\ v(X,Y) \to 0\,,\ \text{as } |X| \to \infty\,. \tag{9.8.17}$$

In (9.8.16), τ is the inverse Bond number given by

$$\tau = \frac{T}{\rho g h^2}\,, \tag{9.8.18}$$

and F is the Froude number which is not known since U_0 is not known.

$$F = \frac{U_0^2}{gh}\,. \tag{9.8.19}$$

Next, we insert the following expansions in series of powers of ε^2 (Nayfeh, Camenschi and Şandru)

$$u = u^{(0)} + \varepsilon^2 u^{(1)} + \ldots,\quad v = v^{(0)} + \varepsilon^2 v^{(1)} + \ldots,$$

$$F = F^{(0)} + \varepsilon^2 F^{(1)} + \ldots,\quad \eta = \eta^{(0)} + \varepsilon^2 \eta^{(1)} + \ldots,$$

into (9.8.12)–(9.8.17) and equate coefficients of like powers of ε. The equations for zero order approximations are

$$u_Y^{(0)} = 0\,,\quad u_X^{(0)} + v_Y^{(0)} = 0\,,\ \text{for } 0 < Y < 1\,, \tag{9.8.20}$$

$$v^{(0)} = 0\,,\ \text{for } Y = 0\,, \tag{9.8.21}$$

$$v^{(0)} = \eta_X^{(0)}\,,\ \text{for } Y = 1\,, \tag{9.8.22}$$

$$\eta^{(0)} + F^{(0)} u^{(0)} = 0\,,\ \text{for } Y = 1\,, \tag{9.8.23}$$

$$\eta^{(0)},\ u^{(0)},\ v^{(0)} \to 0\,,\ \text{as } |X| \to \infty\,. \tag{9.8.24}$$

In these equations $\eta^{(0)}$ is free, being subjected to (9.8.24). The solutions of the equations (9.8.20)–(9.8.23) are given by

$$F^{(0)} = 1,\ u^{(0)}(X,Y) = -\eta^{(0)}(X),\ v^{(0)}(X,Y) = Y\eta_X^{(0)}(X)\,. \tag{9.8.25}$$

From (9.8.25), the equations for the next approximation are

$$u_Y^{(1)} = Y u_{XX}^{(0)}\,,\quad u_X^{(1)} + v_Y^{(1)} = \eta^{(0)}\eta_X^{(0)}\,,\ \text{for } 0 < Y < 1\,, \tag{9.8.26}$$

$$v^{(1)} = 0\,,\ \text{for } Y = 0\,, \tag{9.8.27}$$

$$v^{(1)} = \eta_X^{(1)} - \eta^{(0)}\eta_X^{(0)}\,,\ \text{for } Y = 1\,, \tag{9.8.28}$$

$$\eta^{(1)} - \tau\eta_{XX}^{(1)} - F^{(1)}\eta^{(0)} + u^{(1)} + \frac{1}{2}\eta^{(0)^2} = 0\,,\ \text{for } Y = 1\,, \tag{9.8.29}$$

$$\eta^{(1)},\ u^{(1)},\ v^{(1)} \to 0\,,\ \text{as } |X| \to \infty\,. \tag{9.8.30}$$

We obtain from (9.8.26)–(9.8.28)

$$u^{(1)}(X,Y) = -f(X) + \frac{Y^2}{2}\eta^{(0)}_{XX}(X), \tag{9.8.31}$$

$$V^{(1)}(X,Y) = y[f_X(X) + \eta^{(0)}(X)\eta^{(0)}_X(X)] - \frac{Y^3}{6}\eta^{(0)}_{XXX}(X). \tag{9.8.32}$$

In (9.8.31), the function f is free. Substituting (9.8.31) and (9.8.32) into (9.8.29) and then into (9.8.28) it results that $\eta^{(0)}$ satisfies the equation

$$(\frac{1}{3} - \tau)\eta^{(0)}_{XXX} = F^{(1)}\eta^{(0)}_X(X) - 3\eta^{(0)}\eta^{(0)}_X. \tag{9.8.33}$$

If $\tau = 1/3$, the equation (9.8.33) yields to $\eta^{(0)} = \text{const}$. From (9.8.7) we obtain $H = \text{const.}$, but this result violates the original hypothesis. For $\tau \neq 1/3$, by integrating twice the equation (9.8.33) we have

$$(\frac{1}{3} - \tau)\eta^{(0)2}_X = (F^{(1)} - \eta^{(0)})\eta^{(0)2}. \tag{9.8.34}$$

In this case, the solution of (9.8.34) is not finite for all x. The conclusion is that the solitary waves do not exist for $\tau \geq 1/3$. For $\tau < 1/3$ the solution of (9.8.34) is obtained by using (9.8.24)

$$\eta^{(0)} = \frac{4a^2}{3}(1 - 3\tau)\operatorname{sech}^2(aX). \tag{9.8.35}$$

By neglecting the terms of the order $\varepsilon^4 a^4$, we obtain

$$F = 1 + \frac{4}{3}\varepsilon^2 a^2 (1 - 3\tau), \tag{9.8.36}$$

$$H = h[1 + \frac{4}{3}\varepsilon^2 a^2 (1 - 3\tau)\operatorname{sech}^2 \frac{\varepsilon a X}{h}]. \tag{9.8.37}$$

For $\tau = 0$, the solutions (9.8.35) and (9.8.36) reduce to the solutions with no surface tension. The effect of surface tension is to reduce the quantity $(F - 1)$ by a factor $(1 - 3\tau)$. Also, the amplitude of the soliton is reduced with the same factor. By imposing the condition (9.8.10) to the solution (9.8.35) we obtain $4a^2(1 - 3\tau) = 3$. Introducing this value of a into (9.8.36) and (9.8.37) we obtain

$$F = 1 + \varepsilon^2, \quad H = h[1 + \varepsilon^2 \operatorname{sech}^2 \frac{\varepsilon X\sqrt{3}}{2h\sqrt{1 - 3\tau}}]. \tag{9.8.38}$$

From (9.8.38) we see that for fixed amplitude, the effect of surface tension is to sharpen the crest of the wave by shortening its length by the factor $\sqrt{1 - 3\tau}$.

Chapter 10

ON THE TZITZEICA SURFACES AND SOME RELATED PROBLEMS

10.1 Scope of the chapter

The study of affine differential geometry was initiated by Gheorghe Tzitzeica (1873–1939) in 1907 by studying a particular class of hyperbolic surfaces. Tzitzeica proved that the surfaces for which the ratio K/d^4 (K is the Gaussian curvature and d, the distance from the origin to the tangent plane at an arbitrary point) is constant, are invariants under the group of centroaffine transformations. The Tzitzeica property proves to be invariant under affine transformations, and his surfaces are called Tzitzeica surfaces by Gheorghiu, or affine spheres by Blaschke because they are analogues of spheres in affine differential geometry, or projective spheres by Wilczynski (see Bâlă). Bâlă (1999) applied the symmetry groups theory to study the partial differential equations which arise in Tzitzeica surfaces theory.

In this chapter, a survey of the Tzitzeica surfaces and the application of the symmetry group theory to the Tzitzeica equations, are presented (Bâlă). The chapter explains the Shen–Ling method to construct a Weierstrass elliptic function from the solutions of the Van der Pol's equation. Next, an application in the field of mechanics, closely related to the Tzitzeica equation, is included. In the last section, the results of Rogers and Schief concerning the capability of the Bäcklund transformation to provide an integrable discretization of the characteristic equations associated to the problem of an anisentropic gas, are considered.

We refer the readers to the articles by Bâlă (1999), Crăşmăreanu (2002), Musette *et al.* (2001), Rogers and Schief (1997, 2002), Shen (1967) and Ling (1981).

10.2 Tzitzeica surfaces

Let $D \subset \mathrm{R}^2$ be an open set and consider a surface Σ in R^3 defined by the position vector $r(u,v)$

$$r(u,v) = x(u,v)i + y(u,v)j + z(u,v)k \ , \ (u,v) \in D \ .$$

The vector $r(u,v)$, which satisfies the condition

$$(r, r_u, r_v) \neq 0\,, \tag{10.2.1}$$

can be considered the solution of the second-order partial differential equations system that defines a surface leaving a centroaffinity aside

$$\begin{aligned} r_{uu} &= ar_u + br_v + cr\,, \\ r_{uv} &= a'r_u + b'r_v + c'r\,, \\ r_{vv} &= a''r_u + b''r_v + c''r\,, \end{aligned} \tag{10.2.2}$$

which is completely integrable, that is

$$(r_{uu})_v = (r_{uv})_u\,,\ \ (r_{uv})_v = (r_{vv})_u\,, \tag{10.2.3}$$

where $a,\, a',\, a''\ldots$ are the centroaffine invariant functions of u and v. For $c = c'' = 0$, the surface Σ is related to the asymptotic lines.

THEOREM 10.2.1 (Tzitzeica) *Let Σ be a surface related to the asymptotic lines. The ration $I = \dfrac{K}{d^4}$ is a constant if and only if $a' = b' = 0$.*

Therefore, the Tzitzeica surfaces are defined by the system of equations

$$\begin{aligned} r_{uu} &= ar_u + br_v\,, \\ r_{uv} &= hr\,, \\ r_{vv} &= a''r_u + b''r_v\,, \end{aligned} \tag{10.2.4}$$

where $c' = h$. The integrability conditions (10.2.3) become

$$\begin{aligned} ah &= h_u\,, \\ a_v &= ba'' + h\,, \\ b_v + bb'' &= 0\,, \\ b''h &= h_v\,, \\ a'' + aa'' &= 0\,, \\ b''_u + a''b &= h\,. \end{aligned} \tag{10.2.5}$$

We have met these equations in Section 2.4, in the study of the symmetry group of the Tzitzeica equations of the surface. If h satisfies the Liouville–Tzitzeica equation

$$(\ln h)_{uv} = h\,, \tag{10.2.6}$$

the Tzitzeica surfaces which are not ruled surfaces are defined by

$$r_{uu} = \frac{h_u}{h} r_u + \frac{\varphi(u)}{h} r_v \, , \; r_{uv} = hr \, , \; r_{vv} = \frac{h_v}{h} r_v \, . \tag{10.2.7}$$

If h satisfies the Tzitzeica equation

$$(\ln h)_{uv} = h - \frac{1}{h^2} \, , \tag{10.2.8}$$

the Tzitzeica surfaces which are not ruled surfaces are defined by

$$r_{uu} = \frac{h_u}{h} r_u + \frac{1}{h} r_v \, ,$$

$$r_{uv} = hr \, ,$$

$$r_{vv} = \frac{1}{h} r_u + \frac{h_v}{h} r_v \, . \tag{10.2.9}$$

The system (10.2.4) can be written in the form

$$\theta_{uu} = a\theta_u + b\theta_v \, ,$$

$$\theta_{uv} = h\theta \, , \tag{10.2.10}$$

$$\theta_{vv} = a''\theta_u + b''\theta_v \, ,$$

with the condition that the three independent solutions $x = x(u,v)$, $y = y(u,v)$, $z = z(u,v)$ of (10.2.10) and (10.2.5) define a Tzitzeica surface.

It is known that every linear combination of x, y, z is a solution of (10.2.10).

An equivalent form of (10.2.4) is given by

$$x_{uu} = ax_u + bx_v \, ,$$

$$x_{uv} = hx \, ,$$

$$x_{vv} = a''x_u + b''x_v \, ,$$

$$y_{uu} = ay_u + by_v \, ,$$

$$y_{uv} = hy \, , \tag{10.2.11}$$

$$y_{vv} = a''y_u + b''y_v \, ,$$

$$z_{uu} = az_u + bz_v \, ,$$

$$z_{uv} = hz \, ,$$

$$z_{vv} = a''z_u + b''z_v \, ,$$

with the conditions (10.2.1) and (10.2.5). This form is useful for studying the symmetries of the system (10.2.4).

10.3 Symmetry group theory applied to Tzitzeica equations

The condition (10.2.1) is written as

$$(y_u z_v - z_u y_v)x - (x_u z_v - x_v z_u)y + (x_u y_v - y_u x_v)z = f\ ,\ f(u,v) \neq 0\ . \qquad (10.3.1)$$

Let $D \times U^{(2)}$ be the second-order jet space associated to (10.2.11) and (10.3.1). The independent variables $u = x^1$, $v = x^2$, and the dependent variables $x = u^1$, $y = u^2$, $z = u^3$ are considered. Let $M \subset D \times U$ be an open set and ζ, η, ϕ, λ and ψ the functions of u, v, x, y and z. The infinitesimal generator of the symmetry group G of (10.2.11) and (10.3.1) is

$$X = \zeta \frac{\partial}{\partial u} + \eta \frac{\partial}{\partial v} + \phi \frac{\partial}{\partial x} + \lambda \frac{\partial}{\partial y} + \psi \frac{\partial}{\partial z},$$

denoted by G_1 a subgroup of G on the variables x, y and z. For $\zeta = 0$, $\eta = 0$ and

$$\phi(x,y,z) = a_{11}x + a_{12}y + a_{13}z\ ,$$

$$\lambda(x,y,z) = a_{21}x + a_{22}y + a_{23}z\ ,$$

$$\psi(x,y,z) = a_{31}x + a_{32}y - (a_{11} + a_{22})z\ ,$$

the infinitesimal generator Y of G_1 is

$$Y = a_{11}Y_1 + a_{22}Y_2 + a_{12}Y_3 + a_{13}Y_4 + a_{21}Y_5 + a_{23}Y_6 + a_{31}Y_7 + a_{32}Y_8\ , \qquad (10.3.2)$$

with

$$Y_1 = x\frac{\partial}{\partial x} - z\frac{\partial}{\partial z},$$

$$Y_2 = y\frac{\partial}{\partial y} - z\frac{\partial}{\partial z},$$

$$Y_3 = y\frac{\partial}{\partial x},\ Y_4 = z\frac{\partial}{\partial x},$$

$$Y_5 = x\frac{\partial}{\partial y},\ Y_6 = z\frac{\partial}{\partial y},$$

$$Y_7 = x\frac{\partial}{\partial z},\ Y_8 = y\frac{\partial}{\partial z}. \qquad (10.3.3)$$

The Lie algebra associated to G_1 of the full symmetry group G of (10.2.11) and (10.3.1) is generated by Y_i, $i = 1, 2, ..., 8$, and thus the Lie subgroup G_1 is the unimodular subgroup of the group of centroaffine transformations.

If we consider the subalgebra described by Y_1 and Y_2, then the function F is invariant and satisfies $Y_1(F) = 0$ and $Y_2(F) = 0$. Therefore, $F = \varphi(u, v, xyz)$, and the group invariant solutions are $u = C_1$, $v = C_2$ and $xyz = C_3$. We obtain from here the known surface of Tzitzeica

$$z = \frac{C}{xy}, \quad C \in \mathrm{R}. \tag{10.3.4}$$

THEOREM 10.3.1 (Bâlă) *The general vector field of the algebra of the infinitesimal symmetries associated to the subgroup* G_2, *where* G_2 *is the subgroup of the full symmetry group* G *of* (10.2.11), *which acts on the space of the independent variables, is*

$$Z = \zeta(u)\frac{\partial}{\partial u} + \eta(v)\frac{\partial}{\partial v},$$

where the functions ζ *and* η *satisfy the equations*

$$\zeta a_u + \eta a_v + a\zeta_u + \zeta_{uu} = 0, \quad \zeta b_u + \eta b_v - b\eta_u + 2b\zeta_u = 0,$$

$$\zeta h_u + \eta h_v + h(\zeta_u + \eta_v) = 0, \quad \zeta a''_u + \eta a''_v - a''\zeta_u + 2a''\eta_v = 0, \tag{10.3.5}$$

$$\zeta b''_u + \eta b''_v + b''\eta_v + \eta_{vv} = 0,$$

and the functions a, b, h, a'', b'' *satisfy* (10.2.5).

Consider the cases:

Case 1. Σ is a ruled Tzitzeica surface given by (10.2.7). The completely integrable conditions (10.2.5) turn in

$$a = \frac{h_u}{h}, \quad b = \frac{\varphi(u)}{h}, \quad a'' = 0, \quad b'' = \frac{h_v}{h}, \tag{10.3.6}$$

with h a solution of the Liouville–Tzitzeica equation (10.2.6).

The relations (10.3.5) become

$$\zeta h_u + \eta h_v + h(\zeta_u + \eta_v) = 0,$$

$$\zeta^3 = \frac{k}{\varphi}, \quad hh_{uv} - h_u h_v h^3. \tag{10.3.7}$$

Writing $\zeta = \frac{1}{U'}$, $\eta = -\frac{1}{V'}$, where $U = U(u)$ and $V = V(v)$, then the first equation of (10.3.7) is

$$h = U'V'\mu(U+V)\,. \tag{10.3.8}$$

Substituting (10.3.8) in the last equation (10.3.7) we have the equation

$$\mu\mu'' - \mu'^2 = \mu^3\,. \tag{10.3.9}$$

The general solution of (10.3.9) is

$$\mu(U+V) = \begin{cases} \dfrac{2}{(U+V+C)^2}, & k=0, \\[2ex] \dfrac{l^2}{2\cos^2[0.5l(U+V)+C]}, & k=-l^2, \\[2ex] \dfrac{l^2}{2\sinh^2[0.5l(U+V)+C]}, & k=l^2,\, l>0. \end{cases} \tag{10.3.10}$$

From (10.3.8) and the change of the functions $\tilde{U} = F(U)$, $\tilde{V} = G(V)$, we have

$$\tilde{U} = U + C\,,\ \tilde{V} = V\,,\ \text{for } k = 0\,,$$

$$\tilde{U} = \tanh\frac{l}{2}(U+C)\,,\ \tilde{V} = \tanh\frac{l}{2}V\,,\ \text{for } k = l^2\,, \tag{10.3.11}$$

$$\tilde{U} = \operatorname{cotan}\frac{l}{2}(U+C)\,,\ \tilde{V} = \tan\frac{l}{2}V\,,\ \text{for } k = -l^2\,.$$

Therefore, the general solution of the Liouville–Tzitzeica equation is

$$h(u,v) = \frac{2\tilde{U}'\tilde{V}'}{(\tilde{U}+\tilde{V})^2}\,. \tag{10.3.12}$$

This solution is expressed in terms of solitons for $k = -l^2$, and

$$\tilde{U}' = \frac{l}{2}\operatorname{sech}^2\frac{l}{2}(U+C)\,,$$

$$\tilde{V}' = \frac{l}{2}\operatorname{sech}^2\frac{l}{2}V\,.$$

Case 2. Σ is a Tzitzeica surface which is not a ruled surface, that is (10.2.9). Then (10.2.5) become

$$a = \frac{h_u}{h}\,,\ b = a'' = \frac{1}{h}\,,\ b'' = \frac{h_v}{h}\,, \tag{10.3.13}$$

where h is a solution of the Tzitzeica equation (10.2.8). Substitution of (10.3.13) into (10.3.5) yield

$$\zeta_u = 0\,,\ \eta_v = 0\,,$$

$$\zeta h_u + \eta h_v + h(\zeta_u + \eta_v) = 0\,. \tag{10.3.14}$$

It results $\zeta = C_1$, $\eta = C_2$ and

$$h = \mu(C_1 v - C_2 u),$$

$$Z = C_1 \frac{\partial}{\partial u} + C_2 \frac{\partial}{\partial v}. \tag{10.3.15}$$

Substitution of the first equation of (10.3.15) into the Tzitzeica equation (10.2.8), gives the equation

$$-C_1 C_2 (\mu\mu'' - \mu'^2) = \mu^3 - 1. \tag{10.3.16}$$

For $C_1 C_2 = 0$, we have $\mu = 1$, $h = 1$. This is a case of Tzitzeica solution (Tzitzeica 1924). For $C_1 C_2 \neq 0$, and denoting $k = -\frac{1}{C_1 C_2}$, the equation (10.3.16) becomes

$$\mu\mu'' - \mu'^2 = k(\mu^3 - 1). \tag{10.3.17}$$

For $k = 1$, the equation (10.3.17) is a Weierstrass equation

$$\mu'^2 = 2\mu^3 + C\mu^2 + 1, \; C \in \mathrm{R}. \tag{10.3.18}$$

Using the change of function $\mu = \frac{l}{2} g$, (10.3.18) turns in

$$g'^2 = g^3 + Cg^2 + 4. \tag{10.3.19}$$

If λ is a real solution of the right side polynomials of (10.3.19), then $\lambda \neq 0$ is not a triple solution. Therefore, (10.3.19) yields

$$g'^2 = (g - \lambda)(g^2 - \frac{4}{\lambda^2} g - \frac{4}{\lambda}). \tag{10.3.20}$$

For $\lambda = -1$, then $C = -3$, and (10.3.19) becomes

$$g'^2 = (g+1)(g-2)^2. \tag{10.3.21}$$

Writing $g = \frac{1}{w^2} + 2$, we have

$$w'^2 = \frac{1}{4}(3w^2 + 1), \tag{10.3.22}$$

with the solution

$$w(u+v) = \frac{1}{\sqrt{3}} \sinh[\frac{(u+v)\sqrt{3}}{2} + C_1], \; C_1 \in \mathrm{R}. \tag{10.3.23}$$

From here the other solution of the Tzitzeica equation is obtained, by writing $h = \frac{1}{2w^2} + 1$, and $C_1 = 0$

$$h(u+v) = \frac{3}{2\sinh^2[(u+v)\sqrt{3/2}]} + 1. \qquad (10.3.24)$$

For $\lambda \neq -1$, the polynomial of (10.3.19) admits three distinct real solutions for $\lambda > -1$ ($C < -3$), and one real and two complex for $\lambda < -1$ ($C > -3$). In this case the integral

$$J = \int \frac{dg}{\sqrt{(g-\lambda)(g^2 - \frac{4}{\lambda^2}g - \frac{4}{\lambda})}},$$

is reduced to an elliptical integral of first-order

$$J = \int \frac{d\varphi}{\sqrt{1-k^2\sin\varphi}}.$$

Thus, for $C \neq -3$, the solutions of the Tzitzeica equations with the form $h = \mu(u+v)$ are given in terms of elliptical functions.

PROPOSITION 10.3.1 (Bâlă) *The solution (10.3.24) is a revolution Tzitzeica surface. Moreover it is an associated ruled Tzitzeica surface.*

The proof is given by Bâlă. Tzitzeica studied the revolution surfaces defined by (10.2.10). From the condition that h must verify $h_u = h_v$, he obtains $h = \mu(u+v)$, and the equation (10.3.17) for $k = 1$

$$\mu\mu'' - \mu'^2 = \mu^3 - 1.$$

Tzitzeica found the solution of (10.2.10) by using

$$\frac{\mu'^2 - 2\mu^3 - 1}{4\mu^2} = -k^2.$$

The solution is

$$\theta(u,v) = k_1 \exp\left(\int \frac{h'-1}{2h} d\alpha\right)\cos k\beta + k_2 \exp\left(\int \frac{h'-1}{2h} d\alpha\right)\sin k\beta + k_3 \exp\left(\int \frac{h^2}{h'+1} d\alpha\right), \qquad (10.3.25)$$

for $k \neq 0$, and

$$\theta(u,v) = \exp\left(\int \frac{h'-1}{2h} d\alpha\right)[k_1(\beta^2 + \int \frac{4\mu}{\mu'+1} d\alpha) + k_2\beta + k_3], \qquad (10.3.26)$$

for $k = 0$. Here $\alpha = u+v$, $\beta = u-v$ and k_i, $i = 1,2,3$, are real constants.

On the other hand, it results $k^2 = -\frac{C}{4}$, and then the function (10.3.24) defines the revolution surface (10.3.25). If we consider

$$\tilde{U} = \tanh\frac{\sqrt{3}}{2}(U + C_1)\,,\ \tilde{V} = \tanh\frac{\sqrt{3}}{2}V\,, \tag{10.3.27}$$

the function h is written as

$$h(u,v) = \frac{2\tilde{U}'\tilde{V}'}{(\tilde{U}+\tilde{V})^2} + 1 = H(u,v) + 1\,. \tag{10.3.28}$$

Therefore, the function H is a solution of the Liouville–Tzitzeica equation (10.2.6), which defines a ruled Tzitzeica surface.

PROPOSITION 10.3.2. (Bâlă). *The solution of Tzitzeica* (*10.3.25*) *is invariant under the transformation subgroup of* G_2, *for which the Lie algebra is generated, in the case of not ruled Tzitzeica surfaces, by* $Z = C_1\frac{\partial}{\partial u} + C_2\frac{\partial}{\partial v}$.

In the following we present the Shen method to construct a Weierstrass elliptic function $\wp(t;\, 2\omega_1, 2\omega_3)$, by using the purely imaginary solution, ω_1 and ω_2, of the Van der Pol's differential equation

$$x_{tt} + \mu(x^2 - 1)x_t + k^2 x = 0\,,\ \mu, k \in \mathrm{R}\,, \tag{10.3.29}$$

with the initial condition

$$x(t_0) = x_0\,. \tag{10.3.30}$$

Shen considers

$$-\mu^2 g_3 = \frac{1}{4} g_2^2\,, \tag{10.3.31}$$

where g_2 and g_3 are the invariants of the Weierstrass equation (10.3.18) written under the standard form

$$\dot{\wp}^2 = 4\wp^3 - g_2\,\wp - g_3\,. \tag{10.3.32}$$

The solution requires that g_2 and g_3 must be real, and g_3 must be negative. Shen showed that g_2 is real if μ and k are comparatively smaller than x_0. Introducing the quantity $q = \exp\frac{i\pi\omega_3}{\omega_1}$, Shen showed also that g_3 is negative for

$$q > \frac{1}{\sqrt{505}}\,. \tag{10.3.33}$$

The quantities ω_1 and ω_3 may be computed by using the complete elliptic integrals in condition

$$0 < g_2 < \frac{16}{27}\mu^4\,. \tag{10.3.34}$$

Shen proved that there exists always one and only one cycle on the phase plane for the solution.

The equation (10.3.34) is quadratic and gives two real roots under the restriction imposed on x_0. One root is positive and the other is negative. The positive root leads to Shen's solution.

Ling has shown that there exists a second solution in which ω_3 is no longer purely imaginary, whereas ω_1 remains real, if the condition (10.3.34) is not satisfied. In this case we have

$$g_2 < 0, \text{ or } g_2 > \frac{16}{27}\mu^4. \tag{10.3.35}$$

So, one of the three roots of the cubic equation remains real but the other two roots are complex conjugate. The negative root, which satisfies the first condition in (10.3.35) leads to a second solution.

Ling introduces two invariants σ_4 and σ_6 in place of g_2 and g_3

$$\sigma_4 = \frac{1}{60} g_2 = \sum_{m,n=-\infty}^{\infty} \frac{1}{(2m\omega_1 + 2n\omega_3)^4},$$

$$\sigma_6 = \frac{1}{140} g_3 = \sum_{m,n=-\infty}^{\infty} \frac{1}{(2m\omega_1 + 2n\omega_3)^6}, \tag{10.3.36}$$

where the summation omits the simultaneous zeros of m and n. From the variation of σ_4 and σ_6 it results that we are in the case when g_2 and g_3 are both negative, and the value of q is given by

$$q = \mathrm{i}\exp(-\pi c),\ 0 < c < \frac{1}{2}. \tag{10.3.37}$$

In this solution, the values of ω_1 and q are found from

$$g_2 = \frac{1}{12}\left(\frac{\pi}{\omega_1}\right)^4 (\theta_3^8 - \theta_3^4\theta_4^4 + \theta_4^8),$$

$$g_3 = \frac{1}{432}\left(\frac{\pi}{\omega_1}\right)^6 (2\theta_3^4 - \theta_4^4)(2\theta_4^4 - \theta_3^4), \tag{10.3.38}$$

where

$$\theta_3 = 1 + 2\sum_{n=1}^{\infty} q^{n^2},$$

$$\theta_4 = 1 + 2\sum_{n=1}^{\infty} (-1)^n q^{n^2}. \tag{10.3.39}$$

Elimination of ω_1 leads to an equation with real coefficients which can be solved numerically for a real root $p=-q^2=\exp(-2\pi c)$, less than unity but greater than $\exp(-\pi)$, for the specified g_2 and g_3. Finally, Ling found

$$2\omega_3=\omega_1+\frac{\mathrm{i}\omega_1}{\pi}\pi\log\left(\frac{1}{p}\right). \qquad (10.3.40)$$

If e_1 is a real root of the cubic equation, the others two roots e_2 and e_3 are complex conjugate.

Consider now the Liouville–Tzitzeica equation (10.2.6) and the Tzitzeica equation (10.2.8). If we denote $\ln h=\omega$, these equations become

$$\omega_{uv}=\exp\omega\,, \qquad (10.3.41)$$

and respectively

$$\omega_{uv}=\exp\omega-\exp(-2\omega)\,. \qquad (10.3.42)$$

THEOREM 10.3.2 (Bâlă) *The generator vector field which describes the algebra of infinitesimal symmetries associated to the Liouville–Tzitzeica equation (10.3.41) are*

$$W=f(u)\frac{\partial}{\partial u}+g(v)\frac{\partial}{\partial v}-[f'(u)+g'(v)]\frac{\partial}{\partial\omega}\,. \qquad (10.3.43)$$

THEOREM 10.3.3 (Bâlă) *The vector fields which generate the Lie algebra of infinitesimal symmetries associated to the Tzitzeica equation (10.3.42) are*

$$U_1=u\frac{\partial}{\partial u}-v\frac{\partial}{\partial v},\ U_2=\frac{\partial}{\partial u},\ U_3=\frac{\partial}{\partial v}\,. \qquad (10.3.44)$$

If $\omega=f(u,v)$ is a solution of the Tzitzeica equation (10.3.42), then the functions

$$\omega_1=f[\exp(\varepsilon)u,\ \exp(-\varepsilon)v],\ \omega_2=f[u-\varepsilon,\ v],\ \omega_3=f[u,\ v-\varepsilon],$$

where $\varepsilon\in\mathrm{R}$, are also solutions of the equation.

10.4 The relation between the forced oscillator and a Tzitzeica curve

Crăşmăreanu considers this example. Let consider the equation

$$\ddot f(t)+\varepsilon f(t)=g(t)\,, \qquad (10.4.1)$$

For $\varepsilon=1$, (10.4.1) represents the forced oscillator equation. The driven function $g(t)$ is defined as

$$g(t)=-\frac{1}{a\varepsilon t+b}\,,\ a,b\in\mathrm{R}\,,\ a\neq 0\,,\ \varepsilon=\pm 1\,. \qquad (10.4.2)$$

We can write (10.4.1) under the form

$$\frac{1}{\varepsilon f(t)+\ddot{f}(t)}=-(a\varepsilon t+b)\,. \tag{10.4.3}$$

Differentiating (10.4.3) with respect to time, we obtain

$$\frac{\varepsilon\dot{f}(t)+\dddot{f}(t)}{[\varepsilon f(t)+\ddot{f}(t)]^2}=a\varepsilon\,. \tag{10.4.4}$$

Consider now in R^3 a curve C in the form $r=r(t)$. This curve is called elliptic cylindrical if it has the expression

$$r(t)=(\cos t,\sin t,f(t))\,,\ f\in C^\infty(R)\,. \tag{10.4.5}$$

The curve C is called hyperbolic cylindrical if it has the expression

$$r(t)=(\cosh t,\sinh t,f(t))\,,\ f\in C^\infty(R)\,. \tag{10.4.6}$$

For an elliptic cylindrical curve, the torsion function $\tau(t)$, and the distance from origin to the osculating plane $d(t)$ are defined by

$$\tau(t)=\frac{f'+f'''}{1+f'^2+f''^2}\,, \tag{10.4.7}$$

$$d(t)=\frac{\pm(f+f'')}{\sqrt{1+f'^2+f''^2}}\,. \tag{10.4.8}$$

For a hyperbolic cylindrical curve, the functions $\tau(t)$ and $d(t)$ are

$$\tau(t)=\frac{f'-f'''}{1+(f'^2+f''^2)(\cosh^2 t+\sinh^2 t)-4ff''\cosh t\sinh t}\,, \tag{10.4.9}$$

$$d(t)=\frac{\pm(f-f'')}{\sqrt{1+(f'^2+f''^2)(\cosh^2 t+\sinh^2 t)-4ff''\cosh t\sinh t}}\,. \tag{10.4.10}$$

Consider now that the curve C is Tzitzeica with the constant $a\neq 0$. For a cylindrical elliptic curve we have

$$\frac{\tau(t)}{d^2(t)}=a=\frac{f'(t)+f'''(t)}{(f(t)+f''(t))^2}\,, \tag{10.4.11}$$

and similarly or a hyperbolic cylindrical curve

$$\frac{\tau(t)}{d^2(t)}=a=\frac{f'(t)-f'''(t)}{(f(t)-f''(t))^2}\,. \tag{10.4.12}$$

The equation (10.4.11) is the same with (10.4.4) for $\varepsilon=1$, and the equation (10.4.12) is respectively the same with (10.4.4) for $\varepsilon=-1$.

The solution of (10.4.4) for $\varepsilon=1$ is given by

$$f(t)=f(0)\cos t-\dot{f}(0)\sin t-\int_0^t\frac{\sin(u-t)}{au+b}\mathrm{d}u\,, \tag{10.4.13}$$

and the solution of (10.4.4) for $\varepsilon=-1$ by

$$f(t)=f(0)\cosh t+\dot{f}(0)\sinh t+\int_0^t\frac{\sinh(u+t)}{au+b}\mathrm{d}u\,. \tag{10.4.14}$$

PROPOSITION 10.4.1 *An elliptic cylindrical curve* $r(t)=(\cos t,\sin t,f(t))$ *is a Tzitzeica curve, if and only if* $f(t)$ *is the solution of the forced harmonic oscillator (10.4.13), with the driven force* $g(t)=-\dfrac{1}{a\varepsilon t+b}$, *where the constant* a *is defined as* $a=\dfrac{\tau(t)}{d^2(t)}$, *and* b *is an arbitrary constant.*

PROPOSITION 10.4.2 *A hyperbolic cylindrical curve* $r(t)=(\cosh t,\sinh t,f(t))$ *is a Tzitzeica curve, if and only if (10.4.14) is verified.*

10.5 Sound propagation in a nonlinear medium

A chiral Cosserat material is isotropic with respect to coordinate rotations but not with respect to inversions. A chiral material has nine elastic constants in comparison to the six considered in the isotropic micropolar solids. Materials may exhibit chirality on the atomic scale, as in quartz and in biological molecules. Materials may also exhibit chirality on a larger scale, as in composites with helical or screw-shaped inclusions, solids containing twisted or spiraling fibers, in which one direction of twist or spiral predominate. Such materials can exhibit odd rank tensor properties such as piezoelectric response. Chiral materials include crystalline materials such as sugar which are chiral on an atomic scale, as well as naturally occurring porous material, bones, foams, cellular solids, honeycomb cell structures (Lakes).

Consider the basic equations for an anisotropic Cosserat solid which is isotropic with respect to coordinate rotations but not with respect to inversions (Cosserat, Eringen).

Balance of momentum

$$\sigma_{kl,k}-\rho u_{l,tt}=0\,. \tag{10.5.1}$$

Balance of moment of momentum

$$m_{rk,r}+\varepsilon_{klr}\sigma_{lr}-\rho j\phi_{k,tt}=0\,. \tag{10.5.2}$$

Balance of energy

$$\rho\varepsilon_{,t}=\sigma_{kl}(u_{l,kt}-\varepsilon_{klr}\phi_{r,t})+m_{kl}\phi_{l,kt}\,. \tag{10.5.3}$$

The generalized Christensen's constitutive law which describes the multi-relaxation processes in a chiral material is given by

$$\sigma_{kl}+\sum_{i=1}^{N}a_{iklmn}\frac{\mathrm{d}^i\sigma_{mn}}{\mathrm{d}t^i}=\sum_{i=0}^{\infty}b_{iklmn}\frac{\mathrm{d}^i\varepsilon_{mn}}{\mathrm{d}t^i}+\lambda\varepsilon_{rr}\delta_{kl}+ \\ +(2\mu+\kappa)\varepsilon_{kl}+\kappa\varepsilon_{klm}(r_m-\phi_m)+C_1\phi_{r,r}\delta_{kl}+C_2\phi_{k,l}+C_3\phi_{l,k}, \tag{10.5.4}$$

$$m_{kl}+\sum_{i=1}^{N}c_{iklmn}\frac{\mathrm{d}^i m_{mn}}{\mathrm{d}t^i}=\sum_{i=0}^{\infty}d_{iklmn}\frac{\mathrm{d}^i\phi_{mn}}{\mathrm{d}t^i}+\alpha\phi_{r,r}\delta_{kl}+\beta\phi_{k,l}+\gamma\phi_{l,k}+ \\ +C_1\varepsilon_{rr}\delta_{kl}+(C_2+C_3)\varepsilon_{kl}+(C_3-C_2)\varepsilon_{klm}(r_m-\phi_m), \tag{10.5.5}$$

where N is the number of different relaxation processes which are activated during the loading process, σ_{kl} is the stress tensor which is asymmetric tensor, m_{kl} is the couple stress tensor (or moment per unit area), ε is the internal energy density, $\varepsilon_{kl}=\frac{1}{2}(u_{k,l}+u_{l,k})+\varepsilon_{klm}(r_m-\phi_m)$ is the strain tensor, u is the displacement vector, j is microinertia, ε_{klm} is the permutation symbol, and ρ the density of the material.

The microrotation vector ϕ_k refers to the rotation of points themselves, while the macrorotation vector $r_k=\frac{1}{2}\varepsilon_{klm}u_{m,l}$ refers to the rotation associated with movement of nearby points. The coefficients a_{iklmn}, b_{iklmn}, ... are related to the material properties and to the deformation of the medium. The usual Einstein summation convention for repeated indices is used and the comma denotes differentiation with respect to spatial coordinates and a superposed dot indicates the time rate. We use rectangular coordinates x_k ($k=1,2,3$) or ($x_1=x$, $x_2=y$, $x_3=z$).

In three dimensions there are nine independent elastic constants required to describe an anisotropic Cosserat elastic solid, that is Lamé elastic constants λ, and μ,. the Cosserațt rotation modulus κ, the Cosserațt rotation gradient moduli α,β,γ, and C_i, $i=1,2,3$, the chiral elastic moduli associated with noncentrosymmetry. For $C_i=0$, the equations of isotropic micropolar elasticity are recovered. For $\alpha=\beta=\gamma=\kappa=0$, equations (10.5.2) and (10.5.3) reduce to the constitutive equations of classical isotropic linear elasticity theory. From the requirement that the internal energy must be non-negative, we obtain restrictions on the micropolar elastic moduli $0\le 3\lambda+2\mu+\kappa$, $0\le 2\mu+\kappa$, $0\le\kappa$, $0\le 3\alpha+\beta+\gamma$, $-\gamma\le\beta\le\gamma$, $0\le\gamma$, and any positive or negative C_1,C_2,C_3. Boundary conditions do not depend on assumed material symmetry. One may prescribe the displacements or the surface traction and the microrotations, or the surface couples on the surface.

Consider, now, a slab of chiral material compressed in the z direction, and bounded by surfaces $z=\pm H$ (Lakes). Assume a monorelaxation process, and one component of displacement $u_z(z)$, and one component of microrotation, $\phi_z(z)$ to be nonzero. The motion equations (10.5.1) and (10.5.2) for stress p and the couple stress m are given by

$$(\lambda+2\mu+\kappa)(u_{z,zz}+\tau_\sigma u_{z,zzt})+(C_1+C_2+C_3)(\phi_{z,zz}+\tau_\sigma\phi_{z,zzt})=\rho(u_{z,tt}+\tau_\varepsilon u_{z,ttt}), \quad (10.5.6)$$

$$(1-K_0^2)(u_{z,zz}+\tau_\sigma u_{z,zzt})+\frac{2\kappa}{\alpha+\beta+\gamma}(\phi_z+\tau_\sigma\phi_{z,t})=\rho j(\phi_{z,TT}+\tau_\varepsilon\phi_{z,ttt}), \quad (10.5.7)$$

where τ_ε, τ_σ are the relaxation times under constant strains, and respectively constant stresses, and K_0 with a coupling coefficient defined as

$$K_0^2=\frac{(C_1+C_2+C_3)^2}{(\lambda+2\mu+\kappa)(\alpha+\beta+\gamma)}. \quad (10.5.8)$$

For the case the relaxation times are zero, the motion equations (10.5.6) and (10.5.7) for stress p and the couple stress m are given by

$$(\lambda+2\mu+\kappa)u_{z,zz}+(C_1+C_2+C_3)\phi_{z,zz}=\rho u_{z,tt}, \quad (10.5.9)$$

$$(1-K_0^2)u_{z,zz}+\frac{2\kappa}{\alpha+\beta+\gamma}\phi_z=\rho j\phi_{z,tt}, \quad (10.5.10)$$

with a coupling coefficient defined by (10.5.8).

The traveling wave solutions of (10.5.9) and (10.5.10) can be found assuming $\xi=z-ct$, where c is the wave velocity. Thus, the above equations become

$$(\lambda+2\mu+\kappa-\rho c^2)u_z''+(C_1+C_2+C_3)\phi_z=0, \quad (10.5.11)$$

$$(1-K_0^2)u_z''+\frac{2\kappa}{\alpha+\beta+\gamma}\phi_z-\rho jc^2\phi_z''=0, \quad (10.5.12)$$

where prime means the differentiation with respect to ξ. The solutions for these equations are

$$\phi_z=\phi_0\sinh p\xi, \quad (10.5.13)$$

$$p=\sqrt{\frac{2\kappa(\lambda+2\mu+\kappa-\rho c^2)}{(\alpha+\beta+\gamma)(1-K_0^2)(C_1+C_2+C_3)+\rho jc^2(\lambda+2\mu+\kappa-\rho c^2)}}, \quad (10.5.14)$$

and

$$u_z=e\xi-\frac{C_1+C_2+C_3}{\lambda+2\mu+\kappa-\rho c^2}\phi_0\sinh\xi p, \quad (10.5.15)$$

with e, c and ϕ_0 to be determined from the boundary and initial conditions.

For the case the relaxation times are not zero, the dimensionless generalized equation of sound propagation in a chiral medium, written for the longitudinal displacement wave u, is

$$u_{tt}+\alpha u_t+\beta u_t^2=c^2[u_{xx}+\gamma u_x+\delta u_x^2+f], \quad (10.5.16)$$

involving the functions of nonlinearity α, β, c, γ, δ and f, whose specific expressions depend on u. With a proper change of function, the equation (10.5.6) may be reduced to

$$\omega_{tt} - \omega_{xx} = F(\omega, \omega_x, \omega_t),$$

which plays an important role in the soliton theory. In particular, for $F(\omega) = \sin\omega$, we have the sine-Gordon equation, for $F(\omega) = \omega - \omega^3$, we obtain an equation, which appears in solid state physics and high energy particle physics (Dodd *et al.*), and for $F(\omega) = \mp(\sin\omega + \frac{1}{2}\sin\frac{\omega}{2})$, the double sine-Gordon equation is obtained.

For $\alpha = \gamma = 0$, $c = 1$, $\beta = \delta = -\frac{1}{u}$ and $f = -4au^2$, $a > 0$ a constant, the equation (10.5.16) becomes

$$u_{tt} - \frac{u_t^2}{u} = u_{xx} - \frac{u_x^2}{u} - 4au^2. \tag{10.5.17}$$

By changing the function $u = \exp\omega$, we have

$$\omega_{tt} - \omega_{xx} = -4a\exp\omega,$$

and by changing the variables $x + t = \xi$, $x - t = \eta$, the above equation yields to the Liouville–Tzitzeica equation (10.3.41)

$$\omega_{\xi\eta} - a\exp\omega = 0. \tag{10.5.18}$$

For $f = -4au^2 + b\frac{4}{u}$, $a, b > 0$ constants, the equation (10.5.16) yields the Tzitzeica equation

$$E(\omega) = \omega_{\xi\eta} - a\exp\omega + b\exp(-2\omega) = 0. \tag{10.5.19}$$

Tzitzeica considers the linear system of equations (Tzitzeica)

$$\begin{gathered}
-\tau_{\xi\xi} + U_\xi\tau_\xi + \sqrt{b}\lambda\exp(-U)\tau_\eta = 0, \\
-\tau_{\eta\eta} + U_\eta\tau_\eta + \sqrt{b}\lambda^{-1}\exp(-U)\tau_\eta = 0, \\
-\tau_{\xi\eta} + a\tau\exp U = 0,
\end{gathered} \tag{10.5.20}$$

with $\lambda \in C$, $\lambda \neq 0$, that are verified if U is a solution of (10.5.19), and τ a function related to the ratio of two entire functions. The system (10.5.20) is invariant under the involution (Tzitzeica)

$$(\tau, \exp U, \lambda) \to (\frac{1}{\tau}, -\exp(U) + \frac{2}{a}\frac{\tau_\xi\tau_\eta}{\tau^2}, -\lambda). \tag{10.5.21}$$

From (10.5.21) it results the Darboux transformation that leaves invariant the equations (10.5.20)

$$\exp(\omega) = -\exp(U) + \frac{2}{a}\frac{\tau_\xi \tau_\eta}{\tau^2},$$

or

$$\exp(\omega) = -\frac{2}{a}\partial_\xi \partial_\eta \log \tau + \exp(U),$$

where ω and U are two different solutions of (10.5.19).

By following the results of Musette and his coworkers, Conte and Verhoeven, we start with the Painlevé analysis of the Tzitzeica equation (10.5.19) with one family of movable singularities, in order to obtain the Bäcklund transformation of this equation. We check if the equation verifies the Painlevé condition. That means assume the existence of the transformation

$$\omega = \bar{U} + D \log \tau, \; E(\omega) = 0, \tag{10.5.22}$$

with ω a solution of (10.5.19), D a singular part operator, and $\bar{U}$ an a priori unconstrained function. Then we choose the two, three, etc., order for the unknown Lax pair and represent this Lax pair by an equivalent Riccati pseudopotential $Y_1, Y_2, \ldots$.

The equation (10.5.19) is invariant under the permutation (Lorentz transformation)

$$(\partial_\xi, \partial_\eta) \to (\partial_\eta, \partial_\xi). \tag{10.5.23}$$

This invariance is conserved by the Lax pair. Since τ is invariant under the permutation (10.5.23), we can define the unknown Lax pair on the basis (ψ_x, ψ_t, ψ) as

$$(\partial_\xi - L)\begin{pmatrix}\psi_\xi \\ \psi_\eta \\ \psi\end{pmatrix} = 0, \quad L = \begin{pmatrix} f_1 & f_2 & f_3 \\ g_1 & g_2 & g_3 \\ 1 & 0 & 0 \end{pmatrix}, \tag{10.5.24}$$

$$(\partial_\eta - M)\begin{pmatrix}\psi_\xi \\ \psi_\eta \\ \psi\end{pmatrix} = 0, \quad M = \begin{pmatrix} g_1 & g_2 & g_3 \\ h_1 & h_2 & h_3 \\ 0 & 1 & 0 \end{pmatrix}, \tag{10.5.25}$$

with the link $\tau = \psi$. The nine coefficients f_j, g_j, h_j, $j = 1,2,3$, are functions to be determined. The Riccati components Y_1 and Y_2 are defined as

$$Y_1 = \frac{\psi_\xi}{\psi}, \; Y_2 = \frac{\psi_\eta}{\psi}.$$

The properties of the cross-derivative conditions are written as

$$Y_{1,\eta} = Y_{2,\xi},$$

$$(Y_{1,\xi})_\eta - (Y_{1,\eta})_\xi = X_0 + X_1 Y_1 + X_2 Y_2 = 0, \tag{10.5.26}$$

$$(Y_{2,\xi})_\eta - (Y_{2,\eta})_\xi = X_3 + X_4 Y_1 + X_5 Y_2 = 0,$$

where

$$Y_{1,\xi} = -Y_1^2 + f_1Y_1 + f_2Y_2 + f_3\,,$$

$$Y_{2,\xi} = -Y_1Y_2 + g_1Y_1 + g_2Y_2 + g_3\,,$$

$$Y_{2,\eta} = -Y_2^2 + h_1Y_1 + h_2Y_2 + h_3\,.$$

The first relation (10.5.26) is identically verified, and the rest of relations (10.5.26) yield $X_j = 0$, $j = 0,1,...,5$, or

$$X_0 = f_{3,t} - g_{3,\xi} + f_1g_3 - f_3g_1 + f_3h_3 - g_2g_3 = 0\,,$$

$$X_1 = f_{2,\eta} - g_{2,\xi} + f_1g_2 - f_2g_1 + f_3 - g_2^2 + f_2h_2 = 0\,,$$

$$X_2 = f_{1,\eta} - g_{1,\xi} + f_2h_1 - g_2g_1 - g_3 = 0\,,$$

$$X_3 = h_{3,\xi} - g_{3,\eta} + g_3h_2 - g_2h_3 + f_3h_1 - g_1g_3 = 0\,,$$

$$X_4 = h_{2,\xi} - g_{2,\eta} + f_2h_1 - g_2g_1 - g_3 = 0\,,$$

$$X_5 = -g_{1,\eta} + h_{1,\xi} + g_1h_2 - g_2h_1 + h_3 - g_1^2 + f_1h_1 = 0\,.$$

The next step is to choose the link $D\log\tau = f(\psi)$ between the function τ and the solution ψ of a scalar Lax pair (10.5.24) and (10.5.25). For the link $\tau = \psi$, the truncation of the difference $E(\omega) - E(U)$ is written as

$$\exp(\omega) = D\log\tau + \exp(U)\,, \tag{10.5.27}$$

$$E(\omega) = \sum_{k=0}^{3}\sum_{l=0}^{3-k} E_{kl}(f_j, g_j, h_j, U)Y_1^kY_2^l = 0\,. \tag{10.5.28}$$

The relations (10.5.27) and (10.5.28) yield to ten equations

$$E_{kl}(f_j, g_j, h_j, U) = 0\,,\ \forall k,l\,, \tag{10.5.29}$$

for determining the unknowns U and the nine coefficients f_j, g_j, h_j, $j = 1,2,3$.

This process is repeated until a success occurs, namely $|\partial_\xi - L, \partial_\eta - M|$ is equivalent to $E(U) = 0$. Then, eliminating ψ from the Darboux transformation

$$\exp(\omega) = -\frac{2}{a}\partial_\xi\partial_\eta\log\psi + \exp(U)\,, \tag{10.5.30}$$

and the scalar Lax pair, two equations for the Bäcklund transformation are obtained.

Musette *et al.* have shown that the truncation of $\exp(-\omega)$ has no solution. The truncation of $\exp(\omega)$

$$\exp(\omega) = D\log\tau + \exp(U),$$

$$E(U) = 0, \quad D = -\frac{2}{a}\partial_\xi\partial_\eta, \tag{10.5.31}$$

generates four equations which admit some particular solutions of the form (10.3.23) and (10.3.24). According to the *PROPOSITION* 10.3.1, the solution (10.3.24) is a revolution ruled Tzitzeica surface.

PROPOSITION 10.5.1 *The solution in displacement of the sound propagation in a slab of a chiral material problem is represented by the revolution ruled Tzitzeica surface.*

10.6 The pseudospherical reduction of a nonlinear problem

Consider the one-dimensional problem of an anisentropic gas with its pseudospherical reduction given by Rogers and Schief in 1997. We present in this section the results of Rogers and Schief. The governing equations in a Lagrangian system of coordinates (X,t) are written as

$$\varepsilon_t = v_X, \tag{10.6.1}$$

$$\rho_0 v_t = \sigma_X. \tag{10.6.2}$$

The constitutive law is given by

$$\sigma = -p = -p(\rho, X). \tag{10.6.3}$$

Here, p and ρ are the pressure and respectively, the density of the medium, $\varepsilon = \frac{\rho_0}{\rho} - 1$ is the stretch, ρ_0 is the density of the medium in the underformed state, and $v(X,t)$ is the material velocity. The Lagrangian equations (10.6.1) and (10.6.2) and a constitutive law of the type $\sigma = \sigma(\varepsilon.X)$ may describe also the uniaxial deformation of nonlinear elastic materials with inhomogeneities.

In terms of the Eulerian coordinates $\overline{x} = \overline{x}(X,t)$, we have

$$\mathrm{d}\overline{x} = (\varepsilon + 1)\mathrm{d}X = v\mathrm{d}t, \tag{10.6.4}$$

so that

$$\rho_0 \mathrm{d}X = \rho\mathrm{d}x - \rho v\mathrm{d}t. \tag{10.6.5}$$

In (10.6.5), X corresponds to the particle function ψ of the Martin formulation. The independent variables are chosen to be p and ψ, and we suppose $\rho_0 = 1$. In this case we obtain the Martin formulation (Rogers and Schief 1997) of the Monge–Ampère equation as

$$\xi_{pp}\xi_{\psi\psi} - \xi_{p\psi}^2 = \varepsilon_p, \tag{10.6.6}$$

where

$$t = \xi_p, \; v = \xi_\psi,$$

$$d\bar{x} = \xi_\psi \xi_{pp} + (\xi_\psi \xi_{p\psi} + \varepsilon)d\psi, \tag{10.6.7}$$

$$0 < |\xi_{pp}\varepsilon| < \infty.$$

If a solution $\xi(p,\psi)$ of this equation is specified, then the particle trajectories are calculated from

$$\bar{x} = \int [\xi_\psi \xi_{pp} + (\xi_\psi \xi_{p\psi} + \varepsilon)d\psi], \; t = \xi_p, \tag{10.6.8}$$

in terms of p, for $\psi = \text{const.}$ The isobars are parametrically expressed in terms of ψ, for $p = \text{const.}$ By solving (10.6.8), the solution $p(\psi,t)$ is obtained, and the original solution of (10.6.1)–(10.6.3) is determined in terms of the Lagrangian variables

$$\bar{x} = \bar{x}(\psi,t), \; v = v(\psi,t), \; p = p(\psi,t). \tag{10.6.9}$$

Rogers and Schief made the geometric connection of this problem. To show this, consider a surface Σ in the Monge parametrisation (1.7.7), whose Gaussian curvature is given by (1.7.9b).

Let us introduce new independent variables p and ψ

$$p = z_x, \; \psi = z_y, \tag{10.6.10}$$

and the dependent variable ξ

$$\xi_p = x, \; \xi_\psi = y. \tag{10.6.11}$$

Therefore, we have

$$\xi_{pp} = \frac{z_{yy}}{z_{xx}z_{yy} - z_{xy}^2}, \; \xi_{\psi\psi} = \frac{z_{xx}}{z_{xx}z_{yy} - z_{xy}^2}, \; \xi_{p\psi} = \frac{z_{xy}}{z_{xx}z_{yy} - z_{xy}^2}. \tag{10.6.12}$$

The Gaussian curvature (1.7.9b) yields

$$K = \frac{1}{(1+p^2+\psi^2)^2(\xi_{pp}\xi_{\psi\psi} - \xi_{p\psi}^2)}. \tag{10.6.13}$$

The Gaussian curvature may be set into correspondence with the Martin's Monge–Ampère equation (10.6.6) by

$$\varepsilon_p = \frac{1}{\mathrm{K}(1+p^2+\psi^2)^2}, \tag{10.6.14}$$

and

$$\mathrm{K} = \frac{\rho^2\lambda^2}{(1+p^2+\psi^2)^2}, \tag{10.6.15}$$

where $\lambda^2 = \frac{\partial p}{\partial \rho |_{\psi}}$, with λ the local speed of sound in gas. For the analog nonlinear elastic problem, we have

$$\mathrm{K} = \frac{A^2}{(1+\sigma^2+X^2)^2}. \tag{10.6.16}$$

where $A^2 = \frac{\partial \sigma}{\partial \varepsilon |_X}$, with A the Lagrangian wave velocity. The surface Σ is restricted to be pseudospherical, that is

$$K = -\frac{1}{a^2}, \ a = \text{const}. \tag{10.6.17}$$

In this case the relation (10.6.16) gives

$$\frac{\partial^2 \sigma}{\partial \varepsilon^2 |_X} = \frac{2}{a^2}(1+\sigma^2+X^2)\sigma\frac{\partial \sigma}{\partial \varepsilon}|_X > 0, \ \sigma > 0. \tag{10.6.18}$$

Integrating (10.6.18) with $p \to -\sigma$ and $\psi \to X$, we get

$$\varepsilon = \frac{a^2}{2(1+X^2)^{3/2}}[\arctan(\frac{\sigma}{1+X^2}) + \frac{\sigma\sqrt{1+X^2}}{1+\sigma^2+X^2}] + \alpha(X), \tag{10.6.19}$$

with $\alpha(X)$ arbitrary. For $\sigma|_{\varepsilon=0} = 0$, it results $\alpha(X) = 0$.

Now, we introduce

$$\sigma = \sqrt{1+X^2}\tan[\frac{\sqrt{1+X^2}}{a}(c-c_0)], \tag{10.6.20}$$

into (10.6.19) and obtain

$$\varepsilon = \frac{a^2}{2(1+X^2)}[\frac{c-c_0}{a} + \frac{1}{\sqrt{1+X^2}}]\sin(\frac{2\sqrt{1+X^2}}{a}(c-c_0)). \tag{10.6.21}$$

Relations (10.6.20) and (10.6.21) represent a parametric representation for the constitutive laws of the type $\sigma = \sigma(\varepsilon, X)$, for which the equations (10.6.1) and (10.6.2) are associated to a pseudospherical surface Σ. These equations lead to

$$\sigma_{XX} = \varepsilon_{tt}. \tag{10.6.22}$$

Using (10.6.14), we have

$$\sigma_{XX} = [\frac{a^2}{(1+\sigma^2+X^2)^2}\sigma_{,t}]_{,t}. \tag{10.6.23}$$

This equation has a solitonic behavior, if we take into account that for a pseudosperical surface Σ, the expression (1.7.12) of the Gauss curvature reduces to the sine-Gordon equation.

The mean curvature H of Σ given by (1.7.9a) is an invariant, and then we can write

$$\cot\omega = -a\frac{(1+\sigma^2)X_v - 2\sigma X\sigma_v - (1+X^2)\sigma_t}{2(1+\sigma^2+X^2)^{3/2}}. \qquad (10.6.24)$$

The map

$$(X,t)\to(X'=v,t'\to t),\ 0<|v_X|<\infty, \qquad (10.6.25)$$

yields

$$\frac{\partial}{\partial X}|_t = v_X|_t \frac{\partial}{\partial v}|_t,$$

$$\frac{\partial}{\partial t}|_X = v_t|_X \frac{\partial}{\partial v}|_t + \frac{\partial}{\partial t}|_v, \qquad (10.6.26)$$

or

$$\frac{\partial}{\partial v}|_t = \frac{1}{v_X|_t}\frac{\partial}{\partial X}|_t,$$

$$\frac{\partial}{\partial t}|_v = -\frac{v_t|_X}{v_X|_t}\frac{\partial}{\partial X}|_t + \frac{\partial}{\partial t}|_X. \qquad (10.6.27)$$

From (10.6.24) it results

$$\cot\omega = -\frac{a}{2v_X}a\frac{1+\sigma^2-2\sigma X\sigma_X-(1+X^2)(\sigma_t v_X - v_t\sigma_X)}{(1+\sigma^2+X^2)^{3/2}}. \qquad (10.6.28)$$

The Gauss equations $r=r(t,v)$ of the surface Σ becomes

$$r_{tt} = z_{tt}e_3,\ r_{tv} = z_{tv}e_3 = -\sigma_v e_3 = X_t e_3,$$

$$r_{vv} = z_{vv}e_3 = X_v e_3, \qquad (10.6.29)$$

with the compatibility condition

$$-\sigma_v = X_t. \qquad (10.6.30)$$

This is a restatement of (10.6.2) with $\rho_0 = 1$, by using

$$\sigma_X|_t = v_X|_t\,\sigma_v|_t,$$

$$X_t|_v = -\frac{v_t|_X}{v_X|_t}, \qquad (10.6.31)$$

in the condition $0<|v_X|<\infty$. The equation (10.6.1) is equivalent to the area preserving map expressed in terms of v and t independent variables

$$\left|\frac{\partial(X,\varepsilon)}{\partial(v,t)}\right| = 1. \tag{10.6.32}$$

The expression of the Gaussian curvature (1.7.9b) becomes

$$-\sigma_t X_v + \sigma_v X_t = \mathrm{K}(1+\sigma^2+X^2)^2, \tag{10.6.33}$$

and then

$$\left|\frac{\partial(X,\varepsilon)}{\partial(v,t)}\right| = -\mathrm{K}(1+\sigma^2+X^2)^2 \frac{\partial\sigma}{\partial\varepsilon}\Big|_X^{-1}. \tag{10.6.34}$$

If r is the position of a surface Σ with Gaussian curvature $\mathrm{K} = -\frac{1}{a^2}$, then (1.7.14) or (1.7.15) hold. So, the construction of such pseudospherical surfaces yield Bäcklund transformation (1.7.16) for the sine-Gordon equation.

In terms of the physical variables, the Bäcklund transformation becomes

$$t - t' = \frac{a\cos\zeta(X-X')}{1+\sigma\sigma'+XX'},\quad v - v' = \frac{a\cos\zeta(\sigma-\sigma')}{1+\sigma\sigma'+XX'},$$

$$z - z' = \frac{a\cos\zeta(\sigma' X - \sigma X')}{1+\sigma\sigma'+XX'} = -\sigma(t'-t) + X(v'-v),$$

$$\frac{1+\sigma\sigma'+XX'}{\sqrt{1+\sigma^2+X^2}\sqrt{1+\sigma'^2+X'^2}} = \cos\zeta, \tag{10.6.35}$$

where

$$\mathrm{d}z = -\sigma\mathrm{d}t + X\mathrm{d}v,\ \ \mathrm{d}z' = -\sigma'\mathrm{d}t' + X'\mathrm{d}v'. \tag{10.6.36}$$

The Monge–Ampère equation

$$z_{tt}z_{vv} - z_{tv}^2 = -\frac{1}{a^2}(1+z_t^2+z_v^2)^2, \tag{10.6.37}$$

determines the pseudosperical surface Σ, being invariant to the Bäcklund transformation (10.6.35). An important point of view is to interpret (10.6.35) in terms of the characteristic equations of (10.6.37) (Courant and Hilbert)

$$t_{\tilde\alpha} = \frac{aX_{\tilde\alpha}}{1+\sigma^2+X^2},\quad t_{\tilde\beta} = -\frac{aX_{\tilde\beta}}{1+\sigma^2+X^2},\quad v_{\tilde\alpha} = \frac{a\sigma_{\tilde\alpha}}{1+\sigma^2+X^2},$$

$$v_{\tilde\beta} = -\frac{a\sigma_{\tilde\beta}}{1+\sigma^2+X^2},\quad z_{\tilde\alpha} + \sigma t_{\tilde\alpha} - Xv_{\tilde\alpha} = 0. \tag{10.6.38}$$

If (t,v,z,X,σ) is a solution of (10.6.38), then a solution of the Monge–Ampère equation (10.6.37) is

$$z = z(\tilde\alpha(t,v),\tilde\beta(t,v)), \tag{10.6.39}$$

obtained from the condition of a nonzero Jacobian $\left|\dfrac{\partial(t,v)}{\partial(\tilde{\alpha},\tilde{\beta})}\right|$. Therefore, the Bäcklund transformation (10.6.35) represents an integrable discretization of the nonlinear characteristic equations (10.6.38) of the Monge–Ampère equation.

If we introduce the continuous variables $\hat{\alpha}$ and $\hat{\beta}$

$$\hat{\alpha} = \varepsilon_1 n_1 , \ \hat{\beta} = \varepsilon_2 n_2 , \tag{10.6.40}$$

for small ε_i, $i = 1,2$, we have the Taylor series

$$\omega_1 = \omega + \varepsilon_1 \omega_{\hat{\alpha}} + O(\varepsilon_1^2) , \ \omega_2 = \omega + \varepsilon_2 \omega_{\hat{\beta}} + O(\varepsilon_2^2) , \tag{10.6.41}$$

$$\omega_{12} = \omega + \varepsilon_1 \omega_{\hat{\alpha}} + \varepsilon_2 \omega_{\hat{\beta}} + \varepsilon_1 \varepsilon_2 \omega_{\hat{\alpha}\hat{\beta}} + O(\varepsilon_1^2, \varepsilon_2^2) .$$

For

$$\frac{\mu_1}{\mu_2} = \frac{\varepsilon_1 \varepsilon_2}{4\hat{a}^2} + O(\varepsilon_1^2, \varepsilon_2^2) , \tag{10.6.42}$$

the permutability theorem (1.7.23) reduces in the limit $\varepsilon_i \to 0$, to the sine-Gordon equation

$$\omega_{\hat{\alpha}\hat{\beta}} = \frac{1}{\hat{a}^2} \sin \omega .$$

In terms of the physical variables, the Bäcklund transformation yields the discrete equations

$$t_1 - t = \frac{a \cos \zeta_1 (X_1 - X)}{1 + \sigma\sigma_1 + X X_1} ,$$

$$t_2 - t = \frac{a \cos \zeta_2 (X_2 - X)}{1 + \sigma\sigma_2 + X X_2} ,$$

$$v_1 - v = \frac{a \cos \zeta_1 (\sigma_1 - \sigma)}{1 + \sigma\sigma_1 + X X_1} ,$$

$$v_2 - v = \frac{a \cos \zeta_2 (\sigma_2 - \sigma)}{1 + \sigma\sigma_2 + X X_2} ,$$

$$z_1 - z == -\sigma(t_1 - t) + X(v_1 - v) ,$$

$$z_2 - z == -\sigma(t_2 - t) + X(v_2 - v) , \tag{10.6.43}$$

and

$$\frac{1+\sigma\sigma_1+XX_1}{\sqrt{1+\sigma^2+X^2}\sqrt{1+\sigma_1^2+X_1^2}}=\cos\zeta_1, \tag{10.6.44}$$

$$\frac{1+\sigma\sigma_2+XX_2}{\sqrt{1+\sigma^2+X^2}\sqrt{1+\sigma_2^2+X_2^2}}=\cos\zeta_2. \tag{10.6.45}$$

Using (10.6.42) we write

$$\zeta_1=\varepsilon_1\kappa_1+O(\varepsilon_1^2),\ \zeta_2=\pi+\varepsilon_2\kappa_2+O(\varepsilon_2^2), \tag{10.6.46}$$

with

$$\mu_i=\tan(\frac{1}{2}\zeta_i). \tag{10.6.47}$$

The discrete equations (10.6.43) become

$$t_{\hat\alpha}=\frac{aX_{\hat\alpha}}{1+\sigma^2+X^2},\ t_{\hat\beta}=-\frac{aX_{\hat\beta}}{1+\sigma^2+X^2},$$

$$v_{\hat\alpha}=\frac{a\sigma_{\hat\alpha}}{1+\sigma^2+X^2},\ v_{\hat\beta}=-\frac{a\sigma_{\hat\beta}}{1+\sigma^2+X^2}, \tag{10.6.48}$$

and

$$z_{\hat\alpha}+\sigma t_{\hat\alpha}-Xv_{\hat\alpha}=0,\ z_{\hat\beta}+\sigma t_{\hat\beta}-Xv_{\hat\beta}=0. \tag{10.6.49}$$

The equations (10.6.44) and (10.6.45) become

$$\frac{\sigma_{\hat\alpha}^2+X_{\hat\alpha}^2+(\sigma X_{\hat\alpha}-X\sigma_{\hat\alpha})^2}{(1+\sigma^2+X^2)^2}=\kappa_1^2(\hat\alpha),$$

$$\frac{\sigma_{\hat\beta}^2+X_{\hat\beta}^2+(\sigma X_{\hat\beta}-X\sigma_{\hat\beta})^2}{(1+\sigma^2+X^2)^2}=\kappa_2^2(\hat\beta). \tag{10.6.50}$$

Therefore, the equations (10.6.43)–(10.6.45) can be interpreted as the integrable discretization equations of the characteristic equations (10.6.38).

References

Ablowitz, M. J., Ramani, A. and Segur, H. (1980). A connection between nonlinear evolution equations and ordinary differential equations of P-type I, II, *J. Math. Phys.*, 21, 715–721, 1006–1015.

Ablowitz, M. J. and Segur, H. (1981). Solitons and the inverse scattering transforms, *SIAM*, Philadelphia.

Ablowitz, M. J. and Clarkson, P. A. (1991). *Solitons, nonlinear evolution equations and inverse scattering*, Cambridge Univ. Press, Cambridge.

Abramowitz, M. and Stegun, I. A. (eds.) (1984). *Handbook of mathematical functions*, U. S. Dept. of Commerce.

Achenbach, J. D. (1973). *Wave propagation in elastic solids*, North-Holland Publishing Company, Amsterdam.

Agranovich, Z. S. and Marchenko, V.A. (1964). *The inverse problem of scattering theory*, Gordon and Breach, New York.

Alippi, A., Crăciun, F., Molinari, E. (1988). Stopband edges in the dispersion curves of Lamb waves propagating in piezoelectric periodical structures, *Appl. Phys. Lett.* 53, 19.

Alippi, A. (1982). Nonlinear acoustic propagation in piezoelectric crystals, *Ferroelectrics*, 42, 109–116.

Antman, S. S. (1974). Kirchhoff's problem for nonlinearity elastic rods, *Quart. of Appl. Math.*, 32.

Antman, S. S. and Tai-Ping-Liu (1979). Travelling waves in hyperelastic rods, *Quart. of Appl. Math.*, 36, 4.

Ariman, T., Turk, M. A. and Sylvester, N. D. (1974). On time-dependent blood flow, *Letters in Appl. and Engn. Sci.*, 2, 21–36.

Arnold, V. I. (1976). *Méthodes mathématiques de la mécanique classique*. Mir, Moscow.

Arts, T., Veenstra, P. C. and Reneman, R. S. (1982). Epicardial deformation and left ventricular wall mechanics during ejection in the dog, *Am. J. Physiol.*, 243, H379–H390.

Baker, G. L. and Gollub, J. (1992). *Chaotic dynamics. An introduction*, Cambridge Univ. Press.

Bâlă, N. (1999). Symmetry groups and Lagrangians associated to Tzitzeica surfaces, *Report* DG/9910138.

Balanis, G. N. (1972). The plasma inverse problem, *J. Math. Phys.*, 13, 1001–1005.

Bardinet, E., Cohen, L. and Ayache, N. (1994). Fitting of iso-surfaces using superquadrics and free-form deformations, *Proceed. IEEE Workshop on Biomedical Image Analysis*, Seattle, Washington.

Bardinet, E., Cohen, L. D. and Ayache, N. (1995). Superquadrics and free-form deformations, a global model and track 3D medical data, *Proc. Conf. on Computer Vision, Virtual Reality and Robotics in Medicine*, Nice, France.

Beju, I., Soós, E. and Teodorescu, P. P. (1976). *Tehnici de calcul vectorial cu aplicații* (*Vector calculus with applications*), Editura Tehnică, Bucharest.

Blaschke, W. (1923). *Vorlesungen über Differentialgeometrie und geometrische Grundlagen von Einsteins Relativitätstheorie, II: Affine Differentialgeometrie*, Grundlehren der math. Wissenschaften 7, Springer, Berlin.

Bolotin, V. V. (1963). *Nonconservative problems of the theory of elastic stability*, Macmillan Comp, NY, 274–312.

Born, M. and Huang, K. (1955). *Dynamical theory of crystal lattices*, Oxford at the Clarendon Press.

Bovendeerd, P. H. M., Arts, T., Huyghe, J. M., Van Campen, D. H. and Reneman, R. S. (1992). Dependence of local ventricular wall mechanics on myocardial fiber orientation. A model study, *J. Biomechanics*, 25, 19, 1129–1140.

Brulin, O. (1982). Linear Micropolar Media, in *Mechanics of Micropolar Media* edited by O. Brulin and R. K. T. Hsieh, World Scientific, 87–145.

Camenschi, G. and Şandru, N. (2003). *Modele matematice în prelucrarea metalelor* (*Mathematical models in metal forming*), Editura Tehnică, Bucharest.

Camenschi, G. (2000). *Introducere în mecanica mediilor continue deformabile* (*Introduction in continuum deformable media*), Ed. Univ. of Bucharest.

Caro, C.G., Pedley, T.J., Schroter, R.C. and Seed, W.A. (1978). *The mechanics of the circulation*, Oxford Univ. Press.

Carroll, R. W. (1991). *Topics in Soliton Theory*, North Holland.

Chiroiu, V. and Chiroiu, C. (2003a). *Probleme inverse în Mecanică* (*Inverse Problems in Mechanics*), Publishing House of the Romanian Academy, Bucharest.

Chiroiu, V., Munteanu, L., Chiroiu, C. and Donescu, Şt. (2003b). The motion of a micropolar fluid in inclined open channels, *Proc. of the Romanian Acad., Series A: Mathematics, Physics, Technical Sciences, Information Science*, 4, 2, 129–135.

Chiroiu, V., Chiroiu, C., Rugină, C., Delsanto, P.P. and Scalerandi, M. (2001a). Propagation of ultrasonic waves in nonlinear multilayered media, *J. of Comp. Acoustics*, 9, 4, 1633–1646.

Chiroiu, C., Delsanto, P. P., Scalerandi, M., Chiroiu, V. and Sireteanu, T. (2001b). Subharmonic generation in piezoelectrics with Cantor-like structure, *J. of Physics D: Applied Physics*, 34, 3, 1579–1586.

Chiroiu, V., Chiroiu, C., Badea, T. and Ruffino, E. (2000). A nonlinear system with essential energy influx: the human cardiovascular system, *Proc. of Romanian Acad., Series A: Mathematics, Physics, Technical Sciences, Information Science,* 1, 1, 41–45.

Chiroiu, C., Munteanu, L., Chiroiu, V., Delsanto, P. P., Scalerandi, M. (2000). A genetic algorithm for determining the elastic constants of a monoclinic crystal, *Inverse Problems, Inst. Phys. Publ.*, 16, 121–132.

Christensen, R. M. (1982). *Theory of Viscoelasticity: an Introduction*, Academic Press, New York.

Cole, J. D. (1951). On a quasilinear parabolic equation occurring in aerodynamics, *Quart. Appl. Math.*, 9, 225–236.

Collatz, L. (1966). *The numerical treatment of differential equations*, Berlin, Springer Verlag.

Coley, A., Levi, D., Milson, R., Rogers, C. and Winternitz, P. (eds) (1999). *Bäcklund and Darboux Transformations. The Geometry of Solitons*, Centre de Recherches Mathematiques CRM Proceed. and Lecture Notes, 29.

Courant, R. and Hilbert, D. (1962). *Methods of Mathematical Physics*, II, Interscience, New York.

Costa, K. D., May-Newman, K. D., Farr, D., Otder, W. G., McCulloch, A. D. and Omens, J. H. (1997). Three-dimensional residual strain in mid anterior canine left ventricle. *Am. J. Physiol.*, 273, H1968–H1976.

Cosserat, E. and F. (1909). *Theorie des corps deformables*, Hermann et Fils, Paris.

Crăciun, F., Bettucci, A., Molinari, E., Petri, A. and Alippi, A. (1992). Direct experimental observation of fracton mode patterns in one-dimensional Cantor composites, *Phys. Rev. Lett.*, 68.

Crăşmăreanu, M. (2002). Cylindrical Tzitzeica curves imply forced harmonic oscillators, *Balkan J. of Geometry and Its Applications,* 7, 37–42.Data, E. and Tanaka, M. (1976). Periodic multi-soliton solutions of Korteweg–de Vries equation and Toda lattice, *Supplement of the Progress of Theoretical Physics 59*, 107–125.

Darboux, G. (1887–1896). *Leçons sur la théorie générale des surfaces*, vol. I–IV, Paris.

Davies, M. A. and Moon, F. C. (2001). Solitons, chaos and modal interactions in periodic structures, In *Dynamics with Friction: Modeling, Analysis and Experiment*, part II, Series of Stability, Vibration and Control of Systems, World Scientific Publishing Company, 7, 99–123.

Delsanto, P. P., Provenzano, V., Uberall, H. (1992). Coherency strain effects in metallic bilayers, *J. Phys.: Condens. Matter*, 4, 3915–3928.

Demer L. L. and Yin F. C. P. (1983). Passive biaxial mechanical properties of isolated canine myocardium, *J. Physiol. Lond.*, 339, 615–630.

Dodd, R. K., Eilbeck, J. C., Gibbon, J. D. and Morris, H. C. (1982). *Solitons and Nonlinear Wave Equations*, Academic Press, London, NewYork.

Donescu, Şt. (2003). The existence of the Coulomb vibrations in a nonlinear pendulum *AMSE: Modelling, Measurement and Control, Series B: Mechanics and Thermics*, 72, 6, 29–38.

Donescu, Şt., Chiroiu, V., Badea, T., Munteanu, L. (2003). The hysteretic behavior of nonlinear mesoscopic materials, *AMSE: Modelling, Measurement and Control, Series B: Mechanics and Thermics*, 72, 6, 55-64.

Donescu, Şt. (2000). Nonlinear vibrations in mechanical systems with two degrees of freedom, *Rev. Roum. Sci. Techn. - Méc. Appl.*, 45, 3, 295–308.

Drazin, G. and Johnson, R. S. (1989). *Solitons. An introduction*, Cambridge Univ. Press.

Drazin, P. G. (1983). *Solitons*, Cambridge Univ. Press.

Dressler, R.F. (1949). Mathematical solution of the problem of roll-waves in inclined open channels, *Comm. on Pure and Appl. Math.*, II, 2-3, 149–194.

Dubrovin B. A., Matveev V. B., Novikov S. P. (1976). Nonlinear equations of Korteweg–de Vries type, finite zone linear operators, and abelian varieties, *Russ. Math. Surv.* 31, 1, 59–146.

Duncan, D. B., Eibeck, J. C., Feddersen, H. and Wattis, J.A.D. (1993). *Solitons on lattices*, Physica D, 68, 1-11.

Easwaran, C. V. and Majumdar, R. A. (1990). A uniqueness theorem for incompressible micropolar flows, *Quart. of Appl. Math.*, 48, 2.

Engelbrecht, J. (1991). *An introduction to asymmetric solitary waves*, Longman Scientific & Technical, John Wiley & Sons, New York.

Engelbrecht, J. and Khamidullin, Y. (1988). On the possible amplification of nonlinear seismic waves, *Phys. Earth. Plan. Int.*, 50, 39–45.

Eringen, A. C. (1966). Linear Theory of Micropolar Elasticity, *J. Math. Mech.*, 15, 909.

Eringen, A. C. (1970). Mechanics of Micropolar Continua, in: *Contributions to Mechanics* (ed. David Abir), Pergamon Press, 23–40.

Faddeev, L. D. (1958). On the relation between the *s*-matrix and potential for the one-dimensional Schrödinger operation, *Sov. Phys. Dokl.*, 3, 747–751.

Faddeev, L. D. (1967). Properties of the *s*-matrix of the one-dimensional Schrödinger equation, *Am. Math. Soc. Transl. Ser.* 2, 65, 139–166.

Fermi, E., Pasta, J. R. and Ulam, S. M. (1955). Studies of nonlinear problems, *Technical Report LA-1940*, Los Alamos Sci. Lab.

Fermi, E., Pasta, J. R. and Ulam, S. M. (1965). In *Collected Papers of Enrico Fermi*, vol. 2, E. Fermi, The Univ. of Chicago Press, Chicago.

Fliess, M. (1981). Fonctionnelles causales et indéterminées non-commutatives, *Bull. Soc. Math. de France*, 109, 3–40.

Frederiksen, P. S. (1994). Estimation of elastic moduli in thick composite plates by inversion of vibrational data, *Inverse Problems in Engineering Mechanics*, Bui, Tanaka et al. (eds.), Balkema, Rotterdam, 111–118.

Freeman, N. C. (1984). Soliton solutions of non-linear evolution equations, *IMA Journal of Applied Mathematics*, 32, 125–145.

Fung Y. C. (1981). *Biomechanics, Mechanical properties of living tissues*, Springer-Verlag, New York.

Gardner, C.S. (1971). Korteweg-de Vries equation and generalization IV. The Korteweg-de Vries equation as a Hamiltonian system, *J. Math. Phys.*, 12, 1548–1551.

Gardner, C. S. and Morikawa, G. K. (1960). *Courant Inst. Math. Sci. Rep.* (NYO-9082).

Gardner, C.S., Greene, J.M., Kruskal, M.D. and Miura, R.M. (1974). Korteweg-de Vries equation and generalization. VI. Methods of exact solutions, *Comm. Pure Appl. Math.*, 27, 97–133.

Gauthier, R. D. (1982). *Experimental investigations on micropolar media,* 395–463, in: Mechanics of Micropolar Media, CISM Courses and lectures, edited by O. Brulin and R.K.T. Hsieh, World scientific.

Gelfand, I. M. and Dorfman, I. Y. (1979). Hamiltonian operators and algebraic structures related to them, *Functional Analysis and its Applications*, 13, 4, 13–30.

Gheorghiu, G.T. (1969). O clasă particulară de spaţii cu conexiune afină (A particular class of affine spaces), St. Cerc. Mat., 21, 8, 1157–1168.

Gleick, J. (1989). *Chaos-making: a new science*, Heinemann-London.

Goldberg, D. E. (1989). *Genetic algorithms in search, optimization, and machine learning*, Massachusetts: Addison-Wesley Publishing Co.

Gromak, V. I. (2001). Bäcklund transformations of the higher order Painlevé equations, *Centre de Recherches Mathematiques CRM Proc. and Lecture Notes,* 29, 3–28.

Guccione J. M., McCulloch A. D. and Waldman L. K. (1991). Passive material properties of intact ventricular myocardium determined from a cylindrical model, *ASME J. Biomech. Engn.* 113, 42–55.

Guccione J. M. and McCulloch A. D., *Finite element modeling of ventricular mechanics*, in Glass L., Hunter P.J., McCulloch A. D. (eds.) (1991). *Theory of heart, biomechanics, biophysics and nonlinear dynamics of cardiac function,* New York, Springer-Verlag, 121–144.

Guckenheimer, J. and Holmes, P. (1983). *Nonlinear oscillations, dynamical systems, and bifurcations of vector fields*, Springer-Verlag, NY.

Halanay, A. (1963). *Teoria calitativă a ecuaţiilor diferenţiale* (*The qualitative theory of differential equations*), Editura Academiei, Bucharest.

Hashizume, Y. (1988). Nonlinear pressure wave propagation in arteries, *J. of the Phys. Soc. of Japan*, 57,12.

Helmholtz, H. J. (1887). *Reine Angew, Math.*, Berlin, 100, 137.

Hirota, R. (1980). *Solitons*, (eds. R. K. Bullough and P. J. Caudrey), Springer, Berlin, 157.

Hirota, R. and Satsuma, J. (1981). Soliton solutions of a coupled Korteweg–de Vries equation, *Physics Letters* A, 85, 407–408.

Hoenselaers, C.A. and Micciché (2001). Transcendental solutions of the sine-Gordon equation, *Centre de Recherches Mathematiques CRM Proc. and Lecture Notes,* 29, 261–271.

Holmes, J. W., Nunez, J. A. and Covell, J. W. (1997). Functional implications of myocardial scar structure. *Am. J. Physiol.*, 272, H2123–30.

Hopf, E. (1950). The partial differential equation $u_t + uu_x = \mu u_{xx}$, *Comm. Pure Appl. Math.*, 3, 201-230.

Humphrey, J. D., Strumpf, R. K. and Yin, F. C. P. (1990). Determination of a constitutive relation for passive myocardium, I. A New functional form. *J. Biomech. Engn.* 112, 333–339.

Huyghe, J. M., Arts, T., Van Campen, D. H. and Reneman, R. S. (1992). Porous medium finite element model of the beating left ventricle, *Am. J. Physiol*, 262, H1256–H1267.

Huyghe, J. M., Van Campen, D. H., Arts, T. and Heethaar, R. M. (1991). A two-phase finite element model of the diastolic left ventricle, *J. Biomech.*, 24, 7, 527–536.

Huyghe, J.M., Van Campen, D. H., Arts, T. and Heethaar, R. M. (1991). The constitutive behavior of passive heart muscle tissue. A quasi-linear viscoelastic formulation. *J. Biomech.*, 4, 841–849.

Huyghe, J. M. (1986). *Nonlinear finite element models of the beating left ventricle and the intramyocardial coronary circulation*, PhD thesis, Eindhoven University of Technology, The Netherlands.

Jankowski, A. F. and Tsakalakos, T. (1985). The effect of strain on the elastic constants of noble metals, J. *Phys. F. Met. Phys.*, 15, 1279–1292.

Kamel, A. A. (1970). Perturbation method in the theory of nonlinear oscillations, *Celestial Mechanics*, 3, 90–106.

Kapelewski, J. and Michalec, J. (1991). Soliton-like SAW in nonlinear isotropic piezoelectrics, *Int. J. Engn. Sci.*, 29, 3, 285–291.

Kato, T. (1961). *Perturbation theory for linear operators*, Springer, New York.

Kawahara, T. and Takaoka, M. (1988). Chaotic motions in an Oscillatory Soliton Lattice, *J. of the Phys. Soc. of Japan*, 57, 11, 3714–3732.

Konno, K. and Ito, H. (1987). Nonlinear interaction between solitons in complex *t*-plane. I, *J. of the Phys. Soc. of Japan,* 56, 3, 897–904.

Korteweg, D. J. and de Vries, G. (1895). On the change of form of long waves advancing in a rectangular canal, and on a new type of long stationary waves, *Phil. Mag.*, 39, 422–443.

Kozák, J. and Šileny, J. (1985). Seismic events with non-shear component: I. Shallow earthquakes with a possible tensile source component, *PAGEOPH*, 123, 1–15.

Kranys, M. (1989). Causal theories of evolution and wave propagation in mathematical physics, *AMR*, 42, 11, 305–322.

Krishnan, E. V. (1986). On the Ito-Type Coupled Nonlinear Wave Equation, *J. of the Phys. Soc. of Japan*, 55, 11, 3753–3755.

Krumhansl, J. A. (1991). Unity in the Science of Physics, *Physics Today*, 33.

Kruskal, M. D. (1978). The birth of the soliton, in Calogero, F. (ed.). *Nonlinear evolution equations solvable by the spectral transform*, Research Notes in Mathematics, 26, Pitman, 1–8.

Lakes, R. S. (2001). Elastic and viscoelastic behaviour of chiral materials, *Int. J. of Mech. Sci.*, 43, 1579–1589.

Lamb, G. L. Jr. (1980). *Elements of soliton theory*, John Wiley & Sons, New York.

Lambert, J. W. (1858). *J. Franklin Inst.*, 266, 83.

Landau, L. D. and Lifshitz, E. M. (1968). *Theory of Elasticity*, Moscow.

Lax, P. (1968). Integrals of nonlinear equations of evolution and solitary waves, *Comm. Pure Appl. Math.* 21.

Lax, P. and Phillips, R. (1967). *Scattering Theory*, Academic Press, New York.

Lewis J. T. (1980). The heterogeneous string: coupled helices in Hilbert space, *Quart. of Appl. Math.*, XXXVIII, 4, 461–467.

Ling, C.B. (1981). On a second solution of the Van der Pol's differential equation, *Quart. of Appl. Math.*, XXXVIII, 4, 511–514.

Love, A. E. H. (1926). *A treatise on the mathematical theory of elasticity*, 4th ed., Dover, New York.

Lyapunov, M.A. (1907). *Ann. Fac. Sci.Univ. Toulouse*, 9, 203.

Magnus W., Oberhettinger R., and Soni, P. (1966). *Formulas and Theorems for the Special Functions of Mathematical Physics*, Springer, New York.

Magri, F. (1978). A simple model of integrable Hamiltonian equation, *J. of Math. Phys.*, 19, 5, 1156–1162.

Matsukawa, M. Watanabe, S. and Tanaka, H. (1988). Stability analysis of a soliton by the Hirota method, *J. of the Phys. Soc. of Japan,* 57, 12, 4097–4100.

McDonald, D. A. (1974). *Blood flow in arteries*, Edward Arnold, London, 2nd ed.

Miura, R. M. (1976). *The Korteweg–de Vries equation. A survey of results*, SIAM Review, 18, 412–459.

Miura, R. M., Gardner, C. S. and Kruskal, M. D. (1968). Korteweg–de Vries equation and generalizations. II. Existence of conservation laws and constants of motion, *J. of Math. Phys.*, 9, 8, 1204–1209.

Moodie, T. B. and Swaters, G. E. (1989). Nonlinear waves and shock calculations for hyperelastic fluid-filled tubes, *Quart. of Appl. Math.*, 47, 4.

Munteanu, L. (2003). *Statics and dynamics of the thin elastic rod*, ch. 10, Topics in applied mechanics (eds. V. Chiroiu and T. Sireteanu), Publishing House of the Romanian Academy, Bucharest, 267–300.

Munteanu, L. and Donescu, Şt. (2002). *Introducere în teoria solitonilor. Aplicaţii in Mecanică* (*Introduction to the Soliton Theory. Applications to Mechanics*), Publishing House of the Romanian Academy, Bucharest.

Munteanu, L., Chiroiu, C. and Chiroiu, V. (2002a). Nonlinear dynamics of the left ventricle, *Physiological Measurement*, 23, 417–435.

Munteanu, L., Badea, T. and Chiroiu, V. (2002b). Linear equivalence method for the analysis of the double pendulum's, *Complexity Int. J.*, 9, 26–43.

Munteanu, L., Chiroiu, V. and Scalerandi, M. (2002c). A genetic algorithm for modeling the constitutive laws for the left ventricle, *Proc. of the Romanian Academy, Series A: Mathematics, Physics, Technical Sciences*, Information Science, 3, 1–2, 25–30.

Munteanu, L., Chiroiu, G. and Chiroiu, V. (1998). Transient flow of blood in arteries, *Rev. Roum. Sci. Techn.-Méc. Appl.*, 43, 3, 383–391.

Munteanu, L. and Boştină, M. (1995). Computational strategy in dynamics of solitons at the focusing, *AMSE*: *Fuzzy Systems & Neural Networks*, 2, 52–63.

Musette, M., Conte, R. and Verhoeven, C. (2001). *Bäcklund transformation and nonlinear superposition formula of the Kaup-Kupershmidt and Tzitzeica equations*, *Centre de Recherches Mathematiques CRM Proc. and Lecture Notes*, 29, 345–362.

Nayfeh, A. (1973). *Perturbation methods*, Wiley, NY.

Nettel, S. (1992). *Wave Physics – Oscillations – Solitons – Chaos*, Springer-Verlag.

Newell, A. C. (1985). *Solitons in mathematics and physics*, SIAM, Philadelphia.

Noether, E. (1918). Invariante variansprobleme, *Nachr. v.d. Ges. d. Wiss. zu Göttingen, Math.-Phys.*, 2, 234–257.

Olver, P. J. (1977). Evolution equations possessing infinite symmetries, *J. of Math. Phys.*, 18, 6, 1212–1215.

Osborne, A. R. (1995). Soliton physics and the periodic inverse scattering transform, *Physica* D, 86, 81–89.

Painlevé, P. (1973). *Leçons sur la théorie analytique des équations differentielles. Leçons de Stockholm*, 1895, Hermann, Paris 1897. Reprinted in Oeuvres de Paul Painlevé, vol. 1, editions du CNRS, Paris.

Perring, J. K. and Skyrme, T. H. R (1962). A model unified field equation, *Nucl. Phys.*, 31, 550–555.

Petrov, V. A. (1983). Dilaton model of thermal fluctuation crack nucleation, *Soviet Phys.-Solid State*, 25, 1800–1802.

Poincaré, H. (1892). *Les methodes nouvelles de la mechanique celeste*, Gauthier-Villars, Paris.

Popescu, H. and Chiroiu, V. (1981). *Calculul structurilor optimale* (*The design of optimal structures*), Publishing House of the Romanian Academy, Bucharest.

Převorovská, S., Marsik, F. and Musil, J. (1996). Pressure generation by chemical reaction in the human cardiovascular system, *Grant Agency of the Czech Republic*, no. 101/96/0151 and nr. 101/95/0352.

Rogacheva, N. (1994). *The Theory of Piezoelectric Shells and Plates*, CRC Press, Boca Raton.

Rogers, C., and Schief, W. K. (1997). The classical Bäcklund transformation and integrable discretisation of characteristic equations, *Phys. Letters A*, 232, 217–223.

Rogers, C., and Schief, W. K (2002). *Bäcklund and Darboux transformations. Geometry and modern appli--cations in soliton theory*, Cambridge Univ. Press.

Rudinger, G. (1966). Review of Current Mathematical Methods for the Analysis of Blood Flow, In "Biomedical Fluid Mechanics Symposium", *ASME*, New York.

Russell, Scott J. (1844). *Report on waves*, British Association Reports.

Sadovski, M. A. and Nikolaev, A. V. (1982). New methods of seismic prospecting, perspectives of development, *Vestnik Akad. Nauk SSSR*, 1, 57–64.

Santilli, R. M. (1978). *Foundations of Theoretical Mechanics* I: *The inverse problem in Newtonian mechanics*, Springer Verlag, Berlin.

Santilli, R. M. (1983). *Foundations of Theoretical Mechanics* II: *Birkhoffian Generalizations of Hamilton Mechanics*, Springer Verlag, Berlin.

Satsuma, J. (1987). Explicit solutions of nonlinear equations with density-dependent diffusion, *J. of the Phys. Soc. of Japan*, 947–1950.

Seymour, B. R. and Eric Varley E. (1982). Exact solutions describing soliton-like interactions in a nondispersive medium, *SIAM J. Appl. Math.*, 42, 4, 804–821.

Skalak, R., (1966). Wave Propagation in Blood Flow, in "Biomechanics", ed. Y.C. Fung, *ASME*, New York.

Scalerandi, M., Delsanto, P. P., Chiroiu, C. and Chiroiu, V. (1999). Numerical simulation of pulse propagation in nonlinear 1-D media, *J. of the Acous. Soc. of America*, 106, 2424–2430.

Shen, Z. (1967). A solution of Van der Pol's differential equation, *Quart. of Appl. Math.*, XXV, 299–301.

Shinbrot, M. (1981). The solitary wave with surface tension, *Quart. of Appl. Math.*, XXXIX, 2, 287–291.

Şoós, E. and Teodosiu, C. (1983). *Calcul tensorial cu aplicaţii în mecanica solidelor* (*Tensorial calculus with applications to solid mechanics*), Editura Stiinţ. şi Encicloped., Bucharest.

Şoós, E. (1974). *Modele discrete şi continue ale solidelor* (*Discrete and continuum models of solids*), Editura Stiinţ., Bucharest.

Solomon, L. (1968). *Elasticité Linéaire*, Masson.

Synge J. L. (1981). On the vibrations of a heterogeneous string, *Quart. of App. Math.*, XXXIX, 2, 292–297.

Taber, L. A. (1995). *Biomechanics of growth, remodelling, and morphogenesis*, Appl. Mech. Rev., 48, 8, 487–545.

Tabor, M. (1989). *Chaos and integrability in nonlinear dynamics*, John Wiley & Sons, New York.

Tanaka, M. and Nakamura, M. (1994). *Application of genetic algorithm to plural defects identification*, Inverse Problems in Engineering Mechanics. Bui, Tanaka et al. (eds.), Balkema, Rotterdam, 377–382.

Teodorescu, P. P. (1984–2002). *Sisteme mecanice*: *modele clasice* (*Mechanical systems*: *classical models*), I-IV, Editura Tehnică, Bucharest.

Teodorescu, P. P. and Nicorovici-Porumbaru, N (1985). *Aplicații ale teoriei grupurilor în mecanică și fizică* (*Applications of the theory of groups in mechanics and physics*). Editura Tehnică, Bucharest.

Teodosiu, C. (1982). *Elastic models of crystal defects*, Editura Academiei, Springer-Verlag.

Toda, M. (1989). *Theory of Nonlinear Lattices*, Springer-Verlag, Berlin-Heidelberg.

Toda, M. and Wadati, M. (1973). A soliton and two solitons in an exponential lattice and related equations, *J. of the Phys. Soc. of Japan*, 34, 18–25.

Toma, I. (1995). *Metoda echivalenței lineare și aplicațiile ei* (*The linear equivalence method and its applications*), Editura Flores, Bucharest.

Tricomi, F. G. (1953). *Equazioni differenziali*, Edizioni Scientifiche Einaudi.

Truesdell, C. (1962). Mechanical basis of diffusion, *J. Chem. Phys*., 37, 2336.

Truesdell, C. and Toupin, R. (1960). *The Classical Field Theories*, in S. Flugge (ed.) Enc. of Physics, Springer, Berlin et al. III/1.

Tsuru, H. (1986). Nonlinear dynamics for thin elastic rod, *J. of the Phys. Soc. of Japan*, 55, 7, 2177–2182.

Tsuru, H. (1987). *Equilibrium Shapes and Vibrations of Thin Elastic Rod*, Journal of the Physical Society of Japan, 56, 7, 2309–2324.

Tzitzeica, G. (1924). *Géométrie projective différentielle des réseaux*, Editura Cultura Natională, Bucharest and Gauthier-Villars, Paris.

Tzitzeica, G. (1910). Sur une nouvelle classe des surfaces, *C. R. Acad. Sci.*, Paris 150, 955–956.

Vâlcovici, V., Bălan, Șt. and Voinea, R. (1963). *Mecanica teoretică* (*Theoretical Mechanics*), Editura Tehnică, Bucharest.

Van Campen, D. H., Huyghe, J. M, Bovendeerd, P, H. M. and Arts, T. (1994). Biomechanics of the heart muscle, *Eur. J. Mech., A/Solids*, 13, 4- suppl., 19–41.

Wahlquist, H. D.and Estabrook, F. B. (1973). Bäcklund transformation for solutions of the KdV equation, *Phys. Rev. Lett*., 31, 1386–1390.

Wang, J. P. (1998). *Symmetries and conservation laws of evolution equations*, PhD thesis, Vrije Universiteit, Amsterdam.

Weiss, J., Tabor, M. and Carnevale, G. (1983). The Painlévé property for partial diferential equations, *J. Math. Phys.* 24, 522–526.

Whitham, G. B. (1974). *Linear and Nonlinear Waves*, Wiley, New York

Whitham, G. B. (1984). Comments on periodic waves and solitons, *IMA J. of Appl. Math*., 32, 353–366.

Yang, J. S. (1995). Variational formulations for the vibration of a piezoelectric body, *Quart. of Appl. Math*., LIII, 1.

Yih, C. S. (1994). Intermodal interaction of internal solitary waves, *Quart. of Appl. Math*., LII, 4, 753–758.

Yomosa, S. (1987). Solitary waves in large blood vessels, *J. of the Phys. Soc. of Japan*, 56, 2, 506–520.

Yoneyama, T. (1986). Interacting Toda equations, *J. of the Phys. Soc. of Japan*, 55, 3, 753–761.

Zabusky, N. J. and Kruskal, M. D. (1965). Interaction of solitons in a collisionless plasma and the recurrence of initial states, *Phys. Rev. Lett*., 15, 240–243.

Zachmann, D. W. (1979). Nonlinear analysis of a twisted axially-loaded elastic rod, *Quart. of Appl. Math*., 36, 1.

Zamarreno, G. (1992). The multidimensional inverse Lagrange problem, *Eur. J. Mech., A/Solids*, 11, 1, 49–64.

Zhurkov, S. N. (1983). Dilaton mechanism of the strength of solids, *Soviet Phys.-Solid State*, 25, 1797–1800.

Index

Active constitutive laws 203, 209
Active media 59, 70
Active stress tensor 199, 201, 204, 208, 209
Affine spheres 273
  centroaffinity 274
Amplification 59, 61, 214, 217
Anisentropic gas 291
Anisotropic material 202, 215, 244, 285, 286
Autonomous system 83

Bäcklund transformation 39, 40, 41, 42, 43, 44, 78, 111, 113, 273, 289, 290, 295, 296
  auto-Bäcklund transformation 40
Bilinear form 25, 30
Biological molecules 285
Biological populations 49
Blood coefficients 236
Blood flow 214, 220, 221, 235
  hematocrit 220, 236, 240, 241
  microgyration 220, 239, 240, 241
  peaking 221
  pulmonary artery 214, 220
  steepening 221, 228, 229, 230
  thoracic aorta 230, 231, 232, 233, 234, 238, 239
Bond number 271
Bones 285
Bound-state 9, 11, 185, 189
Boussinesq equation 77, 107, 108, 153
Breaking the waves 84
Breather solution 119
Brillouin zone 192
Burgers equation 56, 74, 86, 107, 110, 111

Cantor-like structure 242, 243, 244
  admittance curve 252
  Cantor set 243, 244
  fracton (localized vibration mode) 242, 243, 244, 253
  phonon (extended vibration mode) 243, 253
  piezoelectricity 244, 245
Cardiovascular system 198, 214, 215, 217, 218
  adenosine triphosphate (ATP) 214, 215
  cardiac cycle 205, 211, 212, 213, 215, 216, 217 218, 219
Cauchy theorem 15
Cauchy–Riemann relations 39
Causality principle 217
Cellular solids 285
Chezy turbulent resisting force 242, 261
Characteristics 54, 85, 110, 166, 167, 220, 258
Chirality 285
Chiral material (medium) 285, 286, 287, 290
Chiral Cosserat material 285
  noncentrosymmetry 286
Chiral elastic moduli 286
Christensen's constitutive law 285
Cosserat material 285, 286
Cnoidal
  functions 18, 19, 20, 139, 187
  method 17, 21, 23, 25, 185, 189, 190, 197, 209, 211, 247
  solutions 1, 17, 78, 137, 138, 173, 185, 190, 206, 285
  vibrations 173, 185, 187, 189, 211
  waves 18, 22, 23, 24, 25, 58, 146, 206, 242
Cole–Hopf transformation 110
Conservation laws 53, 62, 66, 74, 78, 108
Constitutive laws 59, 206, 293
  active 203
  passive 201
Continuity equation 87, 201, 216, 221, 256
Convergence criterion 250, 267
Coulomb wave functions 183, 185
Couple stress 242, 260, 262, 286, 287
Coupled pendulum 173, 175
Crystalline material 285

D'Alembert solution 54, 170, 172
Darboux transformation 39, 288, 290
Darcy's law 201
Dielectric constants 244, 245
Dielectric wires 113
Diffusion equation 75, 86, 110
  diffusion coefficient 110

diffusion system 76
reaction–diffusion 49
Dilaton 59, 61, 213
dilatonic mechanism 59
energy influx 198, 213, 214
Dirac function 12
Direct scattering problem 90
Dispersion 55, 56, 59, 83, 85, 253
relations 21, 55, 56, 150, 156, 157, 158, 193, 225
Dissipation 56, 59, 110, 154
equations 56, 58, 86, 113, 114
DNA 121
DRIP media 149, 159, 160, 163
Dynamic viscosity 261

Eigenfunction 2, 4, 96
Eigenvalue 2, 3, 35, 38, 96, 228, 248
problem 248, 249, 251, 252
Eigenfrequency 249, 251, 252
Elastica 121, 143, 145, 146, 152, 191
Elastic potential 174, 201, 246
Elastic constants 54, 55, 159, 244, 285, 286
third-order 244
Electric field 244, 245, 247, 249, 252
potential 244
induction vector 244
Elliptic cylindrical curve 284
Elliptic integrals 18, 19, 23, 118, 138, 139, 141, 143, 147, 148, 281
Elliptic Weierstrass function 20, 187, 273, 281
Epoxy resin 243, 244, 251, 253
Equipartition of energy 107
Error 24, 250
Euler angles 122, 123, 126, 128, 130, 131, 132, 139, 143, 147, 148, 149, 155
Euler function 91, 92, 93
Euler–Lagrange equations 53, 66, 69, 255
Eulerian coordinates 59, 122, 123, 198, 205, 291
Evolution equations 1, 25, 29, 43, 44, 53, 62, 73, 74, 75, 78, 198
Extended mode of vibration (phonon) 242, 244, 253, 255

Faddeev condition 5, 13
Fermi, Pasta and Ulam problem 82, 107, 159, 173, 191
Fitness function 207, 208, 250, 266, 267
Fliess expansions 34
Flow velocity 220, 221, 224, 228, 229, 234, 235, 238, 239, 240, 241
Flux 84, 216
Foams 160, 285
Focusing of waves 59
Forced oscillator 283, 284
Fourier transform 12, 14
Fractal 243
Fracton (localized mode of vibration) 242, 244, 252, 253, 255
Frequency 55, 60, 146, 157, 193, 216, 231, 234, 237, 242, 245, 249, 253
Froude number 271
Fundamental solutions 5, 7, 8, 10, 12, 17, 90, 95, 100, 101, 103

Gaussian curvature 41, 273, 291, 292, 294
Gauss equation 294
Gaussian function 57
Genetic algorithm 197, 204, 207, 208, 209, 251, 252, 266
Green–Lagrange strain tensor 199, 201
Group of symmetry 53, 62, 63, 64, 65, 66, 273, 274, 275, 276, 277
infinitesimal symmetries 65, 277, 283
Group velocity 55, 56, 150

Hamiltonian 67, 74, 78, 85, 108, 110
Hamilton principle 126
Heat equation 56, 110
Heaviside function 12
Helical inclusions 296
Helical rod 149, 155, 157, 158
Helmholtz conditions 69, 70, 72
Hemodynamics 220
Heterogeneous string 149, 159, 166, 168
Hirota method 25, 29
Hirota operator 25
Honeycomb cell 296
Hydrodynamic resistance 216
Hyperbolic cylindrical curve 284, 285
Hyperbolic functions 1, 18, 122, 143, 152
Hypergeometric series 90, 91

Interaction of waves 102, 163, 242, 258, 259
Internal energy 59, 248, 249, 286
Inverse Lagrange problem 69
Inverse scattering problem 2, 3, 12, 17, 30, 85, 94, 105, 173, 228
Inverse algorithm 207, 249, 250
Involution 288
Irrotational motion 86, 269
Isobars 291
Isotropic energy function 199, 201

Jacobian 295
Jacobi elliptic functions 1, 18, 19, 20, 21, 58, 81, 110, 173
Jacobi elliptic integrals 18
Jacobi matrix 21, 63
Jacobi multiplier 70
Jordan lemma 15

Kadomtsev and Petviashvili equation (K-P) 30
Kinetic energy 129, 130, 173
Kink 42, 51, 113, 152, 153, 154, 155
anti-kink 153
Klein–Gordon equation 56
Konno–Ito method 105, 153
Korteweg–de Vries equation 3, 5, 18, 21, 25, 30, 44, 53, 55, 58, 61, 75, 79, 84, 89, 94, 105, 107, 110, 153, 227

Krishnan solution 187

Lagrange equations 53, 66, 67, 69, 127, 131, 174
Lagrangian 66, 67, 68, 69, 70, 72, 130, 131
coordinates 134, 198, 205, 290
Green strain 202
multipliers 69, 70, 72
Lamé constants 55, 286
Lamb waves 245
Laplace equation 39
Lax pairs 85, 288, 289, 290
Laurent series 50, 51
Left ventricle 197, 198, 199, 205, 207, 209, 211, 214, 216, 220
cardiac cycle 205, 211, 212, 213, 215, 216, 218
endocardium 197, 200, 204
epicardium 197, 200, 204, 209
heart 220, 229, 230
intramyocardial pressure 211, 213
myocardial tissue 200, 201, 215
myocardial deformation 197
porosity 199
sarcomere length 200, 204, 208
super ellipsoid surface 198
Legendre equations 7
Legendre function 93, 97
Lewis method 168
LEM solutions 34, 35, 173, 177, 180, 183, 190
Lie algebra 65, 276, 281, 283
Lie derivative 33, 34
Lie transformation 80. 89
Linear equivalence method (LEM) 31, 32, 35, 37, 190
Liouville–Tzitzeica equation 76, 274, 277, 278, 281, 282, 288

Localized waves 1, 5, 9, 53, 78, 82, 83, 122, 173, 193
Localized mode of vibration (fracton) 242, 244, 252, 253, 255
Lorentz transformation 288

Marchenko equations 17, 98, 101, 104
Martin formulation 291, 292
Material symmetry 286
Maxwell stress tensor 245
McLaurin series 107.
Mean curvature 40, 293
Micropolar coefficients 260, 267, 286
Micropolar elastic moduli 286
Micropolar model 220, 235, 242, 259, 285
Microrotation 235, 242, 259, 260, 269, 286
Microinertia 236, 286
Miura transformation 85, 109
Modal interacting 191
intermodal interaction 242
Modified Korteweg- de Vries equation 74
Monge representation of the surface 40, 291
Monge–Ampère equation 291, 292, 295
Monoclinic symmetry 244

Morse interaction force 192

Navier–Stokes equation 221, 260
Nijenhuis operator (hereditary operator) 74 111, 228
Noether theorem 53, 66, 67, 68
Nonlinear diffusion equation 75, 86, 110
Nonlinear partial differential equations 1, 21, 132

Nonlinear Schrödinger equation 77
Normal mode 107, 252, 253, 254

Objective function 207, 208, 249, 250
Open channels 242, 259
Ordinary differential equations 69, 128
Orthonormal basis 122

Painlevé equations 48
Painlevé property 1, 2, 46, 47, 48, 49, 51, 52, 113, 288
Passive constitutive laws 201
Pendulum 72, 111, 115, 117, 151, 173, 174, 175, 191
Periodic boundary conditions 21, 82
Periodic solutions 105, 109, 110, 111, 191, 193, 238, 266
Periodic structures 191, 193, 242, 253
Permeability tensor 199
Permutability theorem of Bianchi 43, 44, 45, 111, 296
Perturbation method 30, 121
Phase portrait 239, 240, 241
Phase shift 102, 103, 106, 153, 259
Phase velocity 23, 55, 56, 230, 269
Phonon (extended mode of vibration) 242, 244, 253, 255
Piola–Kirchhoff stress tensor 199, 200, 201, 203
Piezoelectric material 242, 243, 244, 246, 248, 251, 253, 285
Piezoelectric constants 244, 245
Poincaré map 218
Potential Burgers equation 75
Potential energy 201, 203, 248
Potential function 2, 5, 9, 12, 13, 17, 90, 96, 99
Potential Korteweg–de Vries equation 74, 75
Pre-hilbertian space 3
Pressure waves 221, 232, 233
Pseudo potential energy 214
Pulsatile flow 220, 228, 235

Quantum mechanics 192

Recursion operators 73, 74, 75
Reflection coefficient 7, 8, 9, 17, 90, 92, 93, 171
Relaxation function 200, 201
time 215, 216, 286, 287
Repulsive energy function 203
repulsive range function 203
Residuum theorem 16
Riccati equations 47, 49

Riccati pseudopotential 288, 289
Rotation vector 259, 304
Runge–Kutta method 216

Saphenous artery 221
Scaling transformation 73, 74
Schrödinger equation 2, 5, 12, 90, 94, 99, 100
Screw-shaped inclusions 285
Screw-symmetric operator 169
Self similarity 243
Seymour–Varley method 149, 160, 161, 163
Shen–Ling method 20, 273, 281
Symmetry transformation 67, 74, 75, 84
Simple pendulum 72, 115, 117, 151
Sine-Gordon equation 41, 42, 43, 76, 111, 112, 113, 293, 294, 296
Singular points 46, 47
Solitary waves 54, 78, 79, 82, 242, 255, 259, 269, 270, 272
Soliton 1, 13, 30, 31, 53, 58, 61, 78, 79, 82, 83, 102, 105, 106, 113, 149, 152, 173, 191, 193, 228, 230, 233, 234, 235, 272, 278
  anti-soliton 153, 154, 217
  asymmetric 61, 198
  solutions 1, 18, 27, 28, 29, 30, 50, 62, 78, 121, 138, 149, 192, 232, 293
  collision (interaction) 105, 153, 154, 193
  multi-soliton solution (N-soliton solution) 29, 94, 100, 103, 113, 193, 228, 266, 269
  two-soliton solution 43, 45, 83, 100, 102, 105, 106, 109, 111, 113, 119, 220, 228, 229, 233, 239, 241
  theory 1, 39, 78, 121, 220, 235
Sound propagation 285, 287, 290
Spiraling fibers 285
Strain energy function 199, 202, 203
Stratified fluid 255
String equation 12, 53, 54, 121, 149, 159, 166
Sturm–Liouville theory 2, 3
Subgroup 65, 276, 277, 281
Subalgebra 277
Super ellipsoid surface 198
Superposition principle
  linear 121, 154
  nonlinear 1, 24, 25, 42, 43, 173, 185, 187, 206, 211, 242
Surface tension 242, 269, 272
Synge method 149, 166

Taylor series 34, 66, 225, 295
Tensors
  asymmetric stress 242
  deformation gradient 199
  Cauchy stress 199
  couple stress 242, 260, 262, 286, 287
  Maxwell stress 245
  Piola–Kirchhoff stress 199, 200, 201, 203
Thermodynamic pressure 259
Theta function 18, 21, 23, 173, 197, 247, 251
Thin elastic rod 121, 122, 125, 128, 130, 132, 149
Toda lattice 191, 192, 242, 299
Toda–Yoneyama equation 173, 194, 196
Torque spring 111
Torsion function 284
Torsional motion 150
Transmission coefficient 7, 8, 9, 11, 15, 93, 94, 95
Tricomi system 115
Troesch problem 37
Turbulent flow 242
Tzitzeica equation 64, 76, 273, 274, 275, 276, 279, 280, 281, 282, 283, 288
Tzitzeica curves 293, 284, 285
Tzitzeica surfaces 64, 273, 274, 275, 277, 278, 280, 281, 290

Ursell number 23

Van der Pol's equation 20, 273, 281
Variational principles 69, 248
Vibrations 107, 149, 156, 166, 168
  cnoidal 173, 185, 187, 189, 211
  Coulomb 183
  fracton 253
  free 249
  linear 149, 150
  nonlinear 149, 173
  phonon 253
  torsional 149
  transverse 54, 149, 155, 157, 158, 159
Viscosity 261
  of blood 247
Volterra integral equations 6, 115

Waves 3, 7, 11, 18, 21, 30, 54, 56, 59, 76, 79, 82, 84, 102, 105, 113, 153, 160, 163, 191, 238, 242, 249, 258, 267, 272, 287
  cnoidal 18, 22, 23, 24, 25, 27, 58, 146, 206, 239, 242, 266
  harmonic 55, 56
  linear 53, 55, 114
  nonlinear 58, 78, 192, 214
  of translation 79
  shock 84, 110, 111
Wavelength 55, 60, 84, 88, 267
Wave equation 31, 59, 108, 159
Wave number 21, 55, 60, 189, 193, 247, 266
Weierstrass equation 21, 23, 49, 58, 81, 110, 135, 173, 185, 186, 187, 279, 281
Wronskian 8, 9

Yang functional 248
Yih analysis 242, 255
Yomosa theory 221, 231
Yoneyama solution 173, 194, 196
Young's elastic modulus 125, 236

Fundamental Theories of Physics

Series Editor: Alwyn van der Merwe, University of Denver, USA

1. M. Sachs: *General Relativity and Matter*. A Spinor Field Theory from Fermis to Light-Years. With a Foreword by C. Kilmister. 1982 ISBN 90-277-1381-2
2. G.H. Duffey: *A Development of Quantum Mechanics*. Based on Symmetry Considerations. 1985 ISBN 90-277-1587-4
3. S. Diner, D. Fargue, G. Lochak and F. Selleri (eds.): *The Wave-Particle Dualism*. A Tribute to Louis de Broglie on his 90th Birthday. 1984 ISBN 90-277-1664-1
4. E. Prugovečki: *Stochastic Quantum Mechanics and Quantum Spacetime*. A Consistent Unification of Relativity and Quantum Theory based on Stochastic Spaces. 1984; 2nd printing 1986 ISBN 90-277-1617-X
5. D. Hestenes and G. Sobczyk: *Clifford Algebra to Geometric Calculus*. A Unified Language for Mathematics and Physics. 1984 ISBN 90-277-1673-0; Pb (1987) 90-277-2561-6
6. P. Exner: *Open Quantum Systems and Feynman Integrals*. 1985 ISBN 90-277-1678-1
7. L. Mayants: *The Enigma of Probability and Physics*. 1984 ISBN 90-277-1674-9
8. E. Tocaci: *Relativistic Mechanics, Time and Inertia*. Translated from Romanian. Edited and with a Foreword by C.W. Kilmister. 1985 ISBN 90-277-1769-9
9. B. Bertotti, F. de Felice and A. Pascolini (eds.): *General Relativity and Gravitation*. Proceedings of the 10th International Conference (Padova, Italy, 1983). 1984 ISBN 90-277-1819-9
10. G. Tarozzi and A. van der Merwe (eds.): *Open Questions in Quantum Physics*. 1985 ISBN 90-277-1853-9
11. J.V. Narlikar and T. Padmanabhan: *Gravity, Gauge Theories and Quantum Cosmology*. 1986 ISBN 90-277-1948-9
12. G.S. Asanov: *Finsler Geometry, Relativity and Gauge Theories*. 1985 ISBN 90-277-1960-8
13. K. Namsrai: *Nonlocal Quantum Field Theory and Stochastic Quantum Mechanics*. 1986 ISBN 90-277-2001-0
14. C. Ray Smith and W.T. Grandy, Jr. (eds.): *Maximum-Entropy and Bayesian Methods in Inverse Problems*. Proceedings of the 1st and 2nd International Workshop (Laramie, Wyoming, USA). 1985 ISBN 90-277-2074-6
15. D. Hestenes: *New Foundations for Classical Mechanics*. 1986 ISBN 90-277-2090-8; Pb (1987) 90-277-2526-8
16. S.J. Prokhovnik: *Light in Einstein's Universe*. The Role of Energy in Cosmology and Relativity. 1985 ISBN 90-277-2093-2
17. Y.S. Kim and M.E. Noz: *Theory and Applications of the Poincaré Group*. 1986 ISBN 90-277-2141-6
18. M. Sachs: *Quantum Mechanics from General Relativity*. An Approximation for a Theory of Inertia. 1986 ISBN 90-277-2247-1
19. W.T. Grandy, Jr.: *Foundations of Statistical Mechanics*. Vol. I: *Equilibrium Theory*. 1987 ISBN 90-277-2489-X
20. H.-H von Borzeszkowski and H.-J. Treder: *The Meaning of Quantum Gravity*. 1988 ISBN 90-277-2518-7
21. C. Ray Smith and G.J. Erickson (eds.): *Maximum-Entropy and Bayesian Spectral Analysis and Estimation Problems*. Proceedings of the 3rd International Workshop (Laramie, Wyoming, USA, 1983). 1987 ISBN 90-277-2579-9
22. A.O. Barut and A. van der Merwe (eds.): *Selected Scientific Papers of Alfred Landé*. [*1888-1975*]. 1988 ISBN 90-277-2594-2

Fundamental Theories of Physics

23. W.T. Grandy, Jr.: *Foundations of Statistical Mechanics.* Vol. II: *Nonequilibrium Phenomena.* 1988 ISBN 90-277-2649-3
24. E.I. Bitsakis and C.A. Nicolaides (eds.): *The Concept of Probability.* Proceedings of the Delphi Conference (Delphi, Greece, 1987). 1989 ISBN 90-277-2679-5
25. A. van der Merwe, F. Selleri and G. Tarozzi (eds.): *Microphysical Reality and Quantum Formalism, Vol. 1.* Proceedings of the International Conference (Urbino, Italy, 1985). 1988 ISBN 90-277-2683-3
26. A. van der Merwe, F. Selleri and G. Tarozzi (eds.): *Microphysical Reality and Quantum Formalism, Vol. 2.* Proceedings of the International Conference (Urbino, Italy, 1985). 1988 ISBN 90-277-2684-1
27. I.D. Novikov and V.P. Frolov: *Physics of Black Holes.* 1989 ISBN 90-277-2685-X
28. G. Tarozzi and A. van der Merwe (eds.): *The Nature of Quantum Paradoxes.* Italian Studies in the Foundations and Philosophy of Modern Physics. 1988 ISBN 90-277-2703-1
29. B.R. Iyer, N. Mukunda and C.V. Vishveshwara (eds.): *Gravitation, Gauge Theories and the Early Universe.* 1989 ISBN 90-277-2710-4
30. H. Mark and L. Wood (eds.): *Energy in Physics, War and Peace.* A Festschrift celebrating Edward Teller's 80th Birthday. 1988 ISBN 90-277-2775-9
31. G.J. Erickson and C.R. Smith (eds.): *Maximum-Entropy and Bayesian Methods in Science and Engineering.* Vol. I: *Foundations.* 1988 ISBN 90-277-2793-7
32. G.J. Erickson and C.R. Smith (eds.): *Maximum-Entropy and Bayesian Methods in Science and Engineering.* Vol. II: *Applications.* 1988 ISBN 90-277-2794-5
33. M.E. Noz and Y.S. Kim (eds.): *Special Relativity and Quantum Theory.* A Collection of Papers on the Poincaré Group. 1988 ISBN 90-277-2799-6
34. I.Yu. Kobzarev and Yu.I. Manin: *Elementary Particles. Mathematics, Physics and Philosophy.* 1989 ISBN 0-7923-0098-X
35. F. Selleri: *Quantum Paradoxes and Physical Reality.* 1990 ISBN 0-7923-0253-2
36. J. Skilling (ed.): *Maximum-Entropy and Bayesian Methods.* Proceedings of the 8th International Workshop (Cambridge, UK, 1988). 1989 ISBN 0-7923-0224-9
37. M. Kafatos (ed.): *Bell's Theorem, Quantum Theory and Conceptions of the Universe.* 1989 ISBN 0-7923-0496-9
38. Yu.A. Izyumov and V.N. Syromyatnikov: *Phase Transitions and Crystal Symmetry.* 1990 ISBN 0-7923-0542-6
39. P.F. Fougère (ed.): *Maximum-Entropy and Bayesian Methods.* Proceedings of the 9th International Workshop (Dartmouth, Massachusetts, USA, 1989). 1990 ISBN 0-7923-0928-6
40. L. de Broglie: *Heisenberg's Uncertainties and the Probabilistic Interpretation of Wave Mechanics.* With Critical Notes of the Author. 1990 ISBN 0-7923-0929-4
41. W.T. Grandy, Jr.: *Relativistic Quantum Mechanics of Leptons and Fields.* 1991 ISBN 0-7923-1049-7
42. Yu.L. Klimontovich: *Turbulent Motion and the Structure of Chaos.* A New Approach to the Statistical Theory of Open Systems. 1991 ISBN 0-7923-1114-0
43. W.T. Grandy, Jr. and L.H. Schick (eds.): *Maximum-Entropy and Bayesian Methods.* Proceedings of the 10th International Workshop (Laramie, Wyoming, USA, 1990). 1991 ISBN 0-7923-1140-X
44. P. Pták and S. Pulmannová: *Orthomodular Structures as Quantum Logics.* Intrinsic Properties, State Space and Probabilistic Topics. 1991 ISBN 0-7923-1207-4
45. D. Hestenes and A. Weingartshofer (eds.): *The Electron.* New Theory and Experiment. 1991 ISBN 0-7923-1356-9

Fundamental Theories of Physics

46. P.P.J.M. Schram: *Kinetic Theory of Gases and Plasmas.* 1991 ISBN 0-7923-1392-5
47. A. Micali, R. Boudet and J. Helmstetter (eds.): *Clifford Algebras and their Applications in Mathematical Physics.* 1992 ISBN 0-7923-1623-1
48. E. Prugovečki: *Quantum Geometry.* A Framework for Quantum General Relativity. 1992 ISBN 0-7923-1640-1
49. M.H. Mac Gregor: *The Enigmatic Electron.* 1992 ISBN 0-7923-1982-6
50. C.R. Smith, G.J. Erickson and P.O. Neudorfer (eds.): *Maximum Entropy and Bayesian Methods.* Proceedings of the 11th International Workshop (Seattle, 1991). 1993 ISBN 0-7923-2031-X
51. D.J. Hoekzema: *The Quantum Labyrinth.* 1993 ISBN 0-7923-2066-2
52. Z. Oziewicz, B. Jancewicz and A. Borowiec (eds.): *Spinors, Twistors, Clifford Algebras and Quantum Deformations.* Proceedings of the Second Max Born Symposium (Wrocław, Poland, 1992). 1993 ISBN 0-7923-2251-7
53. A. Mohammad-Djafari and G. Demoment (eds.): *Maximum Entropy and Bayesian Methods.* Proceedings of the 12th International Workshop (Paris, France, 1992). 1993 ISBN 0-7923-2280-0
54. M. Riesz: *Clifford Numbers and Spinors* with Riesz' Private Lectures to E. Folke Bolinder and a Historical Review by Pertti Lounesto. E.F. Bolinder and P. Lounesto (eds.). 1993 ISBN 0-7923-2299-1
55. F. Brackx, R. Delanghe and H. Serras (eds.): *Clifford Algebras and their Applications in Mathematical Physics.* Proceedings of the Third Conference (Deinze, 1993) 1993 ISBN 0-7923-2347-5
56. J.R. Fanchi: *Parametrized Relativistic Quantum Theory.* 1993 ISBN 0-7923-2376-9
57. A. Peres: *Quantum Theory: Concepts and Methods.* 1993 ISBN 0-7923-2549-4
58. P.L. Antonelli, R.S. Ingarden and M. Matsumoto: *The Theory of Sprays and Finsler Spaces with Applications in Physics and Biology.* 1993 ISBN 0-7923-2577-X
59. R. Miron and M. Anastasiei: *The Geometry of Lagrange Spaces: Theory and Applications.* 1994 ISBN 0-7923-2591-5
60. G. Adomian: *Solving Frontier Problems of Physics: The Decomposition Method.* 1994 ISBN 0-7923-2644-X
61. B.S. Kerner and V.V. Osipov: *Autosolitons.* A New Approach to Problems of Self-Organization and Turbulence. 1994 ISBN 0-7923-2816-7
62. G.R. Heidbreder (ed.): *Maximum Entropy and Bayesian Methods.* Proceedings of the 13th International Workshop (Santa Barbara, USA, 1993) 1996 ISBN 0-7923-2851-5
63. J. Peřina, Z. Hradil and B. Jurčo: *Quantum Optics and Fundamentals of Physics.* 1994 ISBN 0-7923-3000-5
64. M. Evans and J.-P. Vigier: *The Enigmatic Photon.* Volume 1: The Field $\boldsymbol{B}^{(3)}$. 1994 ISBN 0-7923-3049-8
65. C.K. Raju: *Time: Towards a Constistent Theory.* 1994 ISBN 0-7923-3103-6
66. A.K.T. Assis: *Weber's Electrodynamics.* 1994 ISBN 0-7923-3137-0
67. Yu. L. Klimontovich: *Statistical Theory of Open Systems.* Volume 1: A Unified Approach to Kinetic Description of Processes in Active Systems. 1995 ISBN 0-7923-3199-0; Pb: ISBN 0-7923-3242-3
68. M. Evans and J.-P. Vigier: *The Enigmatic Photon.* Volume 2: Non-Abelian Electrodynamics. 1995 ISBN 0-7923-3288-1
69. G. Esposito: *Complex General Relativity.* 1995 ISBN 0-7923-3340-3

Fundamental Theories of Physics

70. J. Skilling and S. Sibisi (eds.): *Maximum Entropy and Bayesian Methods.* Proceedings of the Fourteenth International Workshop on Maximum Entropy and Bayesian Methods. 1996 ISBN 0-7923-3452-3
71. C. Garola and A. Rossi (eds.): *The Foundations of Quantum Mechanics Historical Analysis and Open Questions.* 1995 ISBN 0-7923-3480-9
72. A. Peres: *Quantum Theory: Concepts and Methods.* 1995 (see for hardback edition, Vol. 57) ISBN Pb 0-7923-3632-1
73. M. Ferrero and A. van der Merwe (eds.): *Fundamental Problems in Quantum Physics.* 1995 ISBN 0-7923-3670-4
74. F.E. Schroeck, Jr.: *Quantum Mechanics on Phase Space.* 1996 ISBN 0-7923-3794-8
75. L. de la Peña and A.M. Cetto: *The Quantum Dice.* An Introduction to Stochastic Electrodynamics. 1996 ISBN 0-7923-3818-9
76. P.L. Antonelli and R. Miron (eds.): *Lagrange and Finsler Geometry.* Applications to Physics and Biology. 1996 ISBN 0-7923-3873-1
77. M.W. Evans, J.-P. Vigier, S. Roy and S. Jeffers: *The Enigmatic Photon.* Volume 3: Theory and Practice of the $\boldsymbol{B}^{(3)}$ Field. 1996 ISBN 0-7923-4044-2
78. W.G.V. Rosser: *Interpretation of Classical Electromagnetism.* 1996 ISBN 0-7923-4187-2
79. K.M. Hanson and R.N. Silver (eds.): *Maximum Entropy and Bayesian Methods.* 1996 ISBN 0-7923-4311-5
80. S. Jeffers, S. Roy, J.-P. Vigier and G. Hunter (eds.): *The Present Status of the Quantum Theory of Light.* Proceedings of a Symposium in Honour of Jean-Pierre Vigier. 1997 ISBN 0-7923-4337-9
81. M. Ferrero and A. van der Merwe (eds.): *New Developments on Fundamental Problems in Quantum Physics.* 1997 ISBN 0-7923-4374-3
82. R. Miron: *The Geometry of Higher-Order Lagrange Spaces.* Applications to Mechanics and Physics. 1997 ISBN 0-7923-4393-X
83. T. Hakioğlu and A.S. Shumovsky (eds.): *Quantum Optics and the Spectroscopy of Solids.* Concepts and Advances. 1997 ISBN 0-7923-4414-6
84. A. Sitenko and V. Tartakovskii: *Theory of Nucleus.* Nuclear Structure and Nuclear Interaction. 1997 ISBN 0-7923-4423-5
85. G. Esposito, A.Yu. Kamenshchik and G. Pollifrone: *Euclidean Quantum Gravity on Manifolds with Boundary.* 1997 ISBN 0-7923-4472-3
86. R.S. Ingarden, A. Kossakowski and M. Ohya: *Information Dynamics and Open Systems.* Classical and Quantum Approach. 1997 ISBN 0-7923-4473-1
87. K. Nakamura: *Quantum versus Chaos.* Questions Emerging from Mesoscopic Cosmos. 1997 ISBN 0-7923-4557-6
88. B.R. Iyer and C.V. Vishveshwara (eds.): *Geometry, Fields and Cosmology.* Techniques and Applications. 1997 ISBN 0-7923-4725-0
89. G.A. Martynov: *Classical Statistical Mechanics.* 1997 ISBN 0-7923-4774-9
90. M.W. Evans, J.-P. Vigier, S. Roy and G. Hunter (eds.): *The Enigmatic Photon.* Volume 4: New Directions. 1998 ISBN 0-7923-4826-5
91. M. Rédei: *Quantum Logic in Algebraic Approach.* 1998 ISBN 0-7923-4903-2
92. S. Roy: *Statistical Geometry and Applications to Microphysics and Cosmology.* 1998 ISBN 0-7923-4907-5
93. B.C. Eu: *Nonequilibrium Statistical Mechanics.* Ensembled Method. 1998 ISBN 0-7923-4980-6

Fundamental Theories of Physics

94. V. Dietrich, K. Habetha and G. Jank (eds.): *Clifford Algebras and Their Application in Mathematical Physics.* Aachen 1996. 1998 ISBN 0-7923-5037-5
95. J.P. Blaizot, X. Campi and M. Ploszajczak (eds.): *Nuclear Matter in Different Phases and Transitions.* 1999 ISBN 0-7923-5660-8
96. V.P. Frolov and I.D. Novikov: *Black Hole Physics.* Basic Concepts and New Developments. 1998 ISBN 0-7923-5145-2; Pb 0-7923-5146
97. G. Hunter, S. Jeffers and J-P. Vigier (eds.): *Causality and Locality in Modern Physics.* 1998 ISBN 0-7923-5227-0
98. G.J. Erickson, J.T. Rychert and C.R. Smith (eds.): *Maximum Entropy and Bayesian Methods.* 1998 ISBN 0-7923-5047-2
99. D. Hestenes: *New Foundations for Classical Mechanics (Second Edition).* 1999 ISBN 0-7923-5302-1; Pb ISBN 0-7923-5514-8
100. B.R. Iyer and B. Bhawal (eds.): *Black Holes, Gravitational Radiation and the Universe.* Essays in Honor of C. V. Vishveshwara. 1999 ISBN 0-7923-5308-0
101. P.L. Antonelli and T.J. Zastawniak: *Fundamentals of Finslerian Diffusion with Applications.* 1998 ISBN 0-7923-5511-3
102. H. Atmanspacher, A. Amann and U. Müller-Herold: *On Quanta, Mind and Matter Hans Primas in Context.* 1999 ISBN 0-7923-5696-9
103. M.A. Trump and W.C. Schieve: *Classical Relativistic Many-Body Dynamics.* 1999 ISBN 0-7923-5737-X
104. A.I. Maimistov and A.M. Basharov: *Nonlinear Optical Waves.* 1999 ISBN 0-7923-5752-3
105. W. von der Linden, V. Dose, R. Fischer and R. Preuss (eds.): *Maximum Entropy and Bayesian Methods Garching, Germany 1998.* 1999 ISBN 0-7923-5766-3
106. M.W. Evans: *The Enigmatic Photon Volume 5: O(3) Electrodynamics.* 1999 ISBN 0-7923-5792-2
107. G.N. Afanasiev: *Topological Effects in Quantum Mecvhanics.* 1999 ISBN 0-7923-5800-7
108. V. Devanathan: *Angular Momentum Techniques in Quantum Mechanics.* 1999 ISBN 0-7923-5866-X
109. P.L. Antonelli (ed.): *Finslerian Geometries A Meeting of Minds.* 1999 ISBN 0-7923-6115-6
110. M.B. Mensky: *Quantum Measurements and Decoherence Models and Phenomenology.* 2000 ISBN 0-7923-6227-6
111. B. Coecke, D. Moore and A. Wilce (eds.): *Current Research in Operation Quantum Logic.* Algebras, Categories, Languages. 2000 ISBN 0-7923-6258-6
112. G. Jumarie: *Maximum Entropy, Information Without Probability and Complex Fractals.* Classical and Quantum Approach. 2000 ISBN 0-7923-6330-2
113. B. Fain: *Irreversibilities in Quantum Mechanics.* 2000 ISBN 0-7923-6581-X
114. T. Borne, G. Lochak and H. Stumpf: *Nonperturbative Quantum Field Theory and the Structure of Matter.* 2001 ISBN 0-7923-6803-7
115. J. Keller: *Theory of the Electron.* A Theory of Matter from START. 2001 ISBN 0-7923-6819-3
116. M. Rivas: *Kinematical Theory of Spinning Particles.* Classical and Quantum Mechanical Formalism of Elementary Particles. 2001 ISBN 0-7923-6824-X
117. A.A. Ungar: *Beyond the Einstein Addition Law and its Gyroscopic Thomas Precession.* The Theory of Gyrogroups and Gyrovector Spaces. 2001 ISBN 0-7923-6909-2
118. R. Miron, D. Hrimiuc, H. Shimada and S.V. Sabau: *The Geometry of Hamilton and Lagrange Spaces.* 2001 ISBN 0-7923-6926-2

Fundamental Theories of Physics

119. M. Pavšič: *The Landscape of Theoretical Physics: A Global View*. From Point Particles to the Brane World and Beyond in Search of a Unifying Principle. 2001 ISBN 0-7923-7006-6
120. R.M. Santilli: *Foundations of Hadronic Chemistry*. With Applications to New Clean Energies and Fuels. 2001 ISBN 1-4020-0087-1
121. S. Fujita and S. Godoy: *Theory of High Temperature Superconductivity*. 2001 ISBN 1-4020-0149-5
122. R. Luzzi, A.R. Vasconcellos and J. Galvão Ramos: *Predictive Statitical Mechanics*. A Nonequilibrium Ensemble Formalism. 2002 ISBN 1-4020-0482-6
123. V.V. Kulish: *Hierarchical Methods*. Hierarchy and Hierarchical Asymptotic Methods in Electrodynamics, Volume 1. 2002 ISBN 1-4020-0757-4; Set: 1-4020-0758-2
124. B.C. Eu: *Generalized Thermodynamics*. Thermodynamics of Irreversible Processes and Generalized Hydrodynamics. 2002 ISBN 1-4020-0788-4
125. A. Mourachkine: *High-Temperature Superconductivity in Cuprates*. The Nonlinear Mechanism and Tunneling Measurements. 2002 ISBN 1-4020-0810-4
126. R.L. Amoroso, G. Hunter, M. Kafatos and J.-P. Vigier (eds.): *Gravitation and Cosmology: From the Hubble Radius to the Planck Scale*. Proceedings of a Symposium in Honour of the 80th Birthday of Jean-Pierre Vigier. 2002 ISBN 1-4020-0885-6
127. W.M. de Muynck: *Foundations of Quantum Mechanics, an Empiricist Approach*. 2002 ISBN 1-4020-0932-1
128. V.V. Kulish: *Hierarchical Methods*. Undulative Electrodynamical Systems, Volume 2. 2002 ISBN 1-4020-0968-2; Set: 1-4020-0758-2
129. M. Mugur-Schächter and A. van der Merwe (eds.): *Quantum Mechanics, Mathematics, Cognition and Action*. Proposals for a Formalized Epistemology. 2002 ISBN 1-4020-1120-2
130. P. Bandyopadhyay: *Geometry, Topology and Quantum Field Theory*. 2003 ISBN 1-4020-1414-7
131. V. Garzó and A. Santos: *Kinetic Theory of Gases in Shear Flows*. Nonlinear Transport. 2003 ISBN 1-4020-1436-8
132. R. Miron: *The Geometry of Higher-Order Hamilton Spaces*. Applications to Hamiltonian Mechanics. 2003 ISBN 1-4020-1574-7
133. S. Esposito, E. Majorana Jr., A. van der Merwe and E. Recami (eds.): *Ettore Majorana: Notes on Theoretical Physics*. 2003 ISBN 1-4020-1649-2
134. J. Hamhalter. *Quantum Measure Theory*. 2003 ISBN 1-4020-1714-6
135. G. Rizzi and M.L. Ruggiero: *Relativity in Rotating Frames*. Relativistic Physics in Rotating Reference Frames. 2004 ISBN 1-4020-1805-3
136. L. Kantorovich: *Quantum Theory of the Solid State: an Introduction*. 2004 ISBN 1-4020-1821-5
137. A. Ghatak and S. Lokanathan: *Quantum Mechanics: Theory and Applications*. 2004 ISBN 1-4020-1850-9
138. A. Khrennikov: *Information Dynamics in Cognitive, Psychological, Social, and Anomalous Phenomena*. 2004 ISBN 1-4020-1868-1
139. V. Faraoni: *Cosmology in Scalar-Tensor Gravity*. 2004 ISBN 1-4020-1988-2
140. P.P. Teodorescu and N.-A. P. Nicorovici: *Applications of the Theory of Groups in Mechanics and Physics*. 2004 ISBN 1-4020-2046-5
141. G. Munteanu: *Complex Spaces in Finsler, Lagrange and Hamilton Geometries*. 2004 ISBN 1-4020-2205-0

Fundamental Theories of Physics

142. G.N. Afanasiev: *Vavilov-Cherenkov and Synchrotron Radiation*. Foundations and Applications. 2004 ISBN 1-4020-2410-X
143. L. Munteanu and S. Donescu: *Introduction to Soliton Theory: Applications to Mechanics*. 2004 ISBN 1-4020-2576-9

KLUWER ACADEMIC PUBLISHERS – DORDRECHT / BOSTON / LONDON